The Science and Technology of Particle Accelerators

The Science and Technology of Particle Accelerators

Rob Appleby
Graeme Burt
James Clarke
Hywel Owen

CRC Press
Taylor & Francis Group
Boca Raton London New York

CRC Press is an imprint of the
Taylor & Francis Group, an **informa** business

First edition published 2021
by CRC Press
6000 Broken Sound Parkway NW, Suite 300, Boca Raton, FL 33487-2742

and by CRC Press
2 Park Square, Milton Park, Abingdon, Oxon, OX14 4RN

First issued in paperback 2022

Library of Congress Cataloging-in-Publication Data

Names: Appleby, Rob, author. | Burt, Graeme (Graeme C.), author. | Clarke, J. A. (James A.), author. | Owen, Hywel, author.
Title: The science and technology of particle accelerators / Rob Appleby, Graeme Burt, James Clark, Hywel Owen.
Description: First edition. | Boca Raton : CRC Press, 2020. | Includes bibliographical references and index.
Identifiers: LCCN 2019059103 | ISBN 9781138499874 (hardback) | ISBN 9781351007962 (ebook)
Subjects: LCSH: Particle accelerators. | Particle acceleration--Mathematics.
Classification: LCC QC787.P3 A66 2020 | DDC 539.7/3--dc23
LC record available at https://lccn.loc.gov/2019059103

ISBN 13: 978-1-03-239984-3 (pbk)
ISBN 13: 978-1-138-49987-4 (hbk)
ISBN 13: 978-1-351-00796-2 (ebk)

DOI: 10.1201/9781351007962

Typeset in Minion
by KnowledgeWorks Global Ltd.

Contents

Preface

There are already a number of excellent textbooks which cover the subject of particle accelerators, so why have we decided that there is a need for yet another one? The motivation comes from our experience at the Cockcroft Institute of supervising and teaching scientists and engineers who are new to the subject. We have found that schools, lectures, and textbooks are great at passing on the fundamentals – which are absolutely core to our subject – but that it is the more practical side of particle accelerators that our staff and students sometimes fail to connect with. The aim of this book is therefore to not just explain the principles that underpin our field, but to also pass on some of the experience that we all need to design, build, and operate these marvellous machines. Many of the things we highlight are straightforward to convey, but before they are actually pointed out they can appear a little mysterious. We hope this book will give some useful guidance.

Over the last ten years or so, the Cockcroft Institute of Accelerator Science and Technology in the UK has built up a teaching programme that encompasses the range of skills and methods required to design, construct, and operate particle accelerators across the range of uses to which they can be put. This range covers: the research required to understand and improve particle acceleration methods; the 'traditional' uses of accelerators in particle, nuclear, and atomic physics; and the applications outside of academic research in such areas as medicine, security, and industry. This book tries to reflect that scope; although we cannot cover all the myriad topics our rich subject has to offer, we have tried to cover enough topics to give a core understanding.

Whilst there are a number of excellent reference guides on the subject – Chao and Tigner's excellent handbook being a notable one, since it is found on the desk of nearly every accelerator scientist we have met – we found it difficult to find a practical introduction to the key topics and calculations an early-career professional in our field might encounter. That is the motivation for our textbook: as well as developing a number of topics in accelerator physics (radio-frequency and other acceleration technology, magnetic design, beam dynamics, and radiation), we hope also to provide guidance on how to correctly apply those ideas, and in what practical situations different calculations ought to be used. This book therefore includes numerous worked examples that show the typical numerical quantities that may be encountered. Exercises are also included for the reader on key points, and these can be found at the end of the chapters. The solutions to all the exercises are freely available to download from the publisher's web site page for this book. We have tried to make this book fall somewhere between a traditional textbook and a handbook of formulae.

We hope you enjoy reading it as much as we enjoyed writing it.

Acknowledgements

During our careers, we have all benefited from working closely with a tremendous number of particle accelerator experts who have been very willing to freely pass on their deep knowledge and understanding for which we are extremely grateful. This widespread culture of willingness to openly share with colleagues from across the globe marks out our subject as a truly inspiring field to pursue.

Special thanks are due to Neil Marks and Richard Carter for educating us all in the delights of accelerator magnets and RF. Thanks are also due to Alex Bainbridge and Ben Shepherd for providing some of the background material for the magnet chapter. Finally, special thanks to Andy Wolski for his insight and willingness to teach others.

Rob Appleby
Graeme Burt
James Clarke
Hywel Owen

The Cockcroft Institute of Accelerator Science and Technology,
Sci-Tech Daresbury,
July 2020

1

An Invitation: Acceleration!

This book is about particle accelerators, what they accelerate, how they work, and how (and why) we build them. In this opening chapter we shall take a short tour of a typical accelerator, moving along the beamline and making sense of the different kinds of accelerator elements. This will give a kind of map for the rest of the book and a map of the subject. We hope the reader finds it useful for an overview and orientation. We are scientists and engineers, and as such are concerned with the observation and understanding of the physical world. The first step to any kind of deeper – and if we are lucky, quantitative – understanding of that world is to group and classify those aspects attracting our attention. How do we classify particle accelerators? Before we start, we should first define what a particle accelerator is.

We may define a particle accelerator as a device – often called a 'machine' – that endows subatomic particles with large and *variable* amounts of kinetic energy. 'Large' here is in comparison with the sorts of energies one obtains from a particle source such as a simpler electron gun or ion source that might produce particles of tens of thousands of electron-volts (eV).* Particle accelerators differ from other sources of energetic particles – such as radioactive decay – in that an accelerator allows us (more or less) to freely *choose* the particle energy; for example, alpha particles from a given radionuclide – say, americium-241* – are emitted with only a single energy (of several MeV). We will see in the next chapter that electric fields are the predominant method of providing a particle with kinetic energy, and this demands that the particles we accelerate are *charged* so as to experience an acceleration from that field; the *beams* of particles that travel through an accelerator are therefore often described in terms of the equivalent *current* they carry. However, there are also so-called *secondary sources* of particles, some of which may be electrically neutral; three important examples are the photon, the neutron and the neutrino, all commonly produced by accelerators and used extensively in science and engineering for quite different things.

In the following chapters we will deal with the manner in which particles are produced, accelerated and used – each of which of course depends on the particular particle. But first let us take an overview, and attempt to classify them by type. A first observation is that some accelerators are straight (i.e. *linear*), in which the accelerated species pass through each *element* of the accelerator only once; often the predominant element is the one that

---

*The electron-volt is generally the most appropriate unit to quantify that kinetic energy.
*Americium-241 is chosen as an example here because it is the most commonly-encountered radioisotope; around 1 μCurie activity (about 0.3 mg of AmO_2) is present in virtually every domestic smoke detector.

DOI: 10.1201/9781351007962-1

performs the acceleration. These are usually called *linear accelerators*, or linacs for short. The alternative is the *circular accelerator*, in which particles circulate many times, very often repeatedly through the same elements; this can allow the re-use of accelerating elements (such as accelerating gaps or cavities), or allow repeated production of some secondary species – such as photons, for instance – from the same primary particle.

So we can classify accelerators into two broad categories – those that accelerate particles in a straight line and those that accelerate particles approximately in a circle (usually called a ring). In the straight (or linear) type, the particles start at one end and pass through every element only once (including the accelerating elements), finishing up at the end of the accelerator. This type of linear accelerator (usually abbreviated as 'linac') is very common and is used all over the world, mostly commonly as the device that supplies electrons at ~10 MeV in an X-ray radiotherapy machine. To construct the second type, we imagine bending our linear accelerator into a ring using dipole magnets (also therefore called bending magnets), so that the particle makes many laps (or turns) of the ring, also passing through the elements making up that ring many times. A widely-encountered example of this type of accelerator is the synchrotron; many synchrotrons are used today to produce high-energy photons (with energies typically of a few keV or more) by bending the circulating beam of electrons; these photons are then used in a variety of techniques by researchers. Other synchrotrons are used to accelerate – for example – protons to high energies to hundreds of GeV or more to undergo collisions to study particle physics.

If we visit a particle accelerator ('accelerator' for short) we find that many are composed of several distinct systems, each of which is commonly regarded as an accelerator in its own right. A famous example is CERN's Large Hadron Collider (LHC), a proton accelerator that lies many metres under the ground near Geneva, and which is large enough, with a 27 km circumference, to cross the Swiss-French border twice! The LHC facility is really a number of connected accelerators, and the protons begin their lives within a bottle of hydrogen gas. An *ion source* is used to strip the electrons from the hydrogen atoms and deliver the protons at some (modest) energy of 70 keV to a pre-injector; a chain of further accelerators, first linacs and then circular *synchrotrons*, progressively increases each proton's energy to a final value of 6.5 TeV (tera-electron-volts).* Two independent beams of protons of the same energy travel around the ring, one clockwise and the other counter-clockwise, and these are made to collide head on into each other at specific locations within the *storage ring* to produce reaction products useful for experiments in fundamental particle physics.* Another example is the *free-electron laser* (or FEL). Here, electrons are generated from an *electron gun* (a cathode from which electrons are emitted in the presence of a strong voltage, sometimes with some assistance from a short pulse of laser photons) and then progressively accelerated by a linac; these electrons then pass through a special magnetic device that prompts the electrons to emit light with laser-like properties (the FEL proper) and so generate tailored pulses of photons. An example of an electron gun is shown in Fig 1.1.

The basic building blocks of any accelerator are: the devices that *generate* the particles (the sources); the devices that *accelerate* the particles, which is almost always done with electric fields; and the devices that confine and control the particles, which are commonly built using magnets. For example, electromagnets are used to *deflect* (bend) particles into a curved path – so-called dipole electromagnets are used to construct a circular accelerator. Other electromagnets such as quadrupoles, sextupoles and so on are used to *confine* particles

---

*7 TeV is the anticipated energy in the future.

*A storage ring is a type of synchrotron, but one in which the energy of the circulating particles is constant.

FIGURE 1.1 Here we see one of the electron guns at CLARA, situated at Daresbury Laboratory. This is a typical photo-injector source, in which a pulsed laser is directed (from the left) onto a cathode (on the right of the photograph) and produces an intense, short-duration bunch of electrons that contains tens of picocoulombs of charge. The electrons are then accelerated to the left by a strong oscillating electric field produced in several coupled cavities, up to a kinetic energy of several MeV. ©STFC

into some desired size (envelope). All these devices generate a predetermined magnetic field that is experienced by the passing charged particles, using some sort of *beam-optical* arrangement; an example is shown in Fig 1.2.

The effect of the electric and magnetic fields can be summarised by a single basic law that is the most important equation encountered in this book, and in the field of particle accelerators – the Lorentz equation (also known as the Lorentz force law). This describes the force $\mathbf{F}$ on a particle with charge q from an electric field $\mathbf{E}$ and magnetic field $\mathbf{B}$ and is given by

$$\mathbf{F} = q\left(\mathbf{E} + \mathbf{v} \times \mathbf{B}\right), \tag{1.1}$$

where $\mathbf{v}$ is the particle velocity. The consequences of this seemingly-simple equation – combined with the other laws of electromagnetism – will occupy us in the following chapters, but straight away we seem two very important differences between the way electric and magnetic fields act upon charged particles; electric fields can perform *work* upon the charges, and therefore can impart (kinetic) energy to them, whereas magnetic fields produce a force at right angles to a charge's motion and so do *no work*. A static magnetic field *cannot* change the energy of a charged particle, and particularly can do nothing at all if the charge is stationary; we will discuss this more in the next chapter. We also note now the very important consequence of special relativity, which is that our accelerated particles often significantly increase in mass rather than velocity as they gain energy. This is always considered when calculating the effect of the Lorentz force, and obviously relativity ultimately determines that our particles cannot travel faster than the speed of light, c. Surprisingly, the ideas of quantum physics do not normally have to be considered, although on occasion we will; this

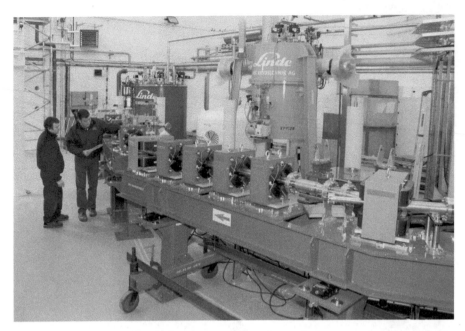

FIGURE 1.2   Part of the ALICE energy-recovery linac electron beam transport system, previously installed at Daresbury Laboratory. Individual beamline magnets are typically mounted onto a girder (usually steel), adjusted so that their magnet centres are aligned and then fixed (with position accuracy of some tens of μm); the girder may then be lifted into its operating position without significant relative movement of the magnets, and they may be adjusted together for efficiency. A typical alignment accuracy from girder to girder of a few tens of μm is also commonly achieved. ©STFC

is most often encountered in the context of photon emission from charges. An important but sometimes overlooked aspect of particle accelerators is that that the *beam pipe* through which the particles pass (and which the electromagnets surround) must be *evacuated* to allow the particles to pass by with little scattering or absorption; *residual gas* within the vacuum system can give rise to such undesirable phenomena as particle loss, emittance degradation (blow-up), ion trapping by electron beams and the analogous electron cloud instability experienced by proton beams.

The most important magnetic devices are the dipole and the quadrupole. Let's look at a dipole first, which can be seen in Fig 1.3, which induces charged particles to follow a curved path (the arc of a circle to be exact); it bends a beam of particles and is composed of two poles (north and south). Linacs often need to utilise dipoles to produce some defined bend angle, for example to steer a produced beam to a precise final location; of course, a circular accelerator requires 360°-worth of total bend, and this is typically achieved using a number of dipoles each contributing a part of the overall deflection. The cyclotron is an example of a circular accelerator that utilises only one dipole, in the form of a single circular magnet. In a dipole a combination of current-carrying copper coils and steel poles produces an almost uniform magnet field, bending the passing charges through some angle determined by the magnetic field and each particle's momentum. This deflection angle θ is proportional to the field strength B; for very large values of B the coil currents must be large, and this may require them to be superconducting. As a general guide, ordinary electromagnetic dipoles generate fields up to 1 or 2 T, and superconducting dipole magnets typically generate fields up to ~8 T (with the prospect for significantly higher fields than this in the future). In Chapter 4 we will see how magnet designers construct dipole magnets to some specified

FIGURE 1.3 A dipole magnet used to deflect (bend) a beam of charged particles. A current-carrying coil – here made of copper channels wound a number of times around each of the two pole pieces – drives the magnetic field. The outer yoke closes the magnetic field lines to maximise efficiency and to limit the stray field away from the magnet so the magnetic field is essentially only present in air between the north and south poles. A vacuum vessel between the poles follows the 60° deflection angle, but also includes extra pieces (here with temporary flanges on) that allow for other uses such as vacuum pumping or for emitted light to be extracted and utilised. ©STFC

strength and field accuracy and we'll see in Chapter 5 how we can compute the motion of particles in a (perhaps non-ideal) dipole field.

The quadrupole – the 'four-pole' magnet – is often the most numerous type of magnet found in an accelerator, and can be seen in Fig 1.2. The magnetic field inside the aperture of a quadrupole has a strength $B_y \propto x$ (the vertical field rises as one moves horizontally from the magnet centre) and also $B_x \propto y$; a quadrupole provides a *gradient* $g = \partial B/\partial x$ in the field with zero field at the magnet centre, so on average provides no deflection at all. As a rule of thumb we typically use gradients of 10 to 100 T/m; smaller magnet apertures make larger gradients easier to achieve. The purpose of quadrupoles is to focus and so basically to confine the beam to within some stable envelope, and in Chapter 5 we will discuss Hill's equation and how it determines if an arrangement of quadrupoles – called a magnetic lattice – gives a stable focusing channel. Often we use a matrix formalism, based on Hill's equation, which allows us to follow – or track – the paths of individual particles. We will see in Chapter 5 how the Courant-Snyder formalism can be used to describe the envelope around those particles using the so-called β-function and the other *Twiss functions*. We will also discuss *higher-order* magnets with more poles, such as the sextupole (in Chapters 4 and 5); these are commonly used to correct beam-optical aberrations and thereby enable magnet lattices to give better stability to the transported particles. Most particle accelerators require magnets that generate these higher-order fields.

At the heart of any accelerator are the devices that generate the accelerating fields, which we will see in Chapter 2 can only be electric fields. A common device is the *accelerating cavity*, within which a time-varying (oscillating) electric field is generated by means of some source of electromagnetic energy. Time-varying fields are often employed as they can more conveniently deliver large electric fields to the particles, albeit with the restriction that we then need to correctly time when those particles pass through the cavity. Particle beams are therefore very often *bunched* so that each bunch arrives within the cavity at the correct phase to see an accelerating field (other phases provide either less acceleration, no acceleration or will decelerate the particles). The fields in the cavity oscillate in time and obey the wave equation; we'll see in Chapter 3 that the longitudinal electric field accelerates particles in the accelerating mode. Note this oscillation in time at a specific frequency and with a characteristic spatial field is typical of resonant structures. Accelerating electric fields are often measured in terms of MV/m; an example of a cavity that makes this sort of field is shown in Fig 1.4.

To generate a large field, cavities are generally *resonators* that store electromagnetic energy, and they can be described also in terms of the voltage gain (which is proportional to the energy gain) a charge sees as it crosses from one end of the cavity to the other; the voltage is the integral of the peak accelerating electric field E, modified by what oscillation phase a particle sees as it travels through that field. Much of modern-day development of cavities and other kinds of accelerating structures is to achieve the highest possible gradient; a larger gradient generally means a smaller – and therefore cheaper – accelerator. Modern-day cavities seek to provide gradients as high as ~ 100 MV/m or more, depending upon how they operate and for which application; superconductivity is again often employed to limit ohmic energy losses in the body of the cavity and thereby to enable more efficient accelerators or to achieve parameters that would otherwise not be possible. However, some types of accelerators – notably cyclotrons and synchrotrons – do not need such high gradients as they can use a smaller voltage repeatedly; in this case, the focus can be more on efficiency and limiting power losses. Large accelerating fields of perhaps 1 GV/m or more can be produced when a plasma has induced within it a significant separation of the electrons from the positive ions; this can be brought about by a variety of means, including either the strong electric field from a focused laser pulse or the passage of a particle bunch. This is an active area of current research, and we will mention it briefly.

The primary particles produced by accelerators are often used directly: for example, the LHC collides very high-energy protons upon each other after accelerating them, whilst a low-energy (several MeV) electron linac can be used to irradiate and sterilise food products and medical equipment. Very often we encounter *targets* onto which these particles are directed. Some of these targets are used to generate *secondary radiation*; an important example is to produce neutrons, where heavy-metal spallation targets taking very intense proton beam powers ~ 1 MW are commonly used. Those neutrons are enormously important for studies of chemistry, physics and engineering studies of novel materials. Particle physics experiments increasingly also call for high-intensity beams to generate such things as muons and neutrinos, the former for future muon colliders and the latter to see signatures of physics beyond the standard model. At lower particle energies around 10 MeV, electron linacs are used to generate bremsstrahlung photons for use in radiotherapy or for scanning cargo; in fact, this is the most likely situation someone will encounter a particle accelerator and there are around 30,000 such accelerators around the world – the majority, in fact. Another medical application is the use of proton cyclotrons to generate radioisotopes; fluorine-18 is the most commonly-produced isotope, made by directing ~ 10 MeV protons onto an enriched water target. Higher-energy protons and other species such as carbon-12 ions are directed into patients to perform particle therapy, another form of radiotherapy. We will

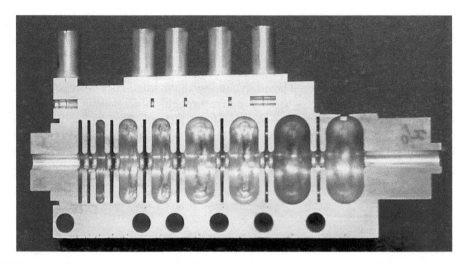

FIGURE 1.4 An accelerating cavity for X-ray cargo scanning, shown cut in half to reveal the internal accelerating cells that oscillate in voltage together at the same frequency. Electrons enter from the left at a low velocity and exit on the right after their acceleration; as they gain velocity they travel further per oscillation period and hence the cavity cells must get longer.

not describe these applications in detail, but will discuss the basics for the operation of the types of accelerator they employ.

From an accelerator point of view the most important secondary particle phenomenon is that of photon production; it's a fundamental behaviour when charges accelerate in electromagnetic fields, and so we discuss it in detail in Chapter 6. We show the basic connection between the bremsstrahlung utilised in radiotherapy and the production of photons via synchrotron radiation. So-called synchrotron light sources are a widespread application of electron accelerators – there are nearly a hundred such facilities around the world now – and they make use of the enormous enhancement of photon production when electrons with a large kinetic energy travel through a specific magnetic field arrangement.

Regardless of the type of accelerator, there are limitations in the accuracy with which it is built, and this must be considered during the design. For example, misalignments of magnets give rise to trajectory errors which must be corrected using additional small steering magnets (also called *corrector magnets*). The effect of the misalignments themselves must be measured using suitable *diagnostic instrumentation* such as beam position monitors (electrostatic pickups, screens and so forth) and measurements of the total beam charge/current and the dimensions of the beam. In virtually every modern accelerator there is a *control system* in which a coordinated set of computers and instrumentation interfaces brings together measurements of the operation and status of each element of the accelerator, to allow adjustment of the operation of it. These are most important during the initial *commissioning* of the accelerator with particle beams.

We see that the field of particle accelerators is very broad, and we have not mentioned many of the possible subsystems that may be encountered; that is the job of more specialist texts. Here, we attempt to give an overview of the principle ideas involved in the practical design and construction of an accelerator system, and discuss the most-often encountered components and phenomena; we therefore restrict ourselves to discussing mostly conventional accelerator components, that is the electromagnets and accelerating cavities used today in the vast majority of particle accelerator facilities.

We divide our discussion into the following chapters. In Chapter 2 we discuss the basic ideas of charges and electromagnetic fields that underpin all the most important ideas in accelerator science – this is our fundamental 'ABC' chapter. Next, in Chapter 3, we shall discuss the methods used to accelerate particles, mainly resonant cavities and their behaviour. In Chapter 4 we meet magnets, and explain the three basic technologies used in their design and construction: electromagnets, permanent magnets, and superconducting systems. The following Chapter 5 introduces beam dynamics, the methodology used to describe and predict the behaviour of particles as they move through an accelerator, including a discussion of non-ideal situations such as machine imperfections. The production of photons by charges is the subject of Chapter 6, including the very important field of synchrotron radiation. Finally, Chapter 7 discusses a few of the most important complexities that arise when particles interact with each other.

2

ABC: Accelerators, Beams, and Charges

The realm of particle accelerators is effectively that of electromagnetism, where we have also to include aspects of the theory of special relativity since in most cases our particles are moving at some significant fraction of the speed of light, $c = 2.99792 \times 10^8$ ms^{-1}. This chapter will explain the key ideas that lay the foundation for the rest of the book. We begin with a brief review of the electromagnetism, assuming that the reader has had some introduction to the topic at the undergraduate level; there are a number of excellent texts on this subject [1, 2, 3, 4]. We then discuss the effect of externally-applied electric and magnetic fields upon charged particles, and discuss when relativistic effects are important. We then discuss the exchange of energy between the electromagnetic field and a set of particles, including the exchange of energy with electromagnetic radiation. We leave the discussion of the production of electromagnetic radiation – by the particles themselves – to Chapter 6. Charges may also exert significant influence upon each other due to their mutually-experienced electric and magnetic fields, and we will introduce that topic ready for a longer discussion in Chapter 7.

It may initially be somewhat surprising that quantum ideas do not often have to be invoked in particle accelerator science; after all, the particles such as electrons and protons with which we are concerned are the basic quantised building blocks in nature. However, we will see that most of the more-or-less classical description of electric and magnetic fields will suffice. For certain phenomena – principally those in which photon emission and absorption is involved – we will occasionally have to resort to quantum ideas.

DOI: 10.1201/9781351007962-2

2.1   The Electromagnetic Field and Its Properties

2.1.1   Maxwell's Equations

Electromagnetism, meaning the motion and dynamics of charges and currents, is governed in the most general way by Maxwell's equations. In the classical description of charged particles, a stationary charge of magnitude q exerts an electric field that is both isotropic and extends to infinity instantaneously; 'instantaneously' means that the presence of a charge can be immediately experienced at some location a distance l away. Straight away we see that this cannot really be true, since the ideas of special relativity tell us that if a charge is moved, it takes some time $t = l/c$ for a distant observer to become aware of that motion; this is the idea of the retarded time; we discuss this later in Chapter 6. Charges are either positive or negative, and one may think of them therefore as being either sources or sinks of field lines; the number of field lines is proportional to the charge.

Field lines must begin and end on charges – there can be no discontinuities in the field lines (a point with a discontinuity implies extra charge there), see Fig 2.1. Therefore, if we surround a charge (or set of charges) with a surface and count up the field lines passing through the surface, the total number is proportional to the charge (we do this correctly by counting field strength perpendicular to the surface, i.e. $\mathbf{E} \cdot d\mathbf{S}$ – the density of field lines is proportional to $\mathbf{E}$). What we have described here is Gauss's Law, which is

$$\oint_s \mathbf{E} \cdot d\mathbf{S} = \frac{q}{\epsilon_0}, \tag{2.1}$$

where S is the surface that encloses the volume V. ϵ_0 is the constant of proportionality between field and charge, and is known as the permittivity of free space. A point charge (such as a single particle) lying at the centre of two concentric spherical surfaces *1* and *2* with radii r_1 and r_2 must have the same number of field lines passing through both surfaces (Fig 2.2). In this case we have spherical symmetry* and so Gauss's Law becomes

$$4\pi r_1^2 E_1 = 4\pi r_2^2 E_2 \tag{2.2}$$

where E_1 and E_2 are the magnitudes of the electric fields passing through each surface (perpendicularly in this case). From this we see that electric field strength around a stationary charge obeys the well-known inverse-square law

$$E(r) = \frac{q}{4\pi\epsilon_0 r^2}. \tag{2.3}$$

The differential form of Gauss's Law is a restatement of the idea that charge creates field, and is

$$\nabla \cdot \mathbf{E} = \frac{\rho}{\epsilon_0}.$$

It is an observed fact that there are no sources or sinks of magnetic field, hence the equivalent to Gauss's Law for magnetic fields is (Fig 2.3)

$$\oint_s \mathbf{B} \cdot d\mathbf{S} = 0. \tag{2.4}$$

The differential form for this equation is

$$\nabla \cdot \mathbf{B} = 0.$$

---

*In other words, we cannot tell which way round a point charge is facing – it has no inherent orientation.

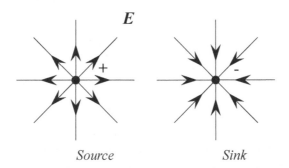

FIGURE 2.1   Charges are sources or sinks of electric field lines; the number of field 'lines' is proportional to the amount of charge q.

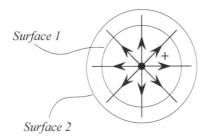

FIGURE 2.2   Two concentric surfaces surrounding a given charge must have the same number of field lines passing through each surface.

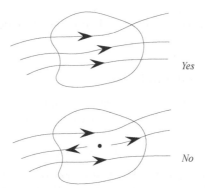

FIGURE 2.3   The total magnetic flux passing through a surface must sum to zero (top figure). If the total flux were non-zero (lower figure) it would imply that there was a source of magnetic field within the volume bounded by that surface; this is not possible for magnetic fields.

Time-varying electric fields can generate a magnetic field, as can a motion (flow) of charges; a flow of charges is a current. The generated magnetic field is described by Ampere's Law, most easily expressed as

$$\nabla \times \mathbf{B} = \mu_0 \mathbf{j} + \mu_0 \epsilon_0 \frac{\partial \mathbf{E}}{\partial t}. \tag{2.5}$$

Similarly, a time-varying magnetic field can generate an electric field; this is Faraday's Law

$$\nabla \times \mathbf{E} = -\frac{\partial \mathbf{B}}{\partial t}. \tag{2.6}$$

(there is no equivalent term to electric current in this equation because there are no such things as magnetic 'charges'). Note the minus sign in Faraday's Law: the induced electric field acts to oppose the change in magnetic field – this is Lenz's Law.

Another important thing we know in electromagnetism is that charge is conserved. Hence, if we have a volume V and consider the charges that may be moving into or out of it, we may relate the current $\mathbf{j}$ flowing through the surface S around V to the change of the total charge Q within it as

$$\oint_S \mathbf{j} \cdot d\mathbf{S} = -\frac{dQ}{dt}. \tag{2.7}$$

Using the divergence theorem we can rewrite the left-hand side of this equation as

$$\oint_S \mathbf{j} \cdot d\mathbf{S} = \int_V \nabla \cdot \mathbf{j} dV, \tag{2.8}$$

and write $Q = \int_V \rho dV$ in terms of the charge density $\rho(\mathbf{r})$ inside V. Hence

$$-\int_V \nabla \cdot \mathbf{j} dV = \frac{d}{dt} \int_V \rho dV = \int_V \frac{\partial \rho}{\partial t} dV, \tag{2.9}$$

where we can make the latter transformation because we choose that the volume V does not change with time. But the chosen volume V may be of any size; we can reduce it to an arbitrarily small volume. We can therefore remove the integrals from this equation to yield

$$\nabla \cdot \mathbf{j} + \frac{d\rho}{dt} = 0. \tag{2.10}$$

This is the Continuity Equation; it's just the differential statement that charge is conserved when charges are moving.

2.1.2   Forces on Charged Particles

The force $\mathbf{F}$ on a charge q moving with velocity $\mathbf{v}$ within the presence of an electric field $\mathbf{E}$ and magnetic field (or more correctly, the magnetic flux density) $\mathbf{B}$ is given by the Lorentz force law

$$\mathbf{F} = q(\mathbf{E} + \mathbf{v} \times \mathbf{B}). \tag{2.11}$$

We see that the electric force points in the direction of the field, and hence the electric field $\mathbf{E}$ can do work upon the charge (or vice versa); the acceleration of a charge due to an electric field is

$$a = \frac{qE}{m}. \tag{2.12}$$

However, magnetic fields do no work since the force exerted on the charge is at right-angles to the direction of $\mathbf{B}$ due to the cross product; also, charges must be moving in order to experience a force due to a magnetic field **B**. It is typical in many accelerators to have charges moving at speeds close to the velocity of light, c. An electromagnet providing a field $B = 1$ T will produce a force $|F| \simeq cB$ equivalent to the force exerted by a electric field of $E = 300$ MV/m. A 1 T electromagnet is routine, but a 300 MV/m electric field is highly challenging to produce and is only really encountered in plasma accelerators. It is for this technological reason that magnetic fields rather than electric fields are predominantly used to guide and focus the charged particles in an accelerator. However, magnetic fields do no work and so cannot change the speed of a charged particle. A charged particle moving through a uniform magnetic field **B** will experience a constant force at right-angles to its motion, and hence move in a circular path due to a transverse acceleration $a_\perp = qvB/m$. The radius of that circle, ρ, mapped out by the particle is dependent upon the momentum of the particle and the magnetic field strength, B, and is

$$\rho = \frac{mv}{qB}. \tag{2.13}$$

Obviously, both accelerations (electric and magnetic) depend upon the charge's mass. At sufficient velocities that mass is no longer the rest mass, but increases due to relativistic effects.

2.2 Relativity

A charged particle may be accelerated to large velocities such that its kinetic energy becomes comparable to or much greater than its rest energy; the effects of relativity must then be taken into account, and this is the case for nearly all the situations encountered in particle accelerator science. The behaviour under the conditions for special relativity will however suffice rather than any effects due to general relativity.

In this book, we adopt the notation conventions m_0 for the rest mass of a particle and E_0 for the rest energy, and we use the convenient units of MeV for energy and MeV/c^2 for mass; 1 eV is equal to about 1.602×10^{-19} J. Since the rest energy of a particle is just $E_0 = m_0 c^2$, we may readily convert for example the rest mass of an electron $m_e = 9.109 \times 10^{-31}$ kg – which is equal to about 0.511 MeV/c^2 – to a rest energy of 0.511 MeV; the numerical value is the same.

The most important variable to determine for a fast-moving, high-energy particle is its relativistic 'gamma' or 'Lorentz' factor, which is the ratio of the total energy of the particle to its rest energy

$$\gamma = \frac{E}{E_0} = \frac{T + E_0}{E_0}, \tag{2.14}$$

where we denote the kinetic energy of the particle as T; a stationary particle has $T = 0$ and so $\gamma = 1$, and γ may be arbitrarily large for a large kinetic energy. The mass of a fast-moving particle increases to $m = \gamma m_0$ and an elapsed time t_0 in its own frame of reference is dilated to $\Delta t = \gamma \Delta t_0$ when viewed by a (stationary) observer; hence an unstable particle with lifetime τ takes longer to decay if it is moving rapidly. The particle velocity is v and so the velocity relative to the velocity of light c is given by β as

$$\frac{v}{c} = \beta = \sqrt{1 - \frac{1}{\gamma^2}}, \tag{2.15}$$

such that β cannot be greater than 1. At very low velocities, this relation reduces to the classical relation $v = \sqrt{2T/m_0} = c\sqrt{2T/E_0}$. Particle momentum is always given as $p = mv$, but it is useful to express it as

$$p = mv = \beta \gamma m_0 c = \beta \gamma E_0/c = \frac{E_0}{c}\sqrt{(\gamma^2 - 1)}. \tag{2.16}$$

Often we are interested in cases where $\gamma \gg 1$, in which case the quantities above have much simpler expressions:

$$\beta \simeq 1, \tag{2.17}$$

$$v \simeq c, \tag{2.18}$$

$$p \simeq \gamma m_0 c, \tag{2.19}$$

$$E \simeq pc. \tag{2.20}$$

A particular example illustrates these concepts. We imagine two particles – a proton and an electron – each with kinetic energy $T = 250$ MeV. The total proton energy is just $E \simeq 1181$ MeV (the proton rest energy is $E_0 \simeq 938$ MeV). The proton has a γ not much more than 1 ($\gamma \simeq 1.27$) such that its velocity is $\beta \simeq 0.614$. The momentum is then just $p \simeq 731$ MeV/c; note that all the numerical values (T, E, and p) are different from each other.

By contrast, an electron with $T = 250$ MeV is 'ultrarelativistic' with $\gamma \simeq 490$, $\beta = 1$ (to a very good accuracy), total energy $E = 250.511$ MeV and $p = 250.511$ MeV/c. Note that $E \simeq T$ so that often these values are interchanged, and when considering ultrarelativistic particles for most practical purposes, the difference between T and E doesn't matter (it's generally far smaller than other uncertainties in other accelerator parameters).

2.3   Particle Motion in Electromagnetic Fields

2.3.1   Curvature in a Magnetic Field

We again consider the separate effect of electric and magnetic forces upon a charge. In the electric field case, as the charge gains in velocity its mass increases and restricts the ultimate velocity to a value less than c. In the magnetic field case, the charge is accelerated transversely without gaining energy, but now

$$\rho = \frac{\gamma m_0 v}{qB}. \tag{2.21}$$

We often write this expression as

$$(B\rho) = \frac{p}{q} \tag{2.22}$$

and call $(B\rho)$ or p/q the beam rigidity. The beam rigidity p/q describes the resistance of a beam of charged particles to being bent into a radius ρ by a given magnetic field B; for a given rigidity $(B\rho)$, doubling the field will halve the bend radius. We may express the beam rigidity equation in several ways. Expressing the momentum in units of GeV/c (1 GeV/c is equivalent to 5.3×10^{-19} kg ms^{-1}), the bending radius is

$$\rho[\text{m}] = 3.33 \frac{p[\text{GeV/c}]}{B[\text{T}]}. \tag{2.23}$$

In general, the bending radius is

$$\rho = \frac{E_0 \sqrt{\gamma^2 - 1}}{qcB} \tag{2.24}$$

for a particle of rest energy E_0. For protons this is

$$\rho[\text{m}] = \frac{3.13 \sqrt{\gamma^2 - 1}}{B[\text{T}]} \tag{2.25}$$

and for electrons

$$\rho[\text{m}] = \frac{1.71 \times 10^{-3}\sqrt{\gamma^2 - 1}}{B[\text{T}]}. \tag{2.26}$$

Note that ultrarelativistic electrons (which practically means any electron with $T > 10$ MeV), $\beta \simeq 1$ and $E \simeq T$ and we may relate the energy as

$$E[\text{GeV}] \simeq 0.3B\rho[\text{Tm}]. \tag{2.27}$$

A 1 GeV electron will move in a 1 T field with a bending radius of 3.33 m.

2.3.2 Conservation of Energy in Electromagnetic Fields

A volume V with electric and/or magnetic fields present within it has associated electric and magnetic field energies U_E and U_B of

$$U = U_E + U_B = \frac{1}{2}\epsilon_0 \int_V E^2 \text{d}V + \frac{1}{2\mu_0} \int_V B^2 \text{d}V. \tag{2.28}$$

The energy density of an electric field (per unit volume) is therefore just $\epsilon_0 E^2/2$ and that of a magnetic field is $B^2/2\mu_0$. Let's now consider the rate of change of energy in a fixed volume. We start by considering a single charge q moving at some velocity v through an electromagnetic field (i.e. composed of both $\mathbf{E}$ and $\mathbf{B}$ components). The charge q may do work upon the field due to the Lorentz force acting upon the charge. The work done on the charge is (as for other forms of work) $\text{d}W = \mathbf{F} \cdot \text{d}\mathbf{l}$, so that the power exerted by the electromagnetic field upon the charge can be determined as

$$\begin{aligned} P &= \frac{\text{d}W}{\text{d}t} \\ &= \mathbf{F} \cdot \frac{\text{d}\mathbf{l}}{\text{d}t} = \mathbf{F} \cdot \mathbf{v}, \tag{2.29} \\ &= q(\mathbf{E} + \mathbf{v} \times \mathbf{B}) \cdot \mathbf{v} \\ &= q\mathbf{E} \cdot \mathbf{v}. \tag{2.30} \end{aligned}$$

(Note that this power P is measured in watts.) We have cancelled out the second term (in $\mathbf{B}$) and we see the well-known fact: magnetic fields never do any work. Only the electric field $\mathbf{E}$ can do work on a charge. We may express this idea in another way: integrating the work done over some path l, we have for the total work

$$W = \int \text{d}W = \int_l q\mathbf{E} \cdot \text{d}\mathbf{l}. \tag{2.31}$$

The work-energy theorem allows us to deduce that an electric field $\mathbf{E}(\mathbf{r})$ (a vector field over some space $\mathbf{r}$) may be equivalently described by a scalar potential $U(\mathbf{r})$ such that

$$\mathbf{E} = -\nabla U. \tag{2.32}$$

It can be shown that U is unique at a particular point $\mathbf{r}$. A charge moving from $\mathbf{r}_1$ to $\mathbf{r}_2$ sees a change in potential ('voltage') U_1 to U_2, and the work done is

$$W = q(U_1 - U_2). \tag{2.33}$$

A change in voltage ΔU gives a change in the kinetic energy of the particle $\Delta T = q\Delta U$. Note that here we adopt the convention U for voltage since we are using V for the volume; however, in the rest of this book we use V for voltage as usual.

If a positive charge is moving in the same direction as the electric field, it gains energy and the field $\mathbf{E}$ does work on the charge; if the charge moves opposite to the $\mathbf{E}$ field direction, then the charge does work on the field. The idea that an electric field $\mathbf{E}$ does work on a charge (or vice versa) implies that energy is exchanged from one to the other. For a distribution of charges moving in an electromagnetic field, we may calculate the power exerted upon (or by) those charges (which have charge density ρ) within an infinitesimal volume dV. This is just

$$P = \rho dV \mathbf{E} \cdot \mathbf{v} \tag{2.34}$$

$$= \mathbf{j} \cdot \mathbf{E}. \tag{2.35}$$

This is easy to see since the total charge in the volume dV is just $dq = \rho dV$, and since the current density (current is just moving charge) is $\mathbf{j} = \rho \mathbf{v} dV$ (if the volume is infinitesimal we may expect that the charges are all moving in the same direction and with the same speed). From this idea, we may now develop an energy conservation law. We start with two equations, Faraday's Law:

$$\mathbf{B} \cdot \frac{\partial \mathbf{B}}{\partial t} = -\mathbf{B} \cdot \nabla \times \mathbf{E} \tag{2.36}$$

and Ampere's Law:

$$\mu_0 \epsilon_0 \mathbf{E} \cdot \frac{\partial \mathbf{E}}{\partial t} = \mathbf{E} \cdot \nabla \times \mathbf{B} - \mu_0 \mathbf{j} \cdot \mathbf{E}. \tag{2.37}$$

Adding these two equations together gives

$$\mathbf{B} \cdot \frac{\partial \mathbf{B}}{\partial t} + \mu_0 \epsilon_0 \mathbf{E} \cdot \frac{\partial \mathbf{E}}{\partial t} = -\mu_0 \mathbf{j} \cdot \mathbf{E} - (\mathbf{B} \cdot \nabla \times \mathbf{E} - \mathbf{E} \cdot \nabla \times \mathbf{B}),$$

$$\frac{1}{2} \frac{\partial}{\partial t} \mathbf{B} \cdot \mathbf{B} + \mu_0 \epsilon_0 \frac{1}{2} \frac{\partial}{\partial t} \mathbf{E} \cdot \mathbf{E} = -\mu_0 \mathbf{j} \cdot \mathbf{E} - \nabla \cdot (\mathbf{E} \times \mathbf{B}). \tag{2.38}$$

Let's define a quantity that will be useful (now and later on)

$$\mathbf{S} = \frac{1}{\mu_0} \mathbf{E} \times \mathbf{B}. \tag{2.39}$$

$\mathbf{S}$ is called the Poynting vector.* With this definition of the Poynting vector we can re-express our equation above as

$$\frac{1}{2} \frac{\partial}{\partial t} (B^2 + \mu_0 \epsilon_0 E^2) = -\mu_0 \mathbf{j} \cdot \mathbf{E} - \nabla \cdot (\mu_0 \mathbf{S}). \tag{2.40}$$

This equation applies to an infinitesimal volume. Let's integrate over some volume V (also dividing through by μ_0 for convenience) which yields

$$\frac{d}{dt} \int_V \left(\frac{1}{2} \epsilon_0 E^2 + \frac{1}{2\mu_0} B^2 \right) dV = - \int_V \mathbf{j} \cdot \mathbf{E} dV - \int_V \nabla \cdot \mathbf{S} dV. \tag{2.41}$$

This is beginning to look a bit like an equation about energy, but we're not quite there yet. Our final step is to apply the divergence term to that last term in $\mathbf{S}$ (and also re-label those terms on the left-hand side) which gives

$$\frac{d}{dt} (U_E + U_B) = - \int_V \mathbf{j} \cdot \mathbf{E} dV - \oint \mathbf{S} \cdot d\mathbf{A}. \tag{2.42}$$

---

*This quantity is named after its inventor Henry Poynting, so don't mis-spell it as 'pointing vector' even though it's quite tempting to.

This is definitely an equation about energy conservation. The left-hand side is the rate of change of energy in the field, which is balanced by the two terms on the right-hand side. The first is the rate of work done on any charges moving in the volume (as we saw above), whilst the second is what power is flowing out of the surface surrounding the volume V. Note that this implies that there is power present in an electromagnetic field, and that the flux of energy (and its direction) is given by the quantity $\mathbf{S}$. The Poynting vector is pointing in the direction of energy flow. We may express our energy conservation equation in differential form as

$$\frac{\partial U}{\partial t} = \nabla \cdot \mathbf{S} + \mathbf{j} \cdot \mathbf{E}. \tag{2.43}$$

This is known as Poynting's Theorem.

2.3.3 Energy in Electromagnetic Plane Waves

We very often have to deal in accelerator physics with the behaviour of plane waves, and so it is instructive to consider the energy embodied in them. Faraday's Law and Ampere's law may be manipulated together to obtain two wave equations, that describe how each vary with position and time in free space (i.e. away from where the currents and charges may have generated them) as

$$\nabla^2 \mathbf{E} - \mu_0 \epsilon_0 \frac{\partial^2 \mathbf{E}}{\partial t^2} = 0, \tag{2.44}$$

$$\nabla^2 \mathbf{B} - \mu_0 \epsilon_0 \frac{\partial^2 \mathbf{B}}{\partial t^2} = 0, \tag{2.45}$$

solutions to which have the form of waves travelling at speed $c = 1/\sqrt{\mu_0 \epsilon_0}$. In a dielectric (non-conducting) material we have a modified permittivity $\epsilon_0 \to \epsilon \epsilon_0$ and permeability $\mu_0 \to \mu \mu_0$ where ϵ and μ are the relative permittivity and permeability characteristic of the particular dielectric; the material modifies the electric and magnetic fields to

$$\mathbf{D} = \epsilon \epsilon_0 \mathbf{E} \tag{2.46}$$

$$\mathbf{H} = \frac{1}{\mu \mu_0} \mathbf{B} \tag{2.47}$$

such that the wave equations become

$$\nabla^2 \mathbf{E} - \mu \mu_0 \epsilon \epsilon_0 \frac{\partial^2 \mathbf{E}}{\partial t^2} = 0, \tag{2.48}$$

$$\nabla^2 \mathbf{B} - \mu \mu_0 \epsilon \epsilon_0 \frac{\partial^2 \mathbf{B}}{\partial t^2} = 0. \tag{2.49}$$

In other words, in a dielectric an electromagnetic wave propagates at a lower speed

$$v = \frac{1}{\sqrt{\mu \mu_0 \epsilon \epsilon_0}} = \frac{c}{\sqrt{\mu \epsilon}} = \frac{c}{n}, \tag{2.50}$$

where the refractive index n of the dielectric is given by $n = \sqrt{\mu \epsilon}$. Very often $\mu \simeq 1$ and can be omitted from equations, and the permittivity can be frequency-dependent – i.e. $\epsilon = \epsilon(f)$ – which leads to the important property of wave dispersion; dispersion is where waves of different frequencies propagate at different velocities. In most accelerator applications we are dealing with electromagnetic waves propagating in a very good vacuum (much less than 1 mbar), so that $\epsilon = \mu = 1$ to a very good accuracy; however, in waveguides (see Chapter 3) the effective wave velocity can be much lower (see below), and in dielectric and plasma accelerators the driving laser will propagate with a modified velocity due to the material being traversed.

A plane wave is a simple solution to the wave equations that has the form

$$E_x = E_x(z,t),$$
$$E_y = E_y(z,t); \tag{2.51}$$

in other words, there is no variation of the field in the x and y directions. Solutions of the form

$$E_x = f(z - ct) \quad \text{or} \quad E_x = g(z + ct) \tag{2.52}$$

are allowed where f and g are arbitrary functions (similar equations may be written for E_y). The first solution describes a disturbance moving in the $+z$ direction whilst the second describes a disturbance moving in the $-z$ direction (see illustration in Fig 2.4). We may build up any function $f(z - ct)$ or $g(z - ct)$ in terms of different-frequency components

$$\mathbf{E} = \mathbf{E_0}e^{i(\omega t \pm kz)} \tag{2.53}$$

where

$$\mathbf{E_0} = \begin{pmatrix} E_{x0} \\ E_{y0} \end{pmatrix}. \tag{2.54}$$

Again, components of the form $(\omega t - kz)$ describe disturbances moving in the $+z$ direction, whilst components of the form $(\omega t + kz)$ describe disturbances moving in the $-z$ direction. We define the dispersion relation for a particular frequency component ω as

$$v = \frac{\omega}{|k|} = \frac{1}{\sqrt{\mu\mu_0\epsilon\epsilon_0}}. \tag{2.55}$$

k is the wavenumber (or 'wavevector') with an associated wavelength $\lambda = 2\pi/k$.

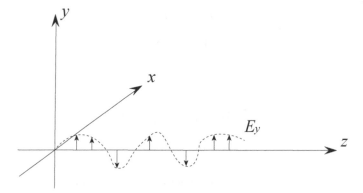

FIGURE 2.4 Illustration of an arbitrary electric field $E_y = E_y(z,t)$, which will satisfy the wave equation if of the form $E_y = f(z - ct)$ or $E_y = g(z + ct)$ for any functions f and g; f and g in turn are described in terms of their frequency components where each frequency ω may propagate at a different velocity according to the particular dispersion relation for that material, leading in general to dispersion (separation over distance) of the different frequency components.

Similar to the electric field, we may define plane wave solutions for the magnetic field as

$$\mathbf{B} = \mathbf{B_0}e^{i(\omega t \pm kz)}. \tag{2.56}$$

However, we already know that varying magnetic and electric fields are coupled together through Faraday's Law as

$$\nabla \times \mathbf{E} = -\frac{\partial \mathbf{B}}{\partial t}. \tag{2.57}$$

Substituting in separately our two solutions for the electric and magnetic fields into Faraday's Law, we obtain

$$i\omega(B_{x0}\hat{\mathbf{x}} + B_{y0}\hat{\mathbf{y}})e^{i(\omega t - kz)} = \begin{vmatrix} \hat{\mathbf{x}} & \hat{\mathbf{y}} & \hat{\mathbf{z}} \\ \frac{\partial}{\partial x} & \frac{\partial}{\partial y} & \frac{\partial}{\partial z} \\ E_{x0} & E_{y0} & 0 \end{vmatrix}$$

$$= \hat{\mathbf{x}}(-ikE_{y0}) + \hat{\mathbf{y}}(-ikE_{x0}), \tag{2.58}$$

where $\hat{\mathbf{x}}$ and $\hat{\mathbf{y}}$ are unit vectors transverse to the direction of motion of the electromagnetic wave. Matching terms in $\hat{\mathbf{x}}$ and $\hat{\mathbf{y}}$, we find

$$B_{x0} = \frac{k}{\omega} E_{y0}, \tag{2.59}$$

$$B_{y0} = \frac{k}{\omega} E_{x0}. \tag{2.60}$$

The electric field in a given direction is coupled to the magnetic field at 90° to it. We may combine these two equations into one as

$$\mathbf{B}_0 = \frac{k}{\omega} \hat{\mathbf{z}} \times \mathbf{E}_0, \tag{2.61}$$

where $\hat{\mathbf{z}}$ is a unit vector along z. We see straightforwardly that

$$\mathbf{E} \cdot \mathbf{B} = -\left(\frac{k}{\omega} E_{y0}\right) E_{x0} + \left(\frac{k}{\omega} E_{x0}\right) E_{y0} = 0. \tag{2.62}$$

Hence $\mathbf{E}$ is perpendicular to $\mathbf{B}$ and

$$B_0 = \frac{k}{\omega} E_0 = \frac{E_0}{c}. \tag{2.63}$$

We see therefore that the two field components in an electromagnetic wave are coupled together as they travel. But what are their typical relative magnitudes? As an example, we consider a radio antenna emitting electromagnetic radiation which at some distance has a peak electric field strength of $E_0 \simeq 3 \times 10^{-3}$ Vm^{-1} = 3 mVm^{-1}. This electric field – which whilst small is still measurable – is much, much bigger than the corresponding magnetic field in the same region, where $B_0 \simeq 10^{-11}$ T. A contrasting situation is that of a high-power laser that may drive a wakefield particle accelerator. It turns out that the typical electric field strength at the laser focus (which then drives the particle acceleration[*]) has values that may readily exceed $E_0 \simeq 10^9$ Vm^{-1} = 1 GVm^{-1}. In this situation the corresponding magnetic field is quite large – with $B_0 \simeq 3$ T. As we will see in Chapter 4 such fields are quite challenging to generate using electromagnets, but arise naturally at the focus of a very strong laser pulse.

We have carried out our derivation linking the electric and magnetic fields by assuming that both are travelling along the z direction. However, it should be obvious that we may

---

[*]A nice example of how the energy in an electromagnetic field can do work on charges and thereby pass energy to them.

equivalently have a plane-polarised wave in an arbitrary direction given by the unit vector $\hat{\mathbf{k}}$. Now, the electric and magnetic fields may be written as

$$\mathbf{E}(\mathbf{r}, t) = \mathbf{E}_0 e^{i(\omega t - \mathbf{k} \cdot \mathbf{r})}, \tag{2.64}$$

$$\mathbf{B}(\mathbf{r}, t) = \frac{1}{c}\hat{\mathbf{k}} \times \mathbf{E} = \frac{1}{\omega}\mathbf{k} \times \mathbf{E}. \tag{2.65}$$

We now consider a plane electromagnetic wave (in a vacuum) travelling in the $\hat{\mathbf{z}}$ direction and assume a linearly-polarised electric field that lies in the x plane, so that

$$\mathbf{E} = E_0\hat{\mathbf{x}}\cos(\omega t - kz), \tag{2.66}$$

$$\mathbf{B} = B_0\hat{\mathbf{y}}\cos(\omega t - kz), \tag{2.67}$$

and where $B_0 = E_0/c$ as shown above; we note that $\mathbf{E}$ and $\mathbf{B}$ oscillate in phase with each other, which we haven't pointed out before now but which is generally true in a vacuum. We see from the definition of $\mathbf{E}$ and $\mathbf{B}$ that the (volumetric) energy density at a given value of z is just given by

$$U_E = \frac{1}{2}\epsilon_0 E_0^2 \cos^2(\omega t - kz), \tag{2.68}$$

$$U_B = \frac{1}{2\mu_0}B_0^2 \cos^2(\omega t - kz). \tag{2.69}$$

It is left as an exercise for the reader to confirm that $U_E = U_B$. In other words, in a plane electromagnetic wave there is equal energy contained in the $\mathbf{E}$ and $\mathbf{B}$ fields, despite the very large disparity in the magnitudes of the actual field strengths. Given that the two energy densities are the same, we may combine them to obtain the total energy in the electromagnetic wave,

$$U = U_E + U_B = \epsilon_0 E_0^2 \cos^2(\omega t - kz). \tag{2.70}$$

U varies both as a function of time t (at a given z) and as a function of position z (at a given t); this is illustrated in Fig 2.5. We may readily calculate the time average of U as

$$\langle U \rangle = \epsilon_0 E_0^2 \left\langle \cos^2(\omega t - kz)\right\rangle = \frac{1}{2}\epsilon_0 E_0^2 = \epsilon_0 E_{rms}^2. \tag{2.71}$$

$\langle U \rangle = \epsilon_0 E_{rms}^2$ is obtained because $E_{rms} = E_0/\sqrt{2}$. Note the various factors of 2 that appear and disappear in these expressions, so care must be taken.

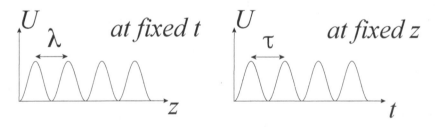

FIGURE 2.5   Illustration showing how the energy density U varies either with position (at a given time) or with time (at a given location); the energy density at a fixed location is not constant, but varies with time.

The average energy density of the electromagnetic wave $\langle U \rangle$ has units of Jm^{-3}. But we also know that, since it's a wave, it is moving at velocity c. Hence the energy flux (rate of energy motion) has units $\text{Jm}^{-3}\times\text{ms}^{-1} = \text{Jm}^{-2}\text{s}^{-1}$. Above, we showed that the Poynting vector $\mathbf{S} = (\mathbf{E} \times \mathbf{B})/\mu_0$ was the energy flux of an electromagnetic wave (as we see, pointing in a direction perpendicular to both $\mathbf{E}$ and $\mathbf{B}$. In the case here of a plane electromagnetic wave we have that

$$\mathbf{S} = \frac{1}{\mu_0} E B \hat{\mathbf{z}} = \frac{1}{\mu_0} E_0 B_0 \cos^2(\omega t - kz) \hat{\mathbf{z}}. \qquad (2.72)$$

$\hat{\mathbf{z}}$ is the direction of energy flow, which is the same as the direction of wave propagation. Averaging over time, we see that

$$\langle S \rangle = \frac{1}{\mu_0} E_0 B_0 \frac{1}{2} \qquad (2.73)$$

where the factor $1/2$ comes from the time average of the $\cos^2$ term. We may then substitute and re-arrange to obtain

$$\langle S \rangle = \frac{1}{2\mu_0} \sqrt{\mu_0 \epsilon_0} E_0^2 = \frac{1}{\sqrt{\mu_0 \epsilon_0}} \frac{1}{2} \epsilon_0 E_0^2 = c \langle U \rangle \qquad (2.74)$$

since $c = 1/\sqrt{\mu_0 \epsilon_0}$ and $\langle U \rangle = \epsilon_0 E_0^2 / 2$. This is a nice result, since it says that an electromagnetic wave with energy density $\langle U \rangle$ transfers that energy to another location at velocity c, which is what we would expect. The energy flux is

$$\langle S \rangle = c \langle U \rangle. \qquad (2.75)$$

2.3.4 Radiation Pressure

We have just derived an expression that relates the energy density of an electromagnetic wave to its energy flux $\mathbf{S}$ (rate of energy flow from one place to another); we did this for a plane electromagnetic wave but it also applies in other situations. Electromagnetic waves of various sorts transfer energy at a speed c, and include such practical devices as TV and radio transmitters, mobile phones (which are of course just miniature transceivers*), microwave ovens (that transmit energy from an electromagnetic wave generator into a target – your dinner), and more esoteric devices such as ray-guns.

We realise that electromagnetic waves carry not only energy but also momentum. Here, we think of the electromagnetic wave as being composed equivalently as a fluence of photons.* Of course, we know that for any particle its energy E is

$$E^2 = p^2 c^2 + m_0^2 c^4, \qquad (2.76)$$

and that for photons $m_0 = 0$ and hence $E = pc$, so that for a given energy E we have $p = E/c$. With this idea, we can consider a volume of space that contains an electromagnetic wave (that has an energy density) and from that define a momentum density which we will label P_a to distinguish it from the other variables also labelled with a 'p'. Momentum has units kgms^{-1}, so that the momentum density P_a must have units kg m s^{-1}/m^3 =kg m^{-2} s^{-1}.

Since $p = E/c$ for an individual photon, we can readily write down that the magnitude of the momentum density is

$$|P_a| = \frac{\langle U \rangle}{c} = \frac{\langle S \rangle}{c^2}. \qquad (2.77)$$

---

*A transceiver is a device that both transmits and receives.
*The fluence of something is the number passing through a given area per unit time, as opposed to the flux which is the total quantity of something such as energy that passes through a given area per unit time.

We can now define a radiation pressure, which must be related to the momentum transferred by the electromagnetic wave when incident upon some area. Pressure = Force/Area (as we know very well); to determine that pressure, we first calculate the total momentum transferred in some time Δt through some surface A (see Fig 2.6). The volume of electromagnetic field that passes through A is just $V = Ac\Delta t$, so that the total momentum p_T (the impulse) transferred through A is

$$p_T = P_a Ac\Delta t \tag{2.78}$$

(momentum density × volume). But the impulse p_T is just $p_T = F\Delta t$, where F is the total force acting over the surface A. In other words

$$F = P_a Ac. \tag{2.79}$$

The radiation pressure P_r may then be simply obtained as

$$P_r = \frac{F}{A} = P_a c = \langle U \rangle. \tag{2.80}$$

The radiation pressure is equal to the energy density – an important result!

$$c\Delta t$$

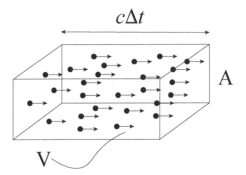

$$A$$

$$V$$

FIGURE 2.6   Illustration showing how a set of photons of momentum $p = E/c$ may impart a radiation pressure on a surface A. The volume traced out by the photons in a time Δt is $V = Ac\Delta t$.

We can summarise the relationship between radiation pressure P_r and the electromagnetic field quantities as

$$P_r = \langle U \rangle = \frac{|\langle \mathbf{S} \rangle|}{c} = \frac{1}{\mu_0 c} |\langle \mathbf{E} \times \mathbf{B} \rangle|. \tag{2.81}$$

Examples of Radiation Pressure

Our first example of radiation pressure is that of an electromagnetic plane wave incident upon a perfectly reflecting mirror. We recall that since the photons bounce off the mirror and then travel backwards, the momentum transferred is twice what it would be if they were just absorbed. Hence the radiation pressure is

$$P_r = 2\frac{|\langle \mathbf{S} \rangle|}{c}. \tag{2.82}$$

This is an important phenomenon in particle accelerators. For example, an accelerating RF cavity will experience a force on its walls due to the electromagnetic waves that are confined within it deforming its shape and changing the resonant frequency, known as Lorentz force detuning; the conducting walls act as a mirror to the photons within the cavity. One counterintuitive – yet true – consequence of this is the radiation pressure within an ordinary microwave oven. With the door closed, there is equal pressure on all the interior surfaces of the oven (including the reflecting door) due to the ~2.45 GHz photons trapped inside. If the door interlock is over-ridden so that the microwave power is still fed in whilst the door is open, then the photons coming out through the door will no longer be reflected and there will be a net 'thrust' on the microwave oven; an open microwave oven in space will move if power is fed into it. The corollary of this is that so-called reactionless rockets (where the idea is of an enclosed cavity providing thrust) cannot possibly work – they would violate conservation of momentum.

Another example is that of an intense laser pulse. We consider a typical CO_2 laser pulse with wavelength $\lambda \sim 10$ μm$= 10^{-5}$ m, pulse length $\tau \sim 10$ ns, beam radius ~1 cm, and pulse energy of 100 J. The energy density in such a pulse $\langle U \rangle$ is

$$\langle U \rangle = \frac{100}{\pi r^2 c\tau} \simeq 10^5 \text{ Jm}^{-3}. \tag{2.83}$$

Therefore

$$\langle S \rangle = c \langle U \rangle \simeq 3.2 \times 10^{13} \text{ Wm}^{-2}, \tag{2.84}$$

and the radiation pressure is

$$P_r = \frac{\langle S \rangle}{c} \simeq 10^5 \text{ Nm}^{-2}. \tag{2.85}$$

This pressure is acting over a 1 cm diameter spot focus, which means the total force is about 32 N. In other words, for 10 ns the laser pushes on that spot focus with the weight of a 3 kg object. This feature of laser pulses is important in laser-driven acceleration, since the intense radiation pressure from the photons falling onto a (thin) target can be sufficient to push the target away from its original position.

2.4 The Basics of Acceleration

The purpose of a particle accelerator is to deliver particles with a chosen amount of kinetic energy; those particles are usually in the form of a beam, i.e. a 'stream' of particles extended over time. We saw that charged particles may have their kinetic energy increased by means of an electric field. The simplest situation is that of a potential difference through which a charge travels; for example, a negatively-charged electron will accelerate towards a positive potential (see Fig 2.7). The electron volt (eV) is defined as that energy gained by a unit charge $e = 1.602 \times 10^{-19}$ coulomb crossing a potential difference of one volt; 1 eV is equal to 1.602×10^{-19} J. The electron-volt is the standard unit of measure in particle accelerators, although we typically work with MeV (million eV) or GeV (billion eV) energies. The kinetic energy gained is $\Delta E = qV$.

Particles can be accelerated with any suitable electric field. In the earliest accelerators a static DC potential difference was used (see Chapter 3), but today the predominant method is to utilise time-varying, oscillatory voltages created in resonant cavities; the requirement for particles to pass at the right time to be in phase with this oscillatory voltage is why most accelerators deliver bunched beams of particles. These cavities typically have resonant frequencies in the radio frequency (RF) part of the electromagnetic spectrum and are therefore known as RF cavities; these are described in detail in the next chapter. RF cavities obtain peak electric fields that are limited to ~200 MV/m, resulting in an average accelerating field of ~100 MV/m, and for higher accelerating fields there has been significant interest in inducing charge separation in plasmas – for example using an intense laser pulse to separate the plasma electrons from the ions – and thereby create a transient electric field exceeding 1 GV/m in some cases. We outline this method too. Very often, however, such large gradients are not required since we may recirculate the particles within a circular accelerator to use a smaller voltage multiple times; for example, in the cyclotron, protons are typically accelerated across a ~100 kV Dee gap and make around 1000 revolutions before being extracted with a kinetic energy of around 100 MeV or more (see Fig 2.8).

The cyclotron was the first circular accelerator – developed originally in the 1930s by Ernest Lawrence and M. Stanley Livingston – and relied on Ernest Lawrence's great insight about the bending radius of a classically-moving charged particle. We saw above that the bending radius ρ (say, of a proton) in a magnetic field is

$$\rho = \frac{mv}{qB}. \tag{2.86}$$

The time taken for one orbit in the cyclotron is $t_r = 2\pi r/v$, so that the proton gyrates in the field B at the cyclotron frequency

$$f_c = \frac{1}{2\pi} \frac{qB}{m}. \tag{2.87}$$

For protons moving transversely to a magnetic field of 1 T, we have $f_c \simeq 15.3$ MHz. We see immediately that the cyclotron frequency is independent of the velocity – as long as the mass of the particle doesn't change; this is highly advantageous as it allows a constant-frequency signal generator to be used to feed the voltage at the cyclotron Dees (Fig 2.8). This, in turn, allows the use of modest accelerating voltages, today typically tens of kilovolts. Another important observation is that the size (i.e. diameter) of a cyclotron scales $\propto 1/B$; larger magnetic fields give a smaller accelerator. Reducing the size of an accelerator is a common aspiration; for a linear accelerator this means maximising electric field gradient or for a circular accelerator a large B is beneficial.

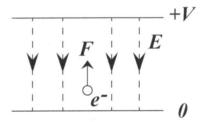

FIGURE 2.7   Illustration of how a charge changes energy due to a voltage difference from 0 to $+V$. Here, an electron (with negative charge q) is accelerated upwards by the force $F = qE$; crossing from one voltage to the other gives an energy gain $\Delta E = qV$.

As a particle is accelerated, its mass increases as $m = \gamma m_0$, and the cyclotron will no longer work isochronously (i.e. with a constant-frequency RF acceleration); indeed, electrons with kinetic energies of even a few hundred keV are moving close to c, and so effectively there is no such thing as an electron cyclotron (although there are such things as ECR – electron cyclotron resonance – ion sources). Above about $\gamma = 1.3$ we must change the accelerating (RF) frequency to maintain synchronism with the accelerating bunches; in a synchrocyclotron the Dee frequency matches the revolution frequency, but only of one accelerated bunch at a time – the maximum bunch extraction rate is therefore the rate at which the Dee frequency can be ramped up and down, typically about 1 kHz. In 1945 Vladimir Veksler and Ed McMillan independently realised the principle of phase stability [5, 6], and this was demonstrated in 1946 on the first synchrocyclotron – adapted from the earlier 37-inch cyclotron at Berkeley.

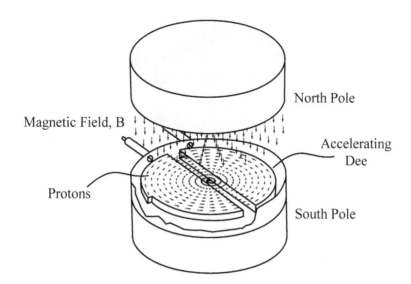

FIGURE 2.8 Illustration of the (classical) cyclotron, where the upper (N) pole has been raised to show the internal layout. A uniform vertical magnetic field B created by the N and S poles confines protons (orbiting horizontally within the vacuum vessel) into a circular path of radius $\rho = mv/qB$. At each Dee gap crossing, the protons gain an energy qV for a Dee voltage V; the voltage polarity must therefore be swapped at each side of the Dee crossing, meaning that the Dee frequency is the same as the cyclotron frequency f_c (or it can be some integer multiple h of it). As the protons accelerate they gain energy and increase in radius ρ, but retain the same f_c as long as their mass does not increase significantly. Many bunches at different energies and radii can co-exist simultaneously in such a cyclotron, each bunch eventually being extracted at the outer radius of the magnet.

The synchrotron improves upon the synchrocyclotron by also varying the magnetic field $B = B(t)$ with time; here, the path of the particles through the magnets is kept constant as the particle energy increases and the RF is matched to be $f_{RF} = hf_r$ where f_r is the (orbital) revolution frequency and the harmonic number h is an integer. An illustration is given in Fig 2.9. The betatron – invented by Donald Kerst also in the 1930s – is similar in that it circulates charged particles (here electrons) at a constant radius, but uses induction acceleration via an e.m.f. generated as the magnetic field itself varies. Frank Goward and D. E. Barnes adapted a betatron to build the first synchrotron in 1946 at Woolwich (London) which accelerated 8 MeV electrons, and the following year an electron synchrotron at General Electric's laboratory demonstrated the production of synchrotron radiation (see Chapter 6). By maintaining a constant beam path that is independent of particle energy, the magnet sizes can be enormously reduced particularly at high energies enabling the very largest colliders such as the LHC to be produced with a realistic cost. The other great advance made around the same time (in 1949) was Nicholas Christofilos's strong-focusing principle, which allows the circulating beam size to be greatly reduced, making the magnets much smaller again; this is discussed later in Chapter 5.

2.5   The Particles Used in Accelerators

Since this book is all about particle accelerators we also need to consider which particles to use. Any charged particle can be accelerated using an electric field; in the broadest sense

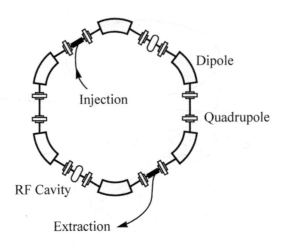

FIGURE 2.9   Illustration of a synchrotron, which uses a number of dipole magnets whose field strength B varies with time; the momentum of the particle $p = q\rho B$ follows the magnetic field strength. Strong focusing – either within the dipoles or here using additional quadrupoles – provides a small beam envelope and thereby a small magnet aperture. Injection and extraction is typically done with pulsed magnetic elements but may also be performed with so-called stripper (charge exchange) foils.

they behave similarly. The differences lie in: the convenience with which they may be obtained; their mass and charge, which determines the acceleration for a given electromagnetic field; and whether they are stable. The most common particle to accelerate therefore is the electron, since it is relatively easy to liberate electrons from a surface by simply heating it and applying a voltage to it; they are the lightest charged particle, and so are also the easiest to make relativistic. We should mention here the positron, the antimatter pair to the electron. From the accelerator's point of view they behave exactly the same, except that since they have the opposite charge ($+e$) they require the opposite polarity for all the fields; this is mostly readily achieved by 'swapping the connections' on all the power supplies. To make protons we can either directly use a radioactive source of β^+ particles (such as sodium-22) or create them in larger numbers using pair production in a suitable target but these methods are not trivial; since positrons for the most part give similar phenomena in accelerators, we rarely use them and instead prefer electrons. The most common use of positrons in an accelerator science is in electron-positron colliders where both particles are accelerated and then made to collide into each other for fundamental particle physics studies.

The second most common particle to accelerate is the proton. As we saw earlier in this chapter, they are much more massive than an electron ($m_p/m_e \simeq 938$ MeV$/0.511$ MeV $= 1836$) and so making them relativistic is harder; we must also account for the varying velocity as described earlier. Protons can be generated in an ion source by ionising hydrogen gas with a suitable large voltage discharge; ion sources are briefly described in Chapter 3. H$^-$ ions are also often used, as they can allow more intense beams to be more efficiently injected or extracted in an accelerator system; a thin stripper foil (perhaps of graphite, aluminium oxide or other robust material) can be placed into the H$^-$ beam causing the electrons to be lost but transmitting most of the remaining protons.

Other particles which may be accelerated include atomic ions, for example, carbon ions for particle radiotherapy or heavy ions such as gold, lead, or uranium for nuclear physics applications. Since the atoms of all the elements, apart from hydrogen, have more than one electron it is possible for ions to have multiple charges. In other words, since lithium has three electrons it is possible to create lithium ions in three different positive charge states (Li^+, Li^{2+}, and Li^{3+}) depending on how many electrons are removed from the atom. Within the same electric field, these three ion states will gain kinetic energy proportional to their charge state. This ability to impart greater kinetic energy to higher charge states is taken advantage of by choosing to work with extreme cases such as $^{238}Ur^{73+}$. It is usual for the kinetic energy gained in an accelerator by an ion to be quoted per nucleon in units of MeV/u (a nucleon is a proton or a neutron, so there are 238 nucleons in this ion). We generally ignore the small difference between the atomic mass unit, u, and the actual nucleon mass for the ion.

Finally, we give an example of exotic particle acceleration: the muon. An elementary particle similar to the electron, with the same charge but about 207 times the mass. At the same kinetic energy, muons radiate far less synchrotron radiation (a factor 207^4 less – see Chapter 6) making a muon-muon collider an attractive prospect. However, at rest, muons have a lifetime of only around 2.2 μs before they decay, and so must be accelerated rapidly to large γ to extend their lifetime via time dilation. No one has yet decided to build such a collider. A possible first step, under consideration, would be to build a muon storage ring to generate intense beams of neutrinos, also for fundamental particle physics measurements.

2.6 The End of ABC

This concludes our introduction to the field of accelerators, our ABC. In the following chapters we discuss the principles of the common elements used in nearly all accelerators – the RF acceleration, the magnet systems, the beam dynamics needed to understand and specify these systems, the radiation the particles may produce and what happens when we have many particles in our bunch. We shall start with the heart of any accelerator, the accelerating structure!

References

1. I. S. Grant and W. R. Phillips. *Electromagnetism (2nd edition)*. Wiley, 1990.
2. G. Bekefi and A. H. Barrett. *Electromagnetic Vibrations, Waves, and Radiation.* Massachusetts Institute of Technology, 1977.
3. P. Lorrain, D. P. Corson, and F. Lorrain. *Electromagnetic Fields and Waves (3rd edition)*. W. H. Freeman and Company, 1988.
4. J. D. Jackson. *Classical Electrodynamics (3rd edition)*. Wiley, 1998.
5. Edwin M. McMillan. The synchrotron—a proposed high energy particle accelerator. *Phys. Rev.*, 68:143–144, Sep 1945.
6. V. I. Veksler. A new method of acceleration of relativistic particles. *J. Phys.*, 9:153–158, 1945.

3

Acceleration

The vast majority of particle accelerators are designed to increase the kinetic energy of charged particles, usually in the form of a particle beam. This is performed by placing those charged particles within a suitable electric field. In this chapter we look at several ways of applying electric fields to charged particles to provide efficient and stable acceleration. First we will examine electrostatic accelerators and their limitations, before moving on to radio-frequency (RF) accelerators.

DOI: 10.1201/9781351007962-3

3.1   Electrostatic Accelerators

The simplest type of particle accelerator can be constructed from two metal plates, an anode and a cathode, separated by a vacuum section and held at different electric potentials by some external voltage source. This simple idea is the foundation of many low-energy or early particle accelerators. There are three possible configurations for creating the very high voltages required for MeV scale accelerators:

1. Van de Graaff Generators (invented by Robert Van de Graaff in 1929), which transfer charge to a high-voltage terminal via a belt;

2. DC power converters which convert a low voltage and high current, to a high voltage at low current;

3. pulsed modulators, which store energy in capacitors or inductors and release it quickly at a higher voltage.

3.1.1   DC Power Converters

DC power converters operating up to 600 kV are readily available at GW of power in the electricity industry and have undergone significant development in recent years for high-voltage DC transmission. Typically, the input to the system is a 3-phase AC input* which is the common method of supplying power from a national electricity grid at large currents and voltages to high-power machinery. The AC signal will be rectified to DC using a full-wave diode rectifier. However, simply using a rectifier would have too much power ripple so a low-pass filter is also necessary to remove the AC frequency and higher harmonics from the output. The size of capacitors and inductors required for the smoothing is inversely proportional to the AC frequency, hence a higher AC frequency is often used. In order to do this, an AC-AC bridge converter is used. In this device switchable diodes or switches such as thyristors are used to first rectify the input AC frequency to a DC signal and then fast switches are used to chop to AC at a higher frequency. In order to create a higher DC voltage than the input voltage, a boost converter is used, as shown in Fig 3.1. Here the load resistance is placed in series with a large inductance. A switch is placed in parallel with the resistance such that, when closed, the current will bypass the load and the inductor will draw a high current. When the switch is opened the load will draw current from both the input supply and from the discharging inductor creating a higher voltage. A capacitor can also be used in parallel with the load to smooth the voltage out so that the load sees a roughly constant voltage. The ratio of the output voltage, V_{out}, to input voltage, V_{in}, is equal to the time the converter is in the off state (when the switch is open), T_{off}, divided by the switching period, T [1],

$$V_{out} = V_{in}\frac{T}{T_{\text{off}}}. \tag{3.1}$$

A DC-DC converter is limited by the maximum voltage that the switch can handle. If particle energies higher than 600 keV are required, then a Cockcroft-Walton voltage multiplier can be utilised [2]. This uses an AC supply or pulsed DC to charge an arrangement of capacitors and diodes, as shown in Fig 3.2. During the first half cycle the first capacitor charges when a negative voltage is applied over it. The diodes ensure that the second capacitor is isolated during this step. In the 2nd half cycle, the polarity is reversed and the 2nd capacitor is charged by the AC supply and the discharging 1st capacitor, thereby

---

*For example, in the UK the 3-phase supply is 415 V at 50 Hz.

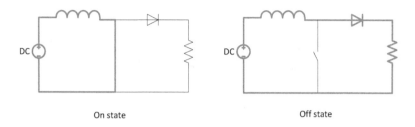

FIGURE 3.1   Circuit diagram of a boost converter showing the 'on' state and the 'off' state.

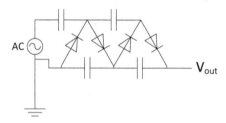

FIGURE 3.2   Circuit diagram of a Cockcroft-Walton voltage multiplier.

providing twice the charging voltage. As the second capacitor only has a positive voltage applied across it, it will have a DC output if the time constant of the capacitor discharging is longer than the switching period. Multiple stages can be provided so that the multiplication increases with each stage. This concept was first utilised in 1932 by John Cockcroft and Ernest Walton in the first nuclear disintegration experiments using 1 MeV beams. They are still in use today in many proton and ion accelerators. The output voltage for a Cockcroft-Walton with n capacitors, with a supply peak-to-peak voltage of V_{pp} is given by [3]

$$V_{out} = \left(\frac{n}{2}\right) V_{pp} - \left[\frac{n^3}{12} + \frac{n^2}{4} - \frac{n}{6}(3D^2 - 3D + 1)\right] \frac{T}{C} I_{out}, \tag{3.2}$$

where T is the switching period, D is the duty cycle ($D = T_{on}/T$), C is the capacitance and I_{out} is the current drawn by the accelerator.

3.1.2   Pulsed Modulators

Another method of generating high voltages is to store energy in a capacitor bank over a long period of time and discharge it over a shorter timescale providing a high voltage and current simultaneously for a short period. Several MW or even GW of power in nanosecond to microsecond pulses can be created using this method. The most common topology for this is the Marx bank generator, invented by Edwin Otto Marx in 1924 [4]. In a Marx bank, a number of capacitors are charged in parallel from a DC supply using the circuit shown

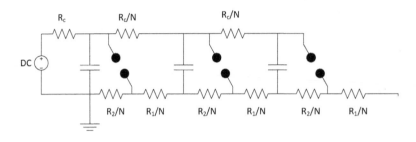

FIGURE 3.3   Circuit diagram of a Marx bank generator, with N stages.

in Fig 3.3. A number of switches are also employed such that, when closed, the capacitors become connected in series rather than parallel so that they can supply a voltage equal to the supply voltage multiplied by the number of capacitors. In many configurations, spark gaps are used instead of switches such that when the capacitors reach a given voltage they automatically conduct providing the series connection. Another type of pulsed modulator is the line type modulator, where the energy is stored in a transmission line, made of a network of capacitors and inductors or coaxial line, such that a square pulse is produced of duration equal to twice the line length divided by the velocity of the pulse on the line.

3.2   Particle Emission

At the start of all accelerators is a source of charged particles, either electrons, protons, ions or negative ions. These sources can provide a continuous or pulsed emission as required.

3.2.1   Electron Emission

In order to accelerate a beam of charged particles we must first obtain charged particles in vacuum. We cannot create charged particles from nothing so they must be either moved from somewhere or created in a nuclear or ionising reaction from neutral particles or in pair production. Electrons most commonly are emitted from a metal or semiconducting cathode. Conducting metals or doped semiconductors contain a number of free electrons in the conduction band; however, these are not able to escape the material due to a finite work function, which is the energy required above the Fermi level (which is the energy of the highest conduction band) to remove an electron from the material to the surrounding vacuum. To create free electrons in a vacuum we must provide enough energy to the electrons to allow them to overcome the work function. This can be achieved by either heating the emitter or with photons via the photoelectric effect. Emission via heating the emitter is known as *thermionic emission* and was first observed by Edmond Becquerel in 1853, and the British physicist Owen Willans Richardson received the Nobel Prize in 1928 for his pioneering work on the subject and the development of the Richardson law, which gives the current density, J, from an emitter as a function of the cathode temperature, T, and work function, ϕ_W, as

$$J = AT^2 e^{-\phi_W/kT}, \tag{3.3}$$

where T is the cathode temperature, ϕ_W is the work function of the material (for example, copper has $\phi_W \simeq 4.7$ eV), and $k = 1.38 \times 10^{-23}$ JK^{-1} is Boltzmann's constant. A is a material-specific constant that has typical values $\sim 3 - 17 \times 10^5 \mathrm{Am}^{-2}\mathrm{K}^{-2}$, and is equal to $\sim 6 \times 10^5 \mathrm{Am}^{-2}\mathrm{K}^{-2}$ for tungsten.

Unfortunately the Richardson equation is only true for low beam currents or high temperatures due to the space charge of the emitted electron beam. Emitted electrons repel electrons near the surface reducing the current that can be emitted. This effect can be overcome, however, by putting the emitter at a negative potential with respect to an anode, thereby creating an electric field which accelerates the electrons from the emitter (cathode) to the anode. By applying the electric field of a continuous electron beam to Maxwell's equations, Child and Langmuir derived an equation for the maximum current density, J (in $\mathrm{A/m^2}$), which can be emitted from a cathode as a function of the applied potential difference, V, and gap, d, between the anode and the cathode [5]. For two parallel plates, the Child-Langmuir law is

$$J = \frac{4}{9}\epsilon_0 \left(\frac{2e}{m_e}\right)^{1/2} \frac{V^{3/2}}{d^2} = 2.33 \times 10^{-6} \frac{V^{3/2}}{d^2}. \tag{3.4}$$

It is common to provide a constant of proportionality between the current and voltage, known as the perveance, P, with units of Perv such that

$$I = PV^{3/2}, \tag{3.5}$$

where for parallel plates, the perveance is given by

$$P = \frac{2.33 \times 10^{-6} A_e}{d^2} \tag{3.6}$$

and A_e is the emission area.

Hence, we arrive at two regimes of electron emission, each limited by one of the two equations above: temperature-limited emission (for high voltages where the space-charge does not limit the emission current density), and space-charge limited emission (for high temperatures where the temperature doesn't limit emission). In space-charge limited emission we can turn the emission of electrons on and off by modulating the applied voltage. Typically for thermionic emission the current density is limited to a few $\mathrm{A/cm^2}$ (typically around 10 $\mathrm{A/cm^2}$) to ensure the cathode doesn't degrade too quickly due to high temperature operation.

Electrons can also be emitted using the photoelectric effect known as photo-emission, where a photon is absorbed and an electron is emitted as a consequence. The process is quantified by the quantum efficiency η of the photocathode which is the average number of electrons emitted for each incident photon; normally $\eta \ll 1$. The quantum efficiency depends on the photocathode material, laser wavelength, accelerating field at the photocathode and the vacuum environment. Photocathodes can be metals or semiconductors. Metal photocathodes have long lifetimes and are very simple but have very low quantum efficiency; for example, copper or molybdenum photocathodes both have $\eta \simeq 0.001$ %. Semiconductor photocathodes such as GaAs or Cs_2Te can have orders of magnitude higher quantum efficiency $\eta \sim 10$ %, but their lifetimes are lower such that their quantum efficiency can drop to a few % in a matter of days [6].

It is also possible to emit electrons from a cathode via quantum tunnelling through the potential barrier created by the work function; this is known as Fowler-Nordheim tunnelling, or more commonly in the accelerator community as field emission. The potential barrier is normally very wide; however, if we apply a potential difference between an anode and a cathode the potential must go linearly from the work function at the cathode to the work function minus the potential difference at the anode. When the potential across the vacuum gap drops below the Fermi level it is possible for electrons to tunnel across the distance to this point from the cathode. The higher the potential difference, or the smaller the gap between the anode and cathode, the smaller the distance the electrons need to tunnel and the higher the probability of an electron tunnelling. Electron emission via this mechanism is known as field emission. If the electric field at the cathode is ~ 100 MV/m or higher, and locally the field can be much larger on the nanometer scale due to surface roughness, then very large currents can be produced by this phenomena. The current density, J_{FN}, in A/m^2 which is produced by a field E_{flat} in V/m, is given by Fowler-Nordheim theory for a triangular barrier as [7]

$$J_{FN}(E) = A_{FN} \frac{(\beta_f E_{flat})^2}{\phi_W} \exp\left(-\frac{B_{FN} \phi_W^{3/2}}{\beta_f E_{flat}} \right), \tag{3.7}$$

where $A_{FN} \simeq 1.54 \times 10^{-6}$ A eV/V^2 and $B_{FN} \simeq 6.83 \times 10^9$ eV$^{3/2}$V/m; β_f is a field enhancement factor, and ϕ_W is the work function in eV. As the emission is dependent on the electric field at the cathode, this emission can be highly dependent on the geometry. Geometries which provide higher electric fields at the cathode for a given potential difference produce more current. One geometry that provides a very high local electric field at the cathode is a whisker or rod which is smaller than the anode cathode gap but with a large ratio of length to radius. Such a geometry can provide a local electric field at the cathode surface, E_{local} several times higher than that of a flat surface, E_{flat}, known as the field enhancement factor β_f so that $E_{local} = \beta_f E_{flat}$. Such whiskers can occur in manufacturing or by damage to a surface on the micron or nanometre scale, giving a higher local electric field on the surface than expected. Field emission can give very high current densities compared to thermionic emission but it is more difficult to produce large emission areas, with high field enhancement factors. In particle accelerators, operating with high electric fields, this effect can be unwanted as the surfaces of RF cavities themselves can emit electrons which can be captured along with the beam in the strong RF fields. These will eventually drift off the beam trajectory and will deposit their energy in whatever they collide with [8]. Field emission can also occur alongside photo-emission in photocathodes degrading the beam quality through unwanted parasitic emission.

Once the electrons have been emitted it is necessary to remove them before they impact the anode so they can be further accelerated. This can be achieved by placing a hole in the aperture connected to a conducting beam tube such that the electrons can travel along this tube to other accelerator components. As particles with *like* charge repel each other the electron bunch will blow up (increase in emittance) between the cathode and the anode. Fortunately, the beam will also have a magnetic field due to the motion of the electrons. The force due to the electron's electric and magnetic fields, known as the space-charge field, cancel each other completely when the beam is travelling at the speed of light c, and partially when the beam is travelling slower than c. In addition, the electrons become heavier due to relativity, hence the effects of space charge on electrons is much more significant at low energy below a few MeV, hence accelerating faster with higher electric fields can minimise the effect; this is covered in more detail in Chapter 7. If the current is well known and the beam is continuous in time, this can be compensated for by curving the cathode and anode to cancel the beam's own space-charge field. This was studied by Pierce who developed the Pierce electrode geometry which is placed at an angle to the cathode to cancel the space-charge field. However, many electron sources (often called electron guns) are required to produce short pulses of electrons in a beam; in this case the beam should require magnetic focusing to compensate and minimise the beam blow-up.

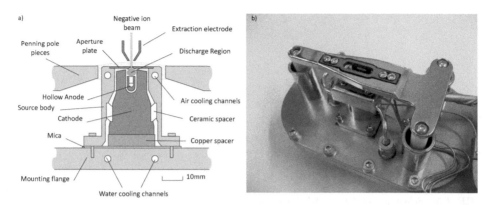

FIGURE 3.4 a) Schematic of the ISIS Penning ion source b) A photograph of the ISIS Penning source. © STFC

3.2.2 Ion Sources

As their name suggests, ion sources provide either positively- or negatively-charged ion species. They are typically composed of two parts: a plasma source (in a chamber) and an extraction system to remove the desired ions from that plasma. The gas can be ionised to create the plasma by either applying a large electric field which creates tunnel ionisation and pulls the positive and negatively charged particles in opposite directions, or by collision with an electron which knocks out a bound electron, known as impact ionisation. It is also possible to ionise a gas via electron capture, to create a negatively-charged ion. Once ionised, a DC accelerator can separate the electrons and the ions. The most common methods used in particle accelerators for generating positive ions are electron bombardment, plasmatrons, microwave, electron beam, laser and vacuum arc [9]. Common methods for producing H^- are surface plasma cold cathodes and multicusp sources. A common example of an ion source is the PIG (Penning Ionisation Gauge) proton source, within which is a small chamber (several millimetres across) that contains hydrogen gas fed in continuously at a known small rate, usually by means of a mass-flow controller. The gas volume has a flat cathode at each end and a cylindrical anode between, with around 2 kV between them. Electrons emitted from these (cold) cathodes take long, helical paths toward the anode due to an additional applied magnetic field applied across the electric field (hence these are cross-field devices), which then create ions via impact ionisation. The PIG source for the ISIS accelerator is shown in Fig 3.4.

3.3 Radio-Frequency Acceleration

The maximum accelerating field of an electrostatic accelerator is limited by the DC Kilpatrick criterion [10] (not to be confused with the RF Kilpatrick criterion given later in this chapter), an empirical formula devised in the 1950s by W.D. Kilpatrick; the maximum voltage V and gradient E satisfy the inequality

$$VE^2 \exp\left(-\frac{1.7 \times 10^7}{E}\right) < 1.8 \times 10^{18}; \tag{3.8}$$

this is an empirical fit, where V is given in V and E in V/m. It shows the accelerating field is dependent on the voltage across the gap between the anode and cathode, and is limited to around 3 MV/m for electrostatic accelerators. The maximum voltage is also limited by this

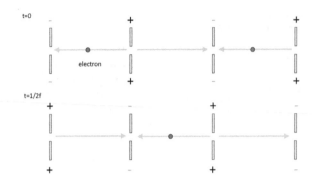

FIGURE 3.5   Schematic of a basic RF linac where the polarity of each electrode alternates along the linac and flips at the RF frequency such that the electron always sees an accelerating field.

criterion for a given size of the accelerator. The charged cathode must be separated from any grounded potentials such that the field does not exceed 3 MV/m for large voltages, as given by the DC Kilpatrick criterion. This means the cathode must be held above the ground and all mechanical supports should be insulated and able to hold off the applied voltage, a limitation that further increases the size of the accelerator. For a 3 GeV machine the cathode would be at least 1 km above ground (likely more) in air and no building or structure could be closer than around 1 km away. This can be reduced to a few hundred metres by using an a pressurised or electronegative gas, such as sulphur hexafluoride, or vacuum to hold off some of the voltage, which can sustain a higher electric field, but the electron/ion path must be in vacuum. This would an unfeasible requirement, and in practice electrostatic accelerators are limited to cathode potentials less than a few tens of MV even when using a pressurised, electronegative gas.

In order to reduce the size of accelerators and allow them to be placed horizontally at ground level it would be ideal to use several gaps in series, with the maximum potential constant along the length. However, as the energy gain is proportional to the difference in potential across the gap, each gap must be at a sequentially increasing potential, thereby negating any benefit of multiple gaps. One option to allow the use of two gaps is to use negative ions and then strip the electrons, making it a positive ion, to allow acceleration in the 2nd gap with the opposite potential difference to the first gap. Such an arrangement is known as a tandem Van de Graaff.

In order to use multiple gaps without increasing the potential at each subsequent electrode we can instead vary the potential in time using a metal drift-tube to shield the particles when the field would be decelerating. Alternatively we can use gaps of alternating potential difference. Here a positive potential – that attracts a negatively-charged particle to it – can be switched to a negative potential when the particle passes, thereby repelling it and giving twice the voltage, as shown in Fig 3.5. The same trick can be used over many hundreds of gaps (or more) allowing the beam to be accelerated to an energy far greater than that given by the potential difference across each gap. As the field varies with time only bunches of charged particles that have a duration much less than the RF period (the time over which the voltage is varying) can be accelerated using this method. Typically the field is alternated at frequencies from tens to thousands of MHz, covering the same frequency band as radio transmissions; hence this is known as radio-frequency (RF) acceleration.

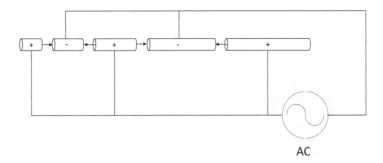

FIGURE 3.6   Schematic of a Wideroe linac, with the polarity switching every drift tube.

3.3.1   The First RF Linacs

The earliest RF accelerator was proposed by Gustav Ising in 1924, and was built by Rolf Widerøe in 1928; it was known as a drift-tube linear accelerator (or linac for short) [11]. Here the positive and negative terminals of an RF oscillator are connected to alternating metal drift-tubes (a hollow metal tube that the electric fields cannot penetrate into, such that the particles drift in a field-free region) such that every drift-tube has the opposite polarity to the drift-tube on either side, as shown in Fig 3.6. Charged particles are accelerated in the gap between each drift-tube. Widerøe's linac was tested on a single drift-tube with only two gaps. As the particles are accelerated they become faster and can travel further in one half RF period hence the gaps increase in length with each successive gap. As the fields oscillate in time the potential difference will change in time as the particles traverse the gap. For this reason for a given potential difference it is optimal to have the particles cross the gap in a finite time period when the field is maximum. However as the particles must be synchronous with the fields, arriving at each gap half an RF period after the previous gap, the drift-tubes need to be sufficiently long to shield the electric fields from the particles for enough time that they enter the next gap at the correct phase. Again the length of the drift tubes increase with particle velocity. Widerøe's original linac used a 1 MHz, 25 kV source to accelerate potassium ions up to 50 keV. The first multi-gap linac was built in 1931 by David Sloan and Ernest Lawrence which produced 1.25 MeV Hg$^+$ ions using an accelerating voltage of 42 kV across 30 gaps, at 10 MHz.

We can generate large potential differences in RF accelerators by storing RF energy over a long period of time and releasing that energy in a shorter time when accelerating the particles. This is achieved by placing the accelerating gap inside a can made of a highly conducting metal which traps the RF fields inside, known as a cavity. The RF fields can then be coupled into the cavity using a small antenna inside it. At certain frequencies, which depends on the size and shape of the cavity, a perfect standing wave is created inside the cavity allowing the energy to be stored for a long time, a few thousand to a few million RF periods dependent on the conductivity of the walls and the coupling. This increases the potential difference across the gap for a given input power compared to the case without a cavity. In circular particle accelerators, where the same cavity can be used for multiple passes of the beam, a single gap cavity can be utilised, but for linear accelerators, in order to minimise the linac length, it is preferred to use multiple gap cavities.

In 1945 Luis Alvarez devised a variant of the Widerøe drift-tube linac (DTL) where several drift-tubes were placed inside a cavity [12]. In this case, the two ends of each drift-tube have opposite potentials and the potential varies along the drift-tube. This means that each gap has the same potential difference and hence the gaps now have to be spaced apart by a full RF period, meaning that the drift tube needs to be almost twice as long. While this means that less of the cavity length is utilised for acceleration, the fact that it utilises a cavity to store the RF energy makes up for this. Alvarez together with Wolfgang Panofsky built a 32 MeV proton DTL operating at 200 MHz in 1947. Currently, Alvarez-type linacs are commonly used to accelerate protons between 50 MeV and 200 MeV.

3.3.2   Disk-Loaded Cavities

In 1933 Jesse Wakefield Beams* developed a method of synchronising the RF in successive cavities by using an artificial lumped-element transmission line with a wave velocity equal to the velocity of the charged particle to be accelerated, with a number of electrodes fed from this line which can accelerate that particle. However, the first electron linac was not produced until 1946. It is often misstated that the first electron linac was at Stanford, but in reality the first electron linac was developed by Donald William Fry at the Telecommunication Research Establishment at Great Malvern in the UK, which was a 0.5 MeV corrugated waveguide linac using a 1 MW, 3 GHz magnetron [13], which was later upgraded to 4 MeV. The reason for the delay was the lack of sufficiently powerful RF sources. The most common RF sources prior to 1937 were magnetrons, similar to those used in microwave ovens today. First invented in 1910 by Harry Boot and John Randall, prior to 1939 magnetrons were not very powerful. In 1937, a new RF source known as the klystron was developed by the Varian brothers, Russell and Sigurd. In both devices the kinetic energy of an electron beam is transferred to an RF wave amplifying either a pre-injected signal or noise in the device. These devices were limited by the maximum RF power that could be generated. During the Second World War many of the world's scientists turned their efforts to helping in the war effort, and many of these were tasked with developing longer range radar systems. During 1939-1945 improved klystrons and magnetrons were rapidly developed that could provide far higher powers in the MW range than their predecessors which were limited to around a kW. From 1945 onwards many of the scientists and engineers working on radar went back to particle and nuclear physics and they brought these new RF sources with them allowing higher particle energies to be reached.

In 1948, unaware of the work of Fry, Bill Hansen improved upon the electron linac design by placing a series of periodic disks inside an RF waveguide, forming a series of small cavities with a potential difference between disks [14], known as a disk-loaded cavity. Each has a small hole for the beam to travel through, and either this hole or other additional holes serve to transfer RF power from cavity to cavity. Each cavity has a slightly different phase of the RF field and the structure will behave like a transmission line with a phase and group velocity which can be altered by changing the coupling of the RF power through the holes in the disk. This device is still the most common type of particle accelerator today. The high RF powers available via the new klystrons and magnetrons allowed accelerating field gradients higher than those available with DC accelerators. The Hansen linac was able to accelerate electrons to 4.5 MeV, and by 1973 the Stanford Linear Accelerator Centre had developed a linac utilising a disk-loaded waveguide that accelerated electrons to 30 GeV.

3.4   Confined Electromagnetic Fields

To have efficient acceleration we must confine the wave in an RF cavity, also known as a resonator. A cavity confines the wave in all 3 directions, which allows a large stored

---

*Jesse Wakefield Beams had probably the most appropriate name of any accelerator scientist.

energy to build up providing large electric fields, while a waveguide confines the wave in 2 directions and allows a power flow in the 3rd direction, which is useful for transporting RF power from the generator or, if slowed down, is also useful for accelerating particles. Here we examine the use of conducting walls to confine the wave. Waveguides are hollow metallic pipes, normally with either rectangular or circular cross section for simplicity. Alternatively, we can instead use concentric cylinders known as coaxial lines to confine the wave. If the two ends are connected to a generator and a load respectively, then power will flow along the waveguide from the generator to the load. A cavity is similar to a waveguide except that both ends have metal walls covering them, causing the wave to reflect between the two ends forming a standing wave inside.

Most waveguides used in accelerators are hollow pipes of rectangular cross section, hence we will begin in Cartesian co-ordinates for simplicity before moving onto cylindrical coordinates as most cavities are cylindrical. In Cartesian co-ordinates the wave equation is

$$\frac{\partial^2 \Phi}{\partial x^2} + \frac{\partial^2 \Phi}{\partial y^2} + \frac{\partial^2 \Phi}{\partial z^2} = \frac{1}{c^2} \frac{\partial^2 \Phi}{\partial t^2} \tag{3.9}$$

where Φ is either the electric or magnetic field in the longitudinal direction, z. If we assume the solution to this equation varies sinusoidally in the longitudinal direction and in time this can be reduced to

$$\frac{\partial^2 \Phi}{\partial x^2} + \frac{\partial^2 \Phi}{\partial y^2} + k_z^2 \Phi - \frac{\omega^2}{c^2} \Phi = 0, \tag{3.10}$$

the solution will be an interference pattern of reflected plane waves travelling at an angle with respect to the propagation direction down the guide. As the walls are parallel to the x and y planes, we expect the solution to have sinusoidal variations in the x and y directions.

From Gauss's law we know that the electric fields, and hence time-varying magnetic fields, must be zero within a conductor. Surface currents can cancel out the magnetic field parallel to the surface, and surface charges can cancel out electric fields perpendicular to the surface, but the other field components must be continuous on both sides of the surface leading to fields inside the conductor and hence losses due to the movement of charges. This leads to the boundary conditions that electric fields parallel to the surface, $E_\parallel$, and magnetic fields perpendicular to the surface, $H_\perp$ should be zero on a perfectly-conducting boundary,

$$E_\parallel = 0,$$
$$H_\perp = 0. \tag{3.11}$$

This implies that these field components must either be zero everywhere or have a variation with distance such that those field components are zero on the walls but finite elsewhere in the waveguide or cavity. Considering these boundary conditions for a waveguide with waveguide width a, waveguide height b, and with metal walls along the x=0 and y=0, we obtain the equations for the transverse variation in the longitudinally directed component of the electric, E_z, and the magnetic, H_z, fields,

$$E_z = E_0(z,t) \sin(k_x x) \sin(k_y y), \tag{3.12}$$

and

$$H_z = H_0(z,t) \cos(k_x x) \cos(k_y y), \tag{3.13}$$

where E_0 and H_0 are the maximum longitudinal electric and magnetic fields, and k_x and k_y are the transverse wavenumbers in the x and y direction respectively where $k_{x,y} = 2\pi/\lambda_{x,y}$.

In order to meet the boundary conditions, it is necessary that the transverse wavenumbers satisfy

$$k_x = \frac{m\pi}{a}, k_y = \frac{n\pi}{b}, \tag{3.14}$$

so that there is an integer number of half wavelengths between the walls; a is the waveguide width, b is the waveguide height, and m and n are arbitrary indices equal to the number of half wavelength variations along the width and height respectively. From the wave equation the wavenumbers should be given by the spatial variation as,

$$\frac{\omega^2}{c^2} = k^2 = k_x^2 + k_y^2 + k_z^2, \tag{3.15}$$

where $\omega = 2\pi f$ is the RF angular frequency, $k = \omega/c$ is the free space wavenumber and $k_z = 2\pi/\lambda_z$ is the wavevector in the longitudinal direction. Often we combine k_x and k_y together into a transverse or cut-off wavenumber k_t,

$$k_t^2 = k_x^2 + k_y^2. \tag{3.16}$$

If we assume the wave varies sinusoidally in the longitudinal direction, z, and in time, t, where longitudinal components point in the direction of the power flow, the electric fields in a waveguide can be given by

$$\mathbf{E}(z,t) = \begin{pmatrix} E_x(x,y) \\ E_y(x,y) \\ E_z(x,y) \end{pmatrix} e^{i(\omega t - k_z z)} \tag{3.17}$$

where E_x, E_y and E_z are the (complex) electric fields in the x, y and z directions, such that each field component may be out of phase with the others.

For each combination of m and n we obtain a different orthogonal mode of the waveguide. In a homogeneous, linear, isotropic and stationary media with a smooth-walled waveguide of constant cross section either the electric or magnetic field in the propagation direction must be zero. It is convenient to split this into two subsets where we calculate the transverse fields from either the longitudinal electric or magnetic fields. Where we have a non-zero longitudinal electric field the magnetic fields are purely transverse, hence we call this a transverse magnetic (TM) mode. Conversely when we have a non-zero longitudinal magnetic field the electric fields are purely transverse hence we call this a transverse electric (TE) mode. There is a third class of mode where both longitudinal electric and magnetic fields are zero, which are called transverse electromagnetic (TEM) modes; however, these can only be supported where there are two electrically isolated conductors, such as a coaxial line where there is an outer and an inner cylinder. Hence we have a set of modes of the waveguide denoted TE_{mn} and TM_{mn} and TEM. As E_z would be parallel to the waveguide walls, m and n cannot be zero for a TM mode, while for TE modes either m or n can be zero but not both.

The electromagnetic waves in a standing-wave cavity can be considered as a superposition of forward- and backward-propagating waves, hence the electric fields are given by

$$\mathbf{E} = \begin{pmatrix} E_x(x,y) \\ E_y(x,y) \\ E_z(x,y) \end{pmatrix} \left[e^{i(\omega t - k_z z)} + e^{i(\omega t + k_z z)} \right]. \tag{3.18}$$

In a cavity k_z can instead only take a finite number of values where the cavity length, L, is an integer number of half wavelengths. Here we provide a third index to define the mode, p, which is the number of half wavelengths along the cavity in the z direction. TM modes can have $p = 0$ and still satisfy the boundary conditions, but TE modes require $p > 1$. The cavity mode is hence defined as TE/TM$_{mnp}$;

$$k_z = \frac{p\pi}{L}.$$

(3.19)

The sum of the wavenumbers squared must still equal the square of the free space wavenumber, and hence each mode can only resonate at a single frequency given by Equation 3.15.

For a waveguide, the mode will propagate if $k > k_t$ and hence k_z has a real component. If $k < k_t$ and therefore k_z is purely imaginary, the wave will decay exponentially and is said to be below cut-off. This implies a minimum frequency at which a mode can propagate, known as the cut-off frequency, $\omega_c = k_t c$. For frequencies below this, k_z is purely imaginary and hence the fields decay exponentially in z. The cut-off frequency is proportional to the waveguide size, hence a low-frequency waveguide is much larger than a high-frequency waveguide. Each mode in the waveguide will have a different cut-off frequency; however, TE and TM modes with the same indices will have the same cut-off frequency in a rectangular waveguide (this is, however, not the case in other waveguide cross-sectional shapes). If a wave propagates in more than one mode, the wave will be distorted due to the different wavenumbers for each mode, hence it is preferred to propagate the RF power in a single mode. Conventionally, the waveguide width is defined as being larger than the height (i.e. $a > b$), hence the lowest frequency mode is the TE$_{10}$ mode and this is the mode typically chosen to transport the power from the RF source to the cavity, although other modes are sometimes used. In order to maximise the frequency band over which the waveguide is single-mode we set $a = 2b$ so that the TE$_{01}$ and the TE$_{20}$ have the same cut-off frequency and the single moded bandwidth is maximised. The dispersion diagram (a plot of ω against k_z) is shown in Fig 3.7. As we will later see, this plot is useful for finding the frequencies of strongest interaction with a beam.

Using Faraday's law and Ampere-Maxwell's law, it can be shown that

$$\begin{aligned}
\mathbf{E}_\perp &= \frac{i}{k_z^2 - k^2}(k_z \nabla_\perp E_z + \omega\mu\nabla \times H_z\hat{\mathbf{z}}), \\
\mathbf{H}_\perp &= \frac{i}{k_z^2 - k^2}(k_z \nabla_\perp H_z - \omega\epsilon\nabla \times E_z\hat{\mathbf{z}}),
\end{aligned}$$

(3.20)

where k is the free-space wavevector ($k = \omega/c$); $\mu = \mu_r\mu_0$ and $\epsilon = \epsilon_r\epsilon_0$ are the permeability and permittivity of the waveguide interior (often we have a vacuum and $\mu = \mu_0$, $\epsilon = \epsilon_0$). This means that once the longitudinal field components have been solved, the transverse field components can then be calculated from them.

The transverse fields for a TM mode are hence

$$\begin{aligned}
E_x &= -\frac{ik_z m\pi}{k_t^2 a}E_0(z,t)\cos\left(\frac{m\pi}{a}x\right)\sin\left(\frac{n\pi}{b}y\right), \\
E_y &= -\frac{ik_z n\pi}{k_t^2 b}E_0(z,t)\sin\left(\frac{m\pi}{a}x\right)\cos\left(\frac{n\pi}{b}y\right), \\
H_x &= \frac{i\omega\epsilon n\pi}{k_t^2 b}E_0(z,t)\sin\left(\frac{m\pi}{a}x\right)\cos\left(\frac{n\pi}{b}y\right), \\
H_y &= -\frac{i\omega\epsilon m\pi}{k_t^2 a}E_0(z,t)\cos\left(\frac{m\pi}{a}x\right)\sin\left(\frac{n\pi}{b}y\right),
\end{aligned}$$

(3.21)

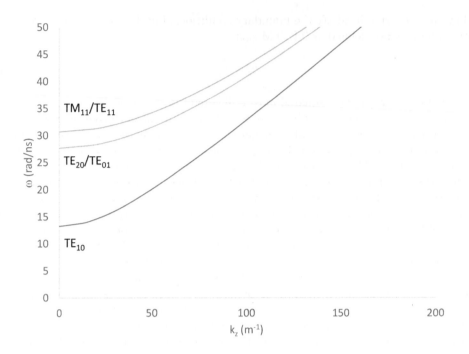

FIGURE 3.7 Dispersion diagram (ω versus k_z) for a waveguide with cross section 71.136 mm × 34 mm (known as WG10 waveguide) for the first five modes. Note that the TM and TE modes with the same indices have the same dispersion in rectangular waveguide.

while for a TE mode they are

$$E_x = \frac{i\omega\epsilon n\pi}{k_t^2 b} H_0(z,t) \cos\left(\frac{m\pi}{a}x\right) \sin\left(\frac{n\pi}{b}y\right),$$

$$E_y = -\frac{i\omega\epsilon m\pi}{k_t^2 a} H_0(z,t) \sin\left(\frac{m\pi}{a}x\right) \cos\left(\frac{n\pi}{b}y\right),$$

$$H_x = \frac{ik_z m\pi}{k_t^2 a} H_0(z,t) \sin\left(\frac{m\pi}{a}x\right) \cos\left(\frac{n\pi}{b}y\right),$$

$$H_y = \frac{ik_z n\pi}{k_t^2 b} H_0(z,t) \cos\left(\frac{m\pi}{a}x\right) \sin\left(\frac{n\pi}{b}y\right). \tag{3.22}$$

The fields in a TE_{10} mode are given by

$$H_z = H_0(z,t) \cos(k_x x) \cos(k_y y),$$

$$E_x = 0,$$

$$E_y = \frac{-i\omega\epsilon\pi}{k_t^2 a} H_0(z,t) \sin\left(\frac{\pi}{a}x\right),$$

$$H_x = \frac{-ik_z \pi}{k_t^2 a} H_0(z,t) \sin\left(\frac{\pi}{a}x\right),$$

$$H_y = 0, \tag{3.23}$$

where E_z is zero everywhere.

The fields of the first two modes (TE_{10} and TE_{20}) in a waveguide where the width is twice the height ($a = 2b$) are shown in Fig 3.8 and Fig 3.9; the fields of the first TM mode (TM_{11}) are shown in Fig 3.10 and Fig 3.11.

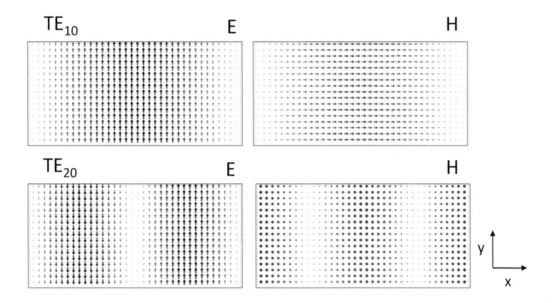

FIGURE 3.8 Electric and magnetic field patterns of the TE_{10} and TE_{20} waveguide modes for a cross-section perpendicular to the propagation direction, where the arrows indicate the direction of the field vector.

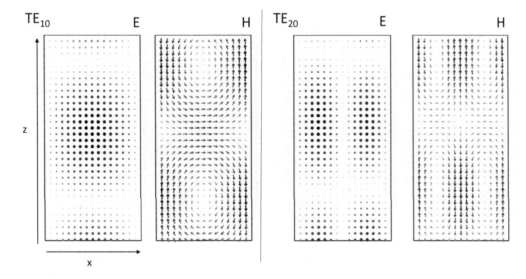

FIGURE 3.9 Electric and magnetic field patterns of the TE_{10} and TE_{20} waveguide modes from a cross-section along the propagation direction and perpendicular to the y direction, where the arrows indicate the direction of the field vector.

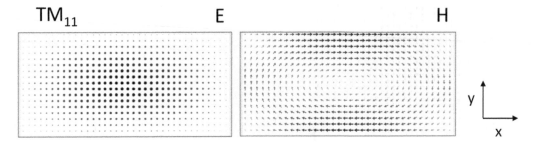

FIGURE 3.10 Electric and magnetic field patterns of the TM_{11} waveguide modes for a cross-section perpendicular to the propagation direction, where the arrows indicate the direction of the field vector.

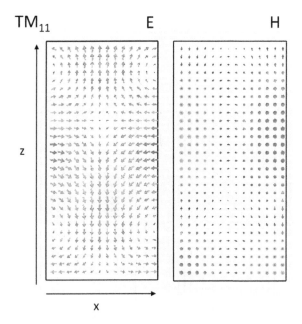

FIGURE 3.11 Electric and magnetic field patterns of the TM_{11} waveguide modes from a cross section along the propagation direction and perpendicular to the y direction, where the arrows indicate the direction of the field vector.

The ratio of the transverse fields $Z = E_\perp/H_\perp$ is known as the wave impedance, which should be real if there are no losses on the cavity walls. For a TM mode the wave impedance is

$$Z_{\mathrm{TM}} = \frac{E_\perp}{H_\perp} = \left(\frac{\mu}{\epsilon}\right)^{1/2}\frac{\lambda}{\lambda_z} = Z_0\frac{\lambda}{\lambda_z}. \tag{3.24}$$

where the constant $Z_0 \simeq 377\ \Omega$ is known as the impedance of free space. For a TE mode the wave impedance is

$$Z_{\mathrm{TE}} = \frac{E_\perp}{H_\perp} = \left(\frac{\mu}{\epsilon}\right)^{1/2}\frac{\lambda_z}{\lambda} = Z_0\frac{\lambda_z}{\lambda}. \tag{3.25}$$

The impedance is useful for calculating reflections from interfaces of different cross section and for the development of equivalent circuit models.

3.4.1 Phase and Group Velocity

As the mode in the waveguide is made up of several plane waves reflecting from the walls and travelling at an angle to the direction of the mode's propagation, the mode will travel slower than the speed of light. However within a pulse the peaks will appear to move at a different velocity, which can be faster than the speed of light. This does not violate causality as it only appears to move faster, the pulse and hence the information always travels slower than the speed of light. The angle at which the plane waves propagate with respect to the z axis is fixed for a given mode and frequency, such that the phase fronts from successive reflections are synchronous, and hence a standing wave is produced in the transverse directions such that the boundary conditions are maintained, shown in Fig 3.12. The distance travelled by the plane wave from one surface to the other and back must be an integer number of free space wavelengths ($\lambda = \omega/c$) so that the wave returns with the same phase. For a waveguide of width a, the distance travelled, l, is related to the angle of propagation, θ, by

$$l = \frac{2a}{\cos\theta} = \lambda. \tag{3.26}$$

The mode has travelled along the waveguide in the longitudinal direction by a distance of only $\lambda\sin\theta$, hence the mode's signal (or group) velocity, which is the velocity component in the longitudinal direction, is given by

$$v_g = c\frac{\lambda\sin\theta}{\lambda} = c\sin\theta. \tag{3.27}$$

If we have a pulse of RF of finite duration, the envelope of the pulse will travel at the group velocity, but the peaks of the wave inside the pulse will move at a different velocity, known as the phase velocity. If we imagine a plane wave propagating at an angle of θ with respect to the z axis, as we have seen, the mode travels in the z direction more slowly; however, as the phase front extends parallel to the direction of propagation to the extents of the waveguide, the phase front seems to have travelled a larger distance, if we look at the distance a peak moves in the direction of propagation in a finite time interval, but in reality the peak at the later time is a different part of the pulse. By considering the phase front as in Fig 3.13 and considering the geometry, we can see that the distance the peak moves in one RF period is

$$l_{phase} = \lambda\sin\theta = \lambda_z, \tag{3.28}$$

hence the phase velocity, v_p, is given as

$$v_p = c\frac{\lambda}{\lambda\sin\theta} = \frac{c}{\sin\theta} = c\frac{k}{k_z} = \frac{\omega}{k_z}, \tag{3.29}$$

and that

$$v_p v_g = c^2. \tag{3.30}$$

The group velocity can also be given by

$$v_g = \frac{\partial \omega}{\partial k_z}. \tag{3.31}$$

The instantaneous directional energy flux at a point in a waveguide is given by the Poynting vector, S, measured in W/m^2, as

$$\mathbf{S} = \mathbf{E} \times \mathbf{H}. \tag{3.32}$$

In a waveguide the Poynting vector points in the direction of propagation. In a cavity the real part of the Poynting vector is zero as there is no net power flow – the forward and backward components cancel; however, there is an imaginary component giving a reactive power back and forth which may have transverse as well as longitudinal components. Later in this chapter we will discuss travelling-wave structures which are a hybrid of a cavity and a waveguide, in which case the Poynting flux has both real and imaginary components. The power contained in the RF wave, P_{av}, is the integral of the time-averaged Poynting vector, $S_{av} = |\mathbf{E} \times \mathbf{H}^*|/2$, where * denotes the complex conjugate,

$$P_{av} = \frac{1}{2} \int_0^a \int_0^b Re|\mathbf{E} \times \mathbf{H}^*| \mathrm{d}x \mathrm{d}y. \tag{3.33}$$

For a rectangular waveguide this is

$$P_{av} = \frac{E_{max}^2}{4 Z_{TE}} ab, \tag{3.34}$$

where E_{max} is the maximum electric field. The maximum power flow in a waveguide is limited by the peak electric field. For a 3 GHz TE$_{10}$ mode in a WG10 standard waveguide ($a = 72.136$ mm, $b = 34.036$ mm), assuming a maximum peak electric field of 3 MV/m (in air), the maximum power flow is 2.26 MW. The group velocity is also related to the power flow in a cavity by

$$v_g = \frac{P_{av} L}{U}, \tag{3.35}$$

where U is the stored energy in a cavity of length L.

In order to have an efficient accelerator we want the electromagnetic fields transported to the accelerating structure to remain in the accelerating structure, only decaying due to ohmic losses in the walls. While a waveguide can be used as an accelerating structure we would have to slow the group velocity down to prevent the power leaving the structure too quickly, while simultaneously reducing the phase velocity to be equal to the particle velocity, which requires the structure to be loaded with either a dielectric lining, a corrugated wall or a series of iris'.

3.4.2   Electromagnetic Fields in Cylindrical Cavities

Cavities are typically cylindrical rather than rectangular. In a real accelerating cavity there will be beampipes with smooth transitions where they meet the cavity; however, it is useful to first understand the fields in a cavity of constant circular cross section but closed at both ends; this is known as a pillbox cavity. In cylindrical coordinates (ϕ, r, z), the wave equation is

$$\frac{1}{r} \frac{\partial}{\partial r} \left(r \frac{\partial \Phi}{\partial r} \right) + \frac{1}{r^2} \frac{\partial^2 \Phi}{\partial \phi^2} + k_z^2 \Phi - \frac{\omega^2}{c^2} \Phi = 0. \tag{3.36}$$

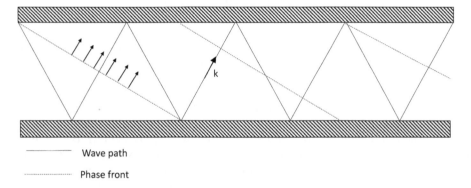

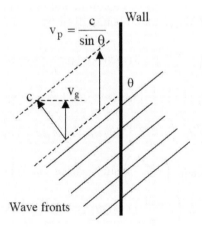

FIGURE 3.12   Wave reflecting inside a waveguide showing wavefront coherence over multiple reflections.

FIGURE 3.13   Group and phase velocities from propagation angles.

TABLE 3.1   The mth root of the
n^{th} Bessel function of the first kind.

$m \backslash n$	0	1	2
1	2.405	3.832	5.136
2	5.520	7.016	8.417
3	8.654	10.173	11.620

Separating variables and applying a periodic boundary condition to the azimuthal component, we find the solution to this equation is a radial varying function, R_m, which satisfies Bessel's equation whose general solution is

$$R_m = A_1 J_m(k_t r) + A_2 N_m(k_t r), \tag{3.37}$$

where J_m is the m_{th} Bessel function of the first kind and N_m is the m_{th} Bessel function of the second kind. Since R_m must be well-behaved at $r = 0$, and $N_m \to -\infty$ at $r = 0$, we set the constant $A_2 = 0$. For TM modes, $E_z = 0$ at the cavity radius, a, due to the boundary conditions, hence $k_t a = \zeta_{mn}$ where ζ_{mn} is the nth root of the mth Bessel function of the first kind. Considering the fields must be sinusoidal in ϕ and z, this leads to

$$E_z = E_0 J_m \left(r \frac{\zeta_{mn}}{a} \right) \cos(m\phi) \cos \left(p\pi \frac{z}{L} \right) \exp(i\omega t). \tag{3.38}$$

The index m is the number of full-wave variations around ϕ and the index n is the number of half-wavelength variations across the cavity diameter. The roots are given in Table 3.1.

As a cavity is fully enclosed by metal walls, the boundary conditions are only satisfied at discrete frequencies, as discussed previously. The resonant frequency of a cavity mode is given by

$$\left(\frac{\omega}{c} \right)^2 = k^2 = k_z^2 + k_t^2 = \left(\frac{\pi p}{L} \right)^2 + \left(\frac{\zeta_{mn}}{a} \right)^2. \tag{3.39}$$

The transverse components of the fields can again be found using Equation 3.20. This leads to the field components of a TM_{mnp} mode being given by

$$E_z = E_0 J_m \left(r \frac{\zeta_{mn}}{a} \right) \cos(m\phi) \cos \left(p\pi \frac{z}{L} \right) \exp(i\omega t),$$

$$E_r = E_0 \frac{k_z}{k_t} J_m' \left(r \frac{\zeta_{mn}}{a} \right) \cos(m\phi) \sin \left(p\pi \frac{z}{L} \right) \exp(i\omega t),$$

$$E_\phi = E_0 \frac{m k_z}{k_t^2 r} J_m \left(r \frac{\zeta_{mn}}{a} \right) \sin(m\phi) \sin \left(p\pi \frac{z}{L} \right) \exp(i\omega t),$$

$$H_r = \frac{i m \omega \epsilon}{k_t^2 r} E_0 J_m \left(r \frac{\zeta_{mn}}{a} \right) \sin(m\phi) \cos \left(p\pi \frac{z}{L} \right) \exp(i\omega t),$$

$$H_\phi = \frac{i \omega \epsilon}{k_t} E_0 J_m' \left(r \frac{\zeta_{mn}}{a} \right) \cos(m\phi) \cos \left(p\pi \frac{z}{L} \right) \exp(i\omega t),$$

$$H_z = 0, \tag{3.40}$$

where $J_m'(x) = \mathrm{d}J_m(x)/\mathrm{d}x$.

In order to accelerate a charged particle beam we need to have an electric field in the direction of the beam's motion; hence, if the beam travels in the z direction, then only a TM mode can accelerate the beam, although some complex structures can distort the fields to give a TE mode an E_z component. Depending on the length of the cavity the lowest

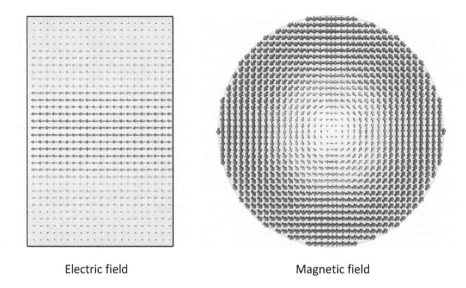

Electric field   Magnetic field

FIGURE 3.14 Electric and magnetic fields of a TM_{010} mode in a pillbox cavity.

resonant frequency will either be the TM_{010} or the TE_{111} mode. In a simple cylindrical cavity used for accelerating relativistic particles, we normally have a short cavity length, hence the TM_{010} will have the lowest resonant frequency; however, some low-energy proton and ion accelerators utilise a TE mode instead. The fields of a TM_{010} mode are shown in Fig 3.14 and are given by the equation

$$E_z \simeq E_0 J_0 \left(\frac{2.405r}{a} \right) \exp(-i\omega t),$$

$$H_\phi \simeq -\frac{iE_0}{Z_0} J_1 \left(\frac{2.405r}{a} \right) \exp(-i\omega t),$$

$$E_r = E_\phi = H_z = H_r = 0, \tag{3.41}$$

where $Z_0 = 377 \ \Omega$ is the impedance of free space, and $J_0'(x) = -J_1(x)$.

It should be noted that while H_ϕ is zero in the cavity centre it doesn't mean that the beam doesn't experience these fields, as the beam will have a finite radius and hence the particles on the outside of the bunch will experience these transverse fields as well as the accelerating field. The effect of this is covered in detail in Chapter 5.

3.4.3 Coaxial Lines

If we have two electrically isolated conductors then the waveguide can also support TEM modes as well as TE and TM modes. TEM modes have no longitudinal field components, $H_z = E_z = 0$, and the electric field parallel to surfaces and magnetic fields perpendicular to surfaces are also zero, i.e. $H_\perp = E_\parallel = 0$. As a consequence, the fields only have variation to conform to the surfaces, and hence the transverse wavenumber, $k_\perp$, is zero. This means that the wave travels longitudinally, and hence propagates at the speed of light in the filling medium; $k = k_z$ and hence has no cut-off frequency. Common waveguides that support TEM modes are parallel plates (two parallel conducting plates separated by some distance), and coaxial lines (two concentric cylinders where the fields propagate in between the inner and

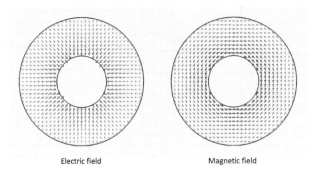

Electric field   Magnetic field

FIGURE 3.15   The electric and magnetic fields of a TEM mode in a coaxial line.

outer conductor). For accelerators operating at low RF frequencies below around 400 MHz, coaxial lines are commonly used to keep the waveguide transverse size down, due to the lack of a cut-off frequency for TEM modes, where a TE mode would require very large dimensions to operate above cut-off. In a coaxial line the field components in cylindrical components are

$$E_r = \frac{E_o(z,t)}{r},$$
$$H_\phi = \frac{E_o(z,t)}{Z_0 r}. \tag{3.42}$$

The fields of a TEM mode in a coaxial line are shown in Fig 3.15. Coaxial lines can also support TE and TM modes, hence it is usual to keep the outer conductor radius within limits to operate below the cut-off of the TE_{11} mode. The transverse wavenumber for the TE_{11} in a coaxial line of outer conductor radius b and inner conductor radius a, is given approximately by

$$k_\perp \approx \frac{2}{a+b}. \tag{3.43}$$

In the limit where the a tends towards b, the cut-off frequency of the TEM mode is 1.8 times lower than in a circular waveguide of radius b.

In a coaxial cavity, with metal walls at both ends connecting the inner and outer conductor, there are an integer number of half wavelengths along the line, and hence the mode is defined as a TEM_{00p} mode.

3.4.4   Walls with Finite Conductivity

Real waveguides and cavities have walls with a finite conductivity and hence work is done to shield the fields inside the metallic walls. Most RF structures are made from good conductors so the charges can redistribute to keep the fields similar to those with a perfect conducting boundary with the electric fields only penetrating a short distance into the conductor known as the skin depth. The fields will decay exponentially between the surface and the skin depth.

The skin depth, δ, is given by

$$\delta = \sqrt{\frac{2}{\sigma\mu\omega}} \tag{3.44}$$

where μ is the permeability of the conductor (where for most RF materials $\mu = \mu_0$) and σ is the electrical conductivity of the conductor. As there is an electric field inside the conductor a current is induced in it, given by Ohm's law $J = \sigma E$, where J is the current density. This in turn leads to a power loss as the current is being driven through a resistance. This surface resistance, R_{surf}, is given by

$$R_{\mathrm{surf}} = \frac{1}{\sigma\delta}. \tag{3.45}$$

Annealed or oxygen-free high conductivity copper, which is a common material for the construction of RF cavities, has a conductivity of around 5.8×10^7 S/m at room temperature, meaning it has a surface resistance of 14.3 $m\Omega$ and a skin depth of 1.2 μm at 3 GHz. The surface current is proportional to the magnetic field in the conductor, so that the RF power loss, P_c, over a surface, S, is given by

$$P_c = \frac{1}{2}R_{surf}\int_S |H|^2 \mathrm{d}S. \tag{3.46}$$

This power is lost directly from the RF field, and in the case of a cavity, reduces the stored energy in the cavity. This power is converted to heat causing the cavity temperature to rise. The RF power loss can be of the order of a few to hundreds of kW for normal conducting cavities, which can raise the cavity temperature to dangerous levels if the cavity isn't sufficiently cooled with circulating water. As the role of the cavity is to store electromagnetic energy, it is desirable to reduce the losses while maximising the stored energy. This leads to the ohmic quality factor, sometimes called the intrinsic Q factor, Q_0, of a cavity, which is proportional to the ratio of stored energy, U, to ohmic losses,

$$Q_0 = \frac{\omega U}{P_c}. \tag{3.47}$$

The higher the Q factor of a cavity, the more energy it can store for a given RF input power. As we will later see, the Q factor is also proportional to the filling time of the cavity and inversely proportional to the cavity bandwidth. A copper cavity will have a $Q_0 \sim 10^4$ at 3 GHz while a superconducting cavity will have $Q_0 \sim 10^9$–10^{10} depending on its operating frequency and operating temperature. In order to compare cavity geometries it is useful to define the geometry factor, G, which is independent of the cavity wall material and is

$$G = R_{\mathrm{surf}}Q_0. \tag{3.48}$$

As an example the superconducting TESLA cavity, operating at 1.3 GHz, has a geometry factor of 250 Ω, providing a $Q_0 \simeq 2.5 \times 10^{10}$ for a surface resistance of 10 nΩ [15]. In the case of a waveguide, as an electromagnetic wave propagates, the energy density of the wave increases as the group velocity of the wave decreases. The losses in a waveguide are dependent on cross-sectional area, waveguide length, group velocity of the wave, the ratio of the operating frequency to the cut-off frequency of the waveguide, and the waveguide conductivity. As the power lost is relative to the incident power, the attenuation is exponential. The attenuation in a waveguide can be represented as an imaginary wavenumber. The loss coefficient, α, is the imaginary part of the axial wavenumber, where β_z is the real part of the axial wavenumber and $k_z = \beta_z - i\alpha$. The transmitted power, P_T, is then given by

$$P_T = P_0 e^{-2\alpha z}, \tag{3.49}$$

where z is the length of the guide and P_0 is the input power. Hence the power lost due to ohmic heating, P_L, is

$$P_L = P_0(1 - e^{-2\alpha z}).\tag{3.50}$$

Differentiating this equation with respect to z and rearranging gives an equation for α;

$$\alpha = \frac{1}{2P_0}\frac{\partial P_L}{\partial z}.\tag{3.51}$$

3.5   Accelerating Modes in Cavities

The main purpose of an RF cavity is to accelerate the beam either to increase the beam's energy or to replace energy lost due to radiation and other processes. The voltage (potential difference), V, between two points A and B is

$$V = \int_A^B E_z(z)\mathrm{d}z;\tag{3.52}$$

however, as the electric field varies in time so will the voltage.

$$V(t) = \int_A^B E_z(z)e^{-i\omega t}\mathrm{d}z,\tag{3.53}$$

where ω is the cavity's resonant frequency. In the case of an accelerating cavity we want to calculate the energy gained by the charged particles. However, the particles cannot travel faster than the speed of light, c, hence it will take a finite time to traverse the cavity's accelerating gap; the electric field will change in strength, meaning the particle will not experience the full voltage of the cavity. Instead we must take into account the particle's position with time, given by $z = vt$ where v is the particle velocity (for convenience we often use the fractional velocity, $\beta = v/c$). The instantaneous accelerating voltage experienced by the beam, V_{acc}, is then

$$V_{acc} = \int_A^B E_z(z)e^{-i\omega z/v}\mathrm{d}z.\tag{3.54}$$

The ratio of the voltage seen by a particle travelling with finite velocity and the voltage seen by a particle travelling with an infinite velocity is known as the transit-time factor, T,

$$T = \frac{|V_{acc}|}{V}.\tag{3.55}$$

It is useful to also express the average accelerating field experienced by the beam, E_{acc}, over a cavity of length, L, known as the cavity accelerating gradient (or gradient for short)

$$E_{acc} = \frac{V_{acc}}{L}.\tag{3.56}$$

The gradient of a pulsed normal conducting cavity can be up to $\sim$100 MV/m for 12 GHz cavities for 200 ns long RF pulse durations, while for 1.3 GHz superconducting cavities it can be up to 35 MV/m*. If we take a pillbox cavity (i.e. a hollow cylinder) such that there

---

*The record is 52 MV/m in a single-cell cavity but this cannot be achieved reliably or in multicell cavities.

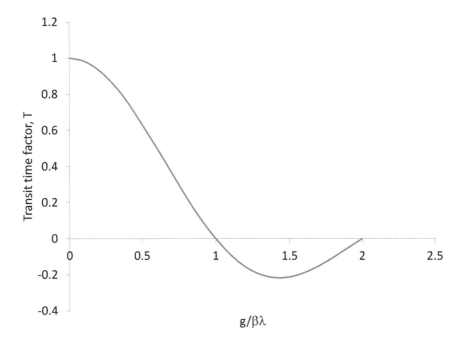

FIGURE 3.16   Transit-time factor as a function of the gap length in an RF cavity.

is no longitudinal variation of the fields ($p = 0$), then the transit-time factor is given by

$$T = \frac{\sin\left(\frac{\omega g}{\beta c}\right)}{\frac{\omega g}{\beta c}}. \tag{3.57}$$

where g is the accelerating gap, which is shorter than the cavity length for multi-cell cavities due to the finite wall thickness. A plot of this equation is shown in Fig 3.16; while it would suggest that we get a larger gradient if we have a chain of very short cavities, in practice this is not the optimum configuration as there must be a finite wall thickness between cavities. If we have more cavities per unit length we also have more walls, and hence more wasted space. More cavities also means more couplers, or if we couple the cavities together, we must synchronise the fields with the beam. In practice the optimum cavity length for a single cell standing wave cavity is roughly given by

$$L_{opt} \simeq \frac{\pi \beta c}{\omega}. \tag{3.58}$$

where the wall thickness is as short as possible. For thin walls, where $g \approx L$, the transit time factor is equal to $2/\pi$ for a pillbox cavity.

The cavity voltage seen by the beam will vary sinusoidally with the beam arrival phase. The maximum voltage of the cavity is often higher than the operating accelerating voltage as we often chose not to inject the beam at a phase corresponding to the peak voltage, as will be discussed in Chapter 5 when we look at beam stability. The ideal cavity should give the maximum voltage for a given dissipated (and hence supplied) RF power. To relate the dissipated power to an accelerating voltage, we use the cavity shunt impedance, $R_{s,circuit}$, defined as

$$R_{s,circuit} = \frac{|V_{acc}|^2}{2P_c}; \tag{3.59}$$

this is equivalent to the power dissipated across a resistor with an AC voltage applied across it, with the factor of 2 in the denominator due to the peak voltage V_{acc} and RMS power P_c being used. Most accelerator physicists use an alternative definition – due to the fact that the particle bunch does not see the sinusoidal variation of the voltage – of

$$R_s = \frac{|V_{acc}|^2}{P_c}. \tag{3.60}$$

For linear accelerators it is often more useful to state the shunt impedance per unit length, r_s.

$$r_s = \frac{|V_{acc}|^2}{P_c L}. \tag{3.61}$$

The CLIC-G accelerating structure, which has an operating frequency of 11.9942 GHz, has a shunt impedance per unit length of 92 MΩ/m [16]. The shunt impedance for a cavity which has all dimensions scaled to a different frequency will have the shunt impedance per cell scaled $\propto 1/\sqrt{f}$; however, as a higher-frequency cavity will have more cells per unit length hence, the shunt impedance per unit length scales $\propto \sqrt{f}$, hence allowing higher-frequency cavities to reach higher gradients for a given input power. However, if you doubled the frequency the aperture would halve for a scaled structure. Normally the aperture of a linac is constrained to a minimum aperture for beam stability and losses, hence when going up in frequency, the ratio of the aperture radius to the wavelength (a/λ) increases reducing the shunt impedance. Fig 3.18 shows the shunt impedance per unit length as a function of aperture for several different frequencies, where it can be seen that for any two frequencies there is a maximum aperture where the higher frequency provides a higher shunt impedance per unit length, and above that aperture the lower frequency is better.

As the shunt impedance per unit length (as well as the maximum gradient as we will see later) is strongly frequency dependent, cavities are grouped by their resonant frequency using the IEEE RADAR RF bands (loosely based on the old NATO bands). L-band (long wave) goes from 1–2 GHz, S-band (short wave) is 2–4 GHz, C-band* from 4–8 GHz, and X-band (short for 'crosshair' as it was used for fire control in World War II) from 8–12 GHz. In each band there are generally two commonly utilised frequencies – either European or US in origin– the difference being whether the wavelength is an integer number of millimetres (European) or integer fractions of an inch (US); for example, X-band structures are either 11.9942 GHz (European) or 11.424 GHz (US). Although many European accelerators now use US frequencies and vice versa.

If a higher shunt impedance is required we can add what are known as nose cones, as shown in Fig 3.17. These are cones around the iris which reduce the accelerating gap, and hence increase the transit time factor, without increasing the magnetic field at the equator, hence the shunt impedance increases.However the peak electric field on the tips of the nose cones increases as the nose cones increase in length meaning this technique is not typically used for very high gradient accelerators. The addition of optimised nose cones will improve the impedance by around 10%.

It is also useful to define the geometric shunt impedance R/Q, which like the geometry constant is independent of the cavity size and material. This impedance also relates the induced cavity voltage to the driving beam current and is a measure of the coupling between the fields in the cavity to the beam.

$$\frac{R}{Q} = \frac{|V_{acc}|^2}{\omega U}. \tag{3.62}$$

---

*The 'C' in C-band stands variously for 'commercial', 'communication', or 'compromise'.

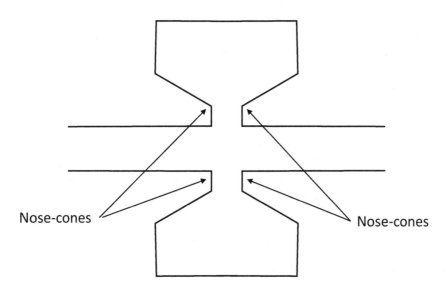

FIGURE 3.17 An RF cavity with nose cones to decrease the gap size while keeping a large cavity volume where the magnetic field is maximum.

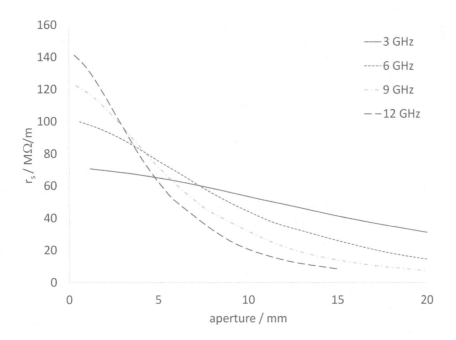

FIGURE 3.18 Shunt impedance per unit length as a function of aperture radius for a disk-loaded cavity for 3, 6, 9 and 12 GHz resonant frequencies where the disk thickness is 0.08 λ.

Each mode in the cavity has a geometric shunt impedance relating the coupling of this mode to the beam. Modes with high geometric shunt impedance can be stimulated within the cavity as a bunch of charged particles traverse the cavity. These unwanted cavity harmonics can lead to detrimental effects on subsequent bunches. Since the effect is induced in the wake of the bunch this type of interaction is known as a *wakefield*. The amount of energy transferred is proportional to the cavity shunt impedance and inversely proportional to the cavity size.

3.5.1 Cavity Equivalent Circuit

An RF cavity can be modelled as an RLC series or parallel circuit, which makes calculations of dynamic behaviour or mode coupling much simpler. The resistance of the circuit is the shunt impedance of the cavity, R_s. The capacitance and inductance are calculated from the cavity resonant frequency, ω_0, and the geometric shunt impedance R/Q using

$$\omega_0 = \frac{1}{\sqrt{LC}}, \quad R/Q = 2\sqrt{\frac{L}{C}} = \frac{2}{\omega C}; \tag{3.63}$$

the Q factor can then be calculated using

$$Q = R_{s,circuit}\sqrt{\frac{C}{L}}. \tag{3.64}$$

3.5.2 Coupling Power into an RF Structure

To connect the RF power supply to the cavity we must construct an antenna that will radiate power into the cavity (see discussion of antenna radiation in Chapter 6); this avoids the power being reflected back up the waveguide. This is normally just a waveguide or coaxial line connected to the cavity via a small hole in the beam pipe or the cavity walls, known as an input or power coupler, which will be discussed later. The strength of the coupling can be represented by defining an external Q factor, Q_e, which relates the stored energy in the cavity to the power that would flow into the coupler if there is no RF power being supplied to the cavity, P_e; this is given as

$$Q_e = \frac{\omega U}{P_e}. \tag{3.65}$$

It is convenient to add the external power lost to the coupler with the input power turned off, P_e, to the cavity ohmic losses, P_c to give the total losses with the RF supply off P_t. Since $P_t = P_e + P_c$ then we can also define a quality factor combining all losses known as the loaded Q factor, Q_L, where for a cavity with a single coupler,

$$\frac{1}{Q_L} = \frac{1}{Q_0} + \frac{1}{Q_e}. \tag{3.66}$$

It is also useful to define the coupling factor, β, which is the ratio of losses through the coupler to the ohmic losses in the cavity walls

$$\beta = \frac{P_e}{P_c} = \frac{Q_0}{Q_e}. \tag{3.67}$$

The cavity will have a finite bandwidth over which power is coupled into the cavity. The impedance of the cavity can then be solved from the equivalent circuit as a function of frequency ω for

$$Z = \frac{R_{s,circuit}}{1 + iQ_L \left(\frac{\omega}{\omega_0} - \frac{\omega_0}{\omega}.\right)}. \tag{3.68}$$

This gives a full width half maximum bandwidth in Z of $2Q_L/\omega$.

At frequencies outside of this bandwidth all the power will be reflected. When we have an RF pulse the rising and falling edges of the pulse will contain a wide range of frequencies, some of which will fall outside the band and will be reflected. For a square pulse, almost all of the power at the rising edge will be outside the band and all of the power will initially be reflected, but over time the bandwidth will reduce, reducing reflections and increasing the power coupled into the cavity. For slower rise times and larger cavity bandwidths the reflections are reduced. To model this, we can consider an equivalent circuit. When RF power is supplied to the cavity there will be a large impedance mismatch between the coupler, which will have an impedance of a few tens of ohms, as it is required to have a high power flow for transport, to the cavity which will have an impedance of several MΩ in order to reach high gradients with minimal power. This means that at the interface between the coupler and the cavity there will be a large reflection in anti-phase to the supplied RF power. This reflected signal from the interface will interfere with the power leaking back into the coupler from the cavity which will be in phase with the supplied RF power. The total power flowing back up the coupler, P_r, when driving the cavity on resonance, will be equal to

$$P_r = \left(\sqrt{P_f} - \sqrt{\frac{\omega U}{Q_e}}\right)^2 \tag{3.69}$$

where P_f is the forward power from the RF source. The reflection from the interface between the cavity and coupler due to the mismatch will be slightly less than 100% in reality. We will refer to the total reverse power going back up the waveguide as the reflected power P_r, the power reflected from the interface between the cavity and coupler when the cavity is empty as the interface reflection, P_i, and the power leaking back up the coupler from the stored energy as the emitted power, P_e. The change in stored energy over time in an RF cavity without beam can be obtained by summing the power flowing into and out of the system as

$$\frac{dU}{dt} = P_f - \left(\sqrt{P_f} - \sqrt{\frac{\omega U}{Q_e}}\right)^2 - \frac{\omega U}{Q_0}. \tag{3.70}$$

Expanding the brackets gives

$$\frac{dU}{dt} = \sqrt{\frac{4P_f \omega U}{Q_e}} - \omega U \left(\frac{1}{Q_0} + \frac{1}{Q_e}\right) \tag{3.71}$$

and inserting the definition of loaded Q factor into this equation gives us

$$\frac{dU}{dt} = \sqrt{\frac{4P_f \omega U}{Q_e}} - \frac{\omega U}{Q_L}. \tag{3.72}$$

We can hence find the steady-state stored energy, U_0, when the stored energy no longer varies with time ($dU/dt = 0$), by solving the quadratic equation for $\sqrt{U}$,

$$U_0 = \frac{4P_f Q_L^2}{Q_e \omega} = \frac{4P_f \beta}{(1+\beta)^2} \frac{Q_0}{\omega}, \tag{3.73}$$

and we can also solve the time dependence of the stored energy, assuming the initial energy is zero, by solving the first-order nonlinear ordinary differential equation as

$$U = U_0 \left(1 - e^{-\omega t / 2 Q_L} \right)^2 . \tag{3.74}$$

It can be seen that the stored energy increases with time as the cavity fills with RF energy, converging to U_0. The time constant for the filling, τ, is

$$\tau = \frac{\omega}{Q_L} \tag{3.75}$$

which is inversely proportional to the loaded Q factor rather than the ohmic Q factor. Having solved for the stored energy we can return to solving the reflected power; inserting Equation 3.73 into Equation 3.69 we obtain the steady-state reflected power

$$P_r = P_f \left(1 - \frac{2 Q_L}{Q_e} \right)^2 \tag{3.76}$$

and inserting the definition of the coupling factor we obtain

$$P_r = P_f \left(1 - \frac{2\beta}{1 + \beta} \right)^2 . \tag{3.77}$$

It can be seen that when $\beta = 1$ the reflected power (flowing back up the coupler) is zero and the cavity is said to be critically-coupled. This can be interpreted as the reflections from the interface – due to the impedance mismatch between the cavity and the waveguide – exactly cancelling out the power emitted from the cavity into the waveguide as they will have equal magnitude but will be 180° out of phase. $\beta = 1$ when the ohmic and external Q factors, and hence the external coupler and ohmic losses, are equal. When $\beta > 1$ the ohmic Q factor is greater than the external Q factor and hence the coupler is said to be over-coupled; when $\beta < 1$ it is said to be under-coupled. This can also be rearranged to find the coupling factor by measuring the steady-state reflections from a cavity to give

$$\beta = \frac{1 \pm \sqrt{P_r / P_f}}{1 \mp \sqrt{P_r / P_f}}, \tag{3.78}$$

with the upper sign used if $\beta > 1$ and the lower sign used if $\beta < 1$. Often $\sqrt{P_r / P_f}$ is referred to as the input port reflection coefficient, S_{11}, which is the first element in the scattering matrix of reflected and transmitted waves from a multiport RF network [17]. By inserting Equation 3.74 into Equation 3.69 we can also solve for the reflected power for the case of the time-dependent reflections

$$P_r = P_f \left[1 - \frac{2\beta}{1 + \beta} \left(1 - e^{-\omega t / 2 Q_L} \right) \right]^2 . \tag{3.79}$$

The first term represents the interface reflection and the second term the emitted power from the cavity. For the case of a critically- or under-coupled cavity, the reflections are initially close to 100% as there is no stored energy in the cavity to cancel the reflections at the interface. As the stored energy builds up in the cavity, so does the power emitted back down the coupler from the cavity, cancelling out some of the power reflected at the interface and reducing the power flowing back up the coupler. For a critically-coupled cavity the reflected

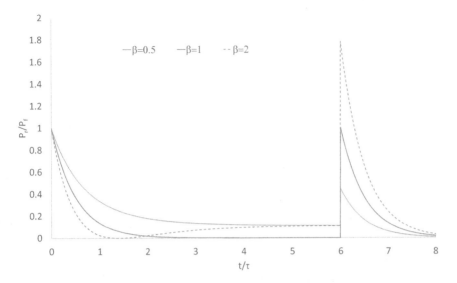

FIGURE 3.19 Transient reflected power for a square wave input pulse, where $t_{pulse} = 6\tau$ for $\beta = 0.5$, 1 and 2.

power tends to zero over the filling time of the cavity, while for an under-coupled case they tend to a finite value. For an over-coupled cavity the behaviour is initially identical but soon the emitted power grows larger than the interface reflection. Hence the superposition of both reverse signals causes the reflected power to reduce to zero and then start increasing again to a finite value, as first the interface reflection dominates the reflected signal, then the emitted power. As these two signals have a 180° phase difference, the phase of the reflected signal also changes by 180° as it crosses zero.

When the RF is switched off suddenly, P_f becomes zero, hence it no longer cancels the emitted power and the reflected power will again spike with the peak reflected power directly after the RF pulse is switched off given by

$$P_r = \frac{\omega U}{Q_e} = P_f \left(\frac{2\beta}{1 + \beta} \right)^2 \tag{3.80}$$

with an over-coupled cavity creating a reflected power spike up to four times the power of the initial forward RF power, a critically-coupled cavity reflected power spike the same size as the forward power and the under-coupled cavity with a smaller spike than the forward power. The reflected signals from a square envelope pulse, of duration t_{pulse}, for each case is shown in Fig 3.19. The stored energy in the cavity will decrease exponentially with the time constant of the cavity

$$U = U_0 e^{-\omega t/Q_L}. \tag{3.81}$$

Hence the stored energy will vary with time as

$$U = U_0 e^{-\omega(t-t_{pulse})/Q_L} = \frac{4 P_f Q_L^2}{Q_e \omega} e^{-\omega(t-t_{pulse})/Q_L}. \tag{3.82}$$

The cavity voltage can then be obtained from the R/Q of the cavity. When the RF is turned off we can set the forward power to zero in Equation 3.69 but maintaining the same stored energy at the moment the RF is turned off; this will then decay exponentially. This yields

$$P_r = P_f \left[\frac{2\beta}{1 + \beta} \left(e^{-\omega(t-t_{pulse})/2Q_L} \right) \right]^2, \tag{3.83}$$

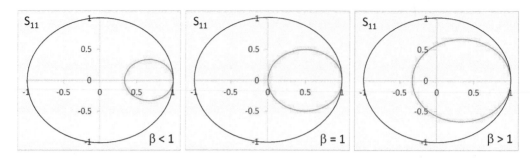

FIGURE 3.20   The reflected signal as a function of frequency on a polar plot for $\beta = 0.5$, 1 and 2.

where P_f is the power before the RF is turned off at time t_{pulse}. If the RF drive frequency, ω, is different than the cavity resonant frequency, ω_0, the steady-state reflected power can be given by [18]

$$P_r = P_f \left(\frac{1 - \beta - iQ_0\delta}{1 + \beta + iQ_0\delta} \right)^2 \tag{3.84}$$

where δ is given by

$$\delta = \frac{\omega}{\omega_0} - \frac{\omega_0}{\omega}. \tag{3.85}$$

Plotting the reflected signal as a function of frequency on a polar plot will give a circle which will not enclose the origin if under-coupled, will cut through the origin if critically-coupled and will enclose the origin if over-coupled; this is shown in Fig 3.20. This allows the coupling to be measured from the reflected power.

3.6   Gradient Limits

The maximum gradient in a normal conducting structure is often limited by a number of phenomena:

- RF breakdown;
- RF heating;
- RF source power limits or operating cost.

RF breakdown is where the high electric fields cause some of the walls to be vaporised, and then ionised causing a plasma to form inside the cavity. RF heating is where the power lost in the cavity walls causes the temperature of the cavity walls to increase causing deformation and stresses which can affect normal operation. The RF sources are also limited by RF breakdown and RF heating and this leads to a limited RF power, which in turn limits the cavity voltage. Using high RF powers also implies a large electricity bill which can also be a limiting factor. If limits of the RF power supply are ignored for now, the physical limits of the cavity gradient are RF breakdown which is dependent on the peak surface electric, E_{pk}, and magnetic, B_{pk}, fields, and RF heating which is related to the surface magnetic field. Hence two important criteria for cavity design are the ratios of the surface fields to the gradient E_{pk}/E_{acc}, and B_{pk}/E_{acc}. Typically values are $E_{pk}/E_{acc} \sim 2 - 4$ and $B_{pk}/E_{acc} \sim 4$ mT/(MV/m).

3.6.1 RF Breakdown

RF breakdown is where a plasma discharge within the accelerating structure grows to an extent where RF operation is not possible. The discharge causes an impedance mismatch which causes the power to be reflected back up the coupler, and absorbs the RF energy inside the cavity, hence stopping the cavity operation. Repeated breakdowns can also cause permanent damage to the structure. RF breakdown – also known as a vacuum arc – requires a gas to ionise but the cavity is initially uder vacuum. This process is initiated by field emission; if the heat created by the current flow and the RF heating becomes large enough there will be vaporisation of material. This gas can then be ionised by the emitted electrons leading to plasma formation. The plasma causes more material vaporisation leading to a growth of the plasma into a runaway current, known as an arc discharge. In 1957 W.D. Kilpatrick devised an empirical formula for the maximum surface electric field that could be sustained before breakdown. This was then reformulated to include a frequency dependence by Boyd [19] to the more common RF Kilpatrick limit

$$f = 1.64 E_{pk}^2 \exp\left(-\frac{8.5}{E_{pk}}\right) \tag{3.86}$$

where the frequency f is in MHz and the electric field E_{pk} is in MV/m. For a 3 GHz cavity the Kilpatrick field limit is 47 MV/m, and rises to 90 MV/m at 12 GHz; this is shown in Fig 3.21. This is not the gradient but the maximum electric field on the surface which is typically twice as large as the gradient. As cavity manufacture and preparation has been improved this limit is now regularly exceeded by a factor of 2, with the CLIC structures demonstrating 100 MV/m gradient (250 MV/m peak surface field) for a 200 ns pulse at 12 GHz [16]. Breakdown is statistical in nature, as the location and time of breakdowns cannot be predicted but rather follows a probability that increases with electric field; so rather than being a hard limit, it is common to refer to the breakdown rate (BDR), given in breakdowns per pulse per metre, at a given field level. The breakdown rate is also dependent on the RF pulse duration, as well as the surface electric field and the cavity frequency. Typically, RF cavities in linacs will aim to operate with less than one breakdown per million pulses per metre to minimise structure damage and disruption to the beam being accelerated. As we saw previously, field emission can be increased on sharp tips, which can serve as breakdown sites, although this is not the only mechanism proposed as the cause of increased field emission. As a cavity is often manufactured with a number of sites which have a higher probability of breakdown it is necessary to condition the cavity. This consists of increasing the RF power very slowly over a number of hours, days or weeks, keeping the breakdown rate below a preset level. If the field is close to the cavities current breakdown limit the breakdown rate will decrease over millions of pulses, allowing the field to be increased. Traditionally this is considered to be due to a number of semi-controlled RF breakdowns, this causes vaporisation of the field emitters just above the breakdown threshold causing a minimum amount of damage. However, this can cause the material to be sputtered elsewhere, creating more field emitters, hence there is a limit to the gains from conditioning. More recently it has been suggested that the conditioning process is dependant on the number of RF pulses rather than the number of breakdowns which may condradict the traditional explanation [20]. After conditioning, the breakdown rate will scale with electric field and pulse duration, t_{pulse}, as

$$\text{BDR} \propto E^{30} t_{pulse}^5. \tag{3.87}$$

Hence the BDR increases very sharply with increasing field producing something close to a hard limit [21].

The causes of the field emitters are not known but several theories exist [22]. One such theory is the electromagnetic field applies a stress to the cavity surface which can give rise to

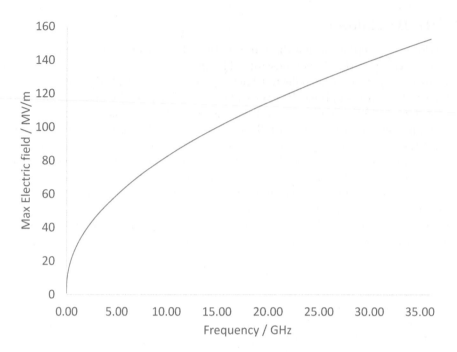

FIGURE 3.21 Maximum surface electric field versus RF frequency from the Kilpatrick criterion.

sharp tips if the stress is applied near a defect under the cavity surface such as a dislocation in the copper atomic lattice [23] and conditioning is a form of work hardening of the surface. As such, the probability of a field emitter appearing, and hence causing a breakdown, would depend on the electric strength and the material properties. The suggested scaling of the BDR with the electric field using the stress is

$$\text{BDR} \propto e^{(\epsilon(\beta_f E_{acc})^2 \Delta V / k_B T)} \tag{3.88}$$

where ΔV is the relaxation volume of the defect, k_B is the Boltzmann constant, and T is the temperature of the defect. More recently CERN [21] has shown good empirical agreement between the BDR of a structure and the peak value on the surface of a modified form of the Poynting vector, known as S_c.

$$S_c = \text{Re}[S] + g_c \text{Im}[S] \tag{3.89}$$

where g_c is a weighting factor due to the different effects of active and reactive power, which is 0.15 to 0.25 depending on the local field enhancement factor, typically taken as 1/6. CERN suggests that for a 12 GHz RF pulse with 200 ns duration, the breakdown rate will be 1 breakdown per million pulses, per metre for an S_c of 5 MW/mm^2. The BDR will scale with S_c and pulse duration as

$$\text{BDR} \propto S_c^{15} t_{pulse}^5. \tag{3.90}$$

3.6.2 Multipactor

Another common cause of electron discharge is multipactor, where the number of free electrons in the cavity vacuum undergoes an exponential growth in time. When an electron strikes a surface it can be absorbed, reflected (elastically backscattered), re-diffused, or create secondary electrons [24]. There may be more than one secondary electron emitted for

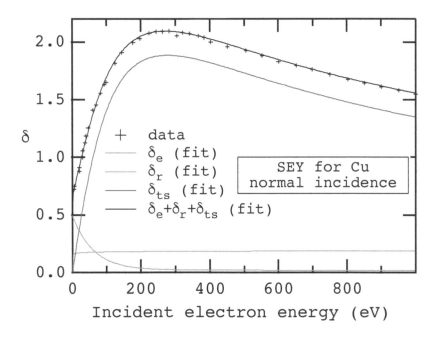

FIGURE 3.22 Secondary emission yield, including true secondaries, δ_{ts}, re-diffused, δ_r, and elastically backscattered electrons, δ_e, as a function of primary electron impact energy for copper [24].

each primary electron impacting the surface depending on the primary electron's impact energy and impact angle. The average number of secondaries per primary is known as the secondary emission yield, and this is shown as a function of impact energy for copper in Fig 3.22. The process is statistical but the average number of secondary electrons per primary electrons, δ, for most metals and ceramics is greater than one for primary impact energies from a few tens of eV up to a few keV, and less than one for other impact energies. These secondaries will experience a force from the RF fields causing it to move from the impact location.

For multipactor to occur, the secondaries created must return to the surface at the correct impact energy over many RF cycles, requiring secondaries to return to the same impact site, at the same phase (although the electron could oscillate between two fixed impact sites, known as two-point multipactor) [25]. This resonance condition will only be met at discrete RF field amplitudes, but when the conditions are met, any stray electrons will cause an exponential growth in the number of secondaries causing RF heating of the surface and absorbing RF power. The number of particles N after a number of impacts, $n_{impacts}$, is given by

$$N(n_{impacts}) = N_0 \langle \delta \rangle^{n_{impacts}}, \tag{3.91}$$

where N_0 is the number of initial electrons, and $\langle \delta \rangle$ is the average number of secondary electrons produced per primary electron impact. Multipactor typically happens in low electric fields either at low cavity voltages, or in locations where the electric field is lower, in order to have the electrons impact the surface at an energy likely to produce more secondaries. A common multipacting trajectory is where electrons make semi-circular cyclotron orbits around high magnetic fields, in the low electric field region, giving a resonant condition

$$B = f\frac{m_e}{e}, \tag{3.92}$$

where m_e is the electron mass. This type of multipactor was a limiting factor in the original superconducting cavities, but this was avoided in later cavities by making the cavity equator elliptical such that the electrons strike the surface at different angles depending on the orbit radii, causing the secondaries to move in successive orbits towards the centre where the electric field is zero and electrons cannot be accelerated to a sufficient energy to create new secondary electrons [26].

3.6.3   RF Heating

At high RF powers, RF heating can be a major issue. Typically, normal conducting RF cavities have their temperature regulated via cooled, turbulent water flowing at high mass flow rates through metal pipes joined to the cavity. For a sufficiently high mass flow rate, the pipes and the cavity surfaces in contact with them can be held at a fixed temperature, although there are limits to flow rate due to cavitation. However, there will always be a temperature gradient between the RF surface where the heat is applied in the skin depth, and the cavity surfaces in contact with the pipes. For a cylinder of internal diameter, d, thickness, t, and length, L, with heat flow, $\dot{Q}$, applied to the inner diameter and the outer surface held at a fixed temperature, the temperature difference between the inner and outer surfaces, ΔT, is given by

$$\Delta T = \frac{\dot{Q}\ln\left(1 + 2t/d\right)}{2\pi L\kappa}, \tag{3.93}$$

where κ is the thermal conductivity of the cylinder. This will result in thermal expansion of the cavity and hence the cavity will detune (change its resonant frequency). For a single-cell cavity, the detuning can be corrected by varying the water, and hence the cavity, temperature or by using a tuner which can change the cavity frequency by perturbing the fields via a moving part or surface deformation. However, for multicell cavities it is likely that all cells will deform differently causing each cell to have a different frequency, and hence causing a cell-to-cell amplitude and phase variation.

For short-pulse RF systems, where the pulse duration, t_{pulse}, is short compared to the time it takes for heat to diffuse into the cavity walls, the temperature rise can be much sharper [27]. When the RF pulse first starts causing ohmic heating, the heat is deposited entirely in the skin depth and hence a very small volume. As the volume heated is small, the temperature rise is large, but will decrease over a few microseconds as the heat diffuses into the bulk. The maximum power density deposited in the wall, P_d, is given by

$$P_d = \frac{R_{surf}H_{max}^2}{2}. \tag{3.94}$$

The temperature rise is given by

$$\Delta T = \frac{2P_d\sqrt{t_{pulse}}}{\sqrt{\pi\rho\kappa c_e}}. \tag{3.95}$$

where ρ is the density, κ is the thermal conductivity and C_e is the specific heat of the wall material. As the surface is at an elevated temperature compared to the bulk, the thermal expansion of the surface layer will be constrained creating a high stress on the surface. The yield strength of copper is exceeded for temperature rises of around 50 K. As the stress is cyclic, surface cracking can occur due to fatigue, which in turn can cause increased RF losses, hence surface heating, and/or field emission.

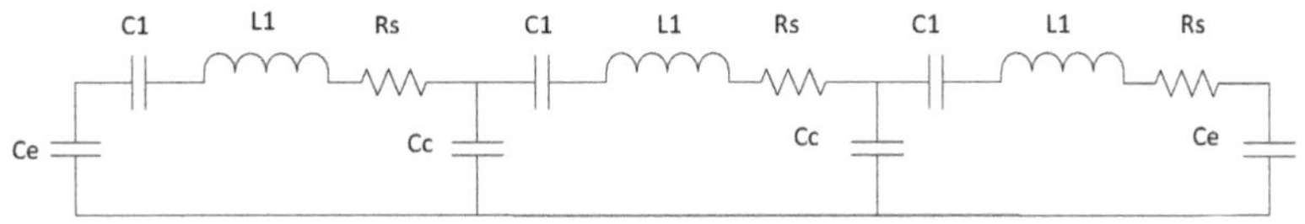

FIGURE 3.23 Circuit diagram of a three-cell RF structure.

3.7 Multi-cell Cavities

For practical accelerators, what is important is not the gradient but the real-estate gradient, which is the accelerating voltage divided by linac length including all ancillaries and drift tubes. Rather than having individual cavities, each with their own power couplers, vacuum pumps, and water cooling, it is preferable to have a series of RF resonators, which we will refer to as cells and for a multi-cell structure the term cavity refers to the entire structure of cells, coupled together such that a single power coupler is needed for each group of cells and the number of cavities/cells that can be fitted per metre of accelerator is increased. This increases the real-estate gradient as the spacing between cells is very short. Typically, for high-energy accelerators, a cavity is made of a circular waveguide with each cell separated by a metal disk with a hole for the beam to pass through. This hole, referred to as the beam aperture, can also be used to provide coupling between the cells. If we add more cells to a cavity but keep the field in each cell the same, both the voltage and dissipated power increase proportional to the number of cells hence the shunt impedance of a multi-cell cavity, $R_{s,cavity}$ is the impedance of a single cell, $R_{s,cell}$ multiplied by the number of cells, N_{cells}

$$R_{s,cavity} = R_{s,cell} N_{cells}. \tag{3.96}$$

In order to understand the behaviour of a chain of coupled RF cells it is convenient to analyse the equivalent circuit. Each cell can be represented as a resonant series RLC circuit as above but with the addition of an additional capacitive or inductive coupling between cells, as shown in Fig 3.23 for a three-cell cavity with capacitive coupling, C_c. The coupling can be via the electric or magnetic fields. If we chose to couple through the beam aperture in a TM_{110} mode, then the coupling will be via the electric field, represented by a parallel capacitance shared between the cell and its neighbour. Another possibility is to have holes in the walls between cells near the equator where the magnetic field is strongest, which is represented by a parallel inductance again shared by both cells.

The effect of this coupling on the frequency of the cavity modes can be found by solving the eigenmodes of the circuit. Applying Kirchoff's loop law to cell n, we obtain the following equation for the current in cell n, I_n,

$$I_{n-1}Z_c - I_n\left(Z + 2Z_c\right) + I_{n+1}Z_c = 0 \tag{3.97}$$

where Z is the impedance of the RLC circuit of the cell and Z_c is the impedance of the coupling capacitor or inductor. The solution to this equation is

$$\cos\phi_a = 1 + \frac{Z}{2Z_c} \tag{3.98}$$

where ϕ_a is the phase difference between the voltage in cell n and cell $n+1$, known as the phase advance. If we solve for a cavity with a finite number of cells we find there

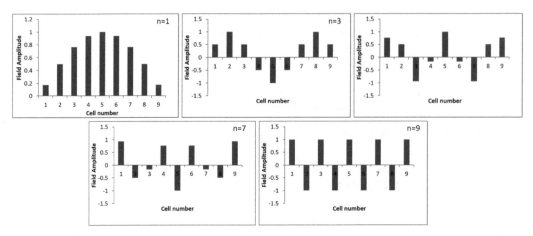

FIGURE 3.24 Field patterns for modes with different phase advances ($\phi_a = n\pi/9$), in a 9-cell cavity, where each bar represents the amplitude of the voltage in each cell.

are a number of possible eigenmodes of the system, where the number of eigenmodes for each cavity is equal to the number of cells. The variation in the field amplitude in each cell for a range of standing-wave phase advances is shown in Fig 3.24. The set of eigenmodes of the multi-cell cavity for each eigenmode of a single-cell cavity is known as that mode's passband. The frequencies of the eigenmodes for N identical cells are not all at the same frequency, as the different currents flowing through the coupling reactance for each mode provides a separation in frequency. For example, in the mode with zero phase advance, the currents flowing in each cell cancel in the coupling reactance, while in a mode with a π phase advance the currents sum together. If we expand Z and Z_c in terms of capacitance and inductance for an N cell cavity, assuming a coupling capacitance, and define a coupling factor $k = C/C_c$ we can find the frequency of each eigenmode as

$$\omega^2 = \omega_{\pi/2}^2 \left(1 + 2k \left[1 - \cos\left(\frac{n\pi}{N}\right)\right]\right), \qquad (3.99)$$

where n is the eigenmode number (an integer from 1 to N) of each mode with phase advance $n\pi/N$. The frequency versus phase advance for a multi-cell cavity with coupling factor, $k = 0.3$ and $f_{\pi/2} = 1$ is shown in Fig 3.25. Having a larger coupling factor, hence a larger coupling capacitance or inductance, provides a larger separation between the modes in the passband. A larger separation reduces the coupling to more than one mode at a given operating frequency, and hence the perturbation of the cavity fields from those other modes.

It is necessary to ensure that the beam arrives at each cell at the same phase, hence the length of each cell, L_{cell} can be calculated so that the phase change during the transit time is equal to the phase advance, based on the beam velocity, the cavity frequency and the phase advance of the operating mode.

$$L_{cell} = \frac{\beta\phi_a c}{\omega}. \qquad (3.100)$$

It is possible to design an accelerating structure to operate in a cavity at any phase advance, but in order to maintain synchronism the cells must get shorter as the phase advance decreases. In a standing-wave cavity, any phase advance that isn't a multiple of π (radians) will result in empty or partially-filled cells, reducing the gradient. The transit-time factor is also affected by the change in cell length as the phase advance is varied, resulting in a reduced gradient as the cells get longer with a transit-time factor of zero for a phase advance

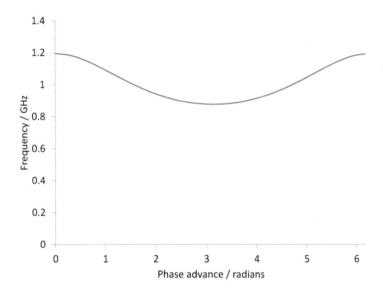

FIGURE 3.25 Resonant frequency of a multicell RF structure as a function of the phase advance between cells for electric coupling, where the $\pi/2$ mode is at 1 GHz.

of 2π. This means the ideal phase advance for maximum shunt impedance in a standing wave cavity is π, hence these structures are known as π-mode structures. If a mode with a 2π phase advance is to be used then the beam must be shielded from the RF fields for at least half of the RF cycle, using drift tubes as mentioned earlier. This reduces their shunt impedance but for low particle velocities the cell lengths for other phase advances become too short to make a practical structure.

3.7.1 Standing-Wave Cavities

All the cells in a π-mode standing-wave cavity should have the same field amplitude, however if one cell has a resonant frequency different from the other cells, that cell will have a different amplitude. The resulting eigenmode will be a hybrid of the required mode and its nearest neighbours. No physical structure can be made to infinite precision so in practice, all cavities will have finite variations in each cell's frequency due to manufacturing or alignment tolerances. In addition the phase or amplitude of each cell may vary due to the finite resistance of each cell, depending on the phase advance.

The spacing between the eigenmodes in the cavity varies sinusoidally with phase advance with minimum spacing at 0, π, and 2π modes, hence these modes are the most affected by manufacturing tolerances. In contrast the $\pi/2$ mode has the largest modal separation, and the modes are symmetric around it, meaning that this mode is the least affected by mechanical tolerances. The mode separation is also proportional to the cell-to-cell coupling, hence more coupling is preferred. As the number of eigenmodes in a cavity is proportional to the number of cells, longer cavities have larger variations in amplitude and phase over the structure, hence the cell-to-cell coupling is often increased for longer structures.

The cell-to-cell coupling can be increased by using larger apertures in the disks either at the beam hole for electric coupling, or at the equator for magnetic coupling. Increasing the aperture however increases the peak electric and magnetic fields on that aperture, which will reduce the shunt impedance and increase peak fields, as can be seen in Fig 3.26. For higher-frequency cavities, there are more cells per unit length and hence more coupling is

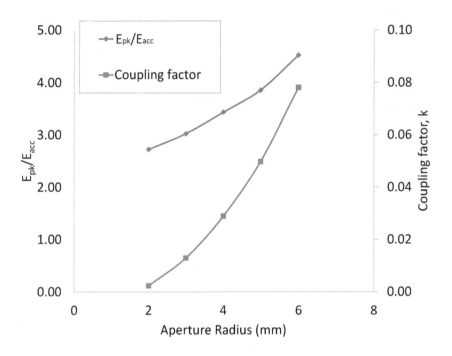

FIGURE 3.26   Peak field versus coupling constant for different aperture coupling [28].

required for a given structure length.

For high-gradient applications, π modes are preferred as they have a higher shunt impedance; however, they are the most sensitive to mechanical tolerances due to the smaller mode separation and the fact that the other modes are not symmetric in frequency around the π mode, requiring higher-cost precision machining or shorter structures. For long standing-wave cavities, or for industrial, medical or security accelerators where larger tolerances are required to keep costs low, a $\pi/2$ mode cavity is often preferred. The $\pi/2$ mode, however, has every second cell unfilled, reducing the cavity shunt impedance by a factor of two if all cells are identical. In order to restore the shunt impedance, the unfilled cells can be modified so that their gap is made small to the beam. The two most common approaches to achieving this are the side-coupled cavity where the unfilled cells are placed offset from the beam axis, such that the filled cells become the same length as a π mode structure, hence restoring the shunt impedance, or bi-periodic structures [28] where the un-filled cells are made very short and the filled cells are lengthened to maintain synchronism, as shown in Fig 3.27. The side-coupled cell designed for the PROBE project [29] is shown in Fig 3.28, where the side-coupled cells have a small capacitive gap to reduce the cavity frequency for a given radius, allowing compact side-coupled cells. It can be seen that nose cones are added around the beam aperture in each cell. As discussed earlier the nose cones allow smaller gaps, and hence transit time factors, without losing synchronism with the beam or increasing the capacitance of the cells. This allows the cavities to have a higher shunt impedance.

It is critical that the accelerating and coupling cells have the same resonant frequency in order for the coupling cells to provide a resonant coupling. If there is a frequency difference between the two cell types, the coupling cells will be capacitive or inductive instead limiting the coupling between any two adjacent accelerating cells, providing two separate passbands with fields either entirely in the accelerating or coupling cells, and the $\pi/2$ modes becoming

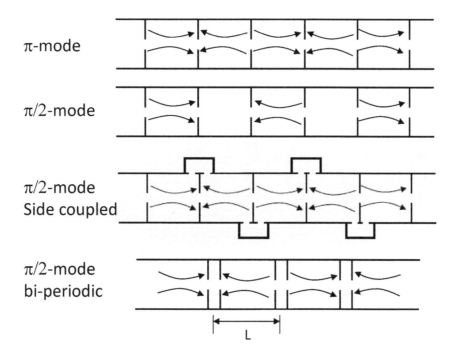

FIGURE 3.27 Geometries of π and $\pi/2$ modes in a cavity with all cells the same, as well as $\pi/2$ modes in side-coupled and bi-periodic cavities.

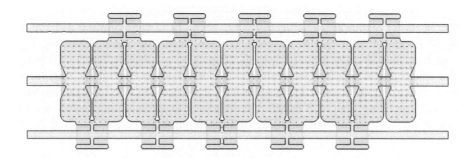

FIGURE 3.28 The side-coupled linac for the PROBE project [29].

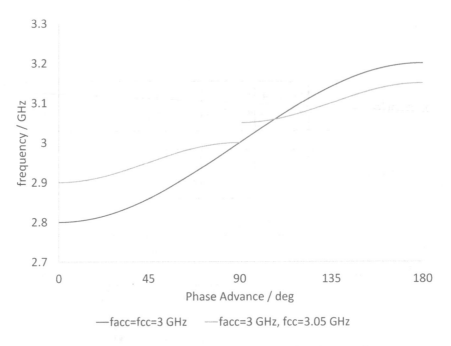

FIGURE 3.29   Dispersion curve for side-coupled linacs, with and without confluence.

π modes losing all the advantages described above. When the frequencies of the two cell types are brought together, known as confluence, all cells are resonantly coupled. In this case the mode with fields only in the accelerating cells becomes a $\pi/2$ mode, with a second $\pi/2$ mode with fields only in the coupling cells, and all other modes having fields in both types of cells. The Brillouin diagram (a plot of frequency against phase advance) for a 3 GHz side-coupled cavity for the case of confluence, and the case of the side-coupled cells being off frequency are shown in Fig 3.29.

For manufacturing errors where the coupling cells are accidentally all at a different frequency from the accelerating cells, the amplitude, A_{2n}, in the accelerating cell $2n$, with the coupler in cell $2m$, is given by [30]

$$A_{2n} = (-1)^{n-m} A_{2m} \left[1 - \frac{2(m^2 - n^2)}{k^2 Q_a Q_c} \right] e^{\left(i \frac{4(m^2 - n^2)}{k^2 Q_a} \frac{\Delta\omega}{\omega_a} \right)} \tag{3.101}$$

where Q_a is the Q of the accelerating cells, Q_c is the Q of the coupling cells, and $\Delta\omega$ is a single cell frequency shift due to mechanical errors such that the accelerating cells have a different frequency from the coupling cells. Typically, a coupling factor of 1% to 5% is chosen to ensure the fields are not significantly perturbed by machining tolerances, with a coupling factor of 2.1 % used in the PROBE structure. As the field deviation from a perfect cavity increases along the length of a structure, larger coupling factors are required for longer cavities, which provide an ultimate limit in number of cells of around 20–30 cells.

3.7.2   Travelling-Wave Structures

As mentioned previously, phase advances that are not integer multiples of 180° result in partially-filled or unfilled cells for standing-wave cavities, as the fields from the forward and backward waves destructively interfere in some cells and constructively in others. This destructive interference can be avoided by using a travelling-wave instead, where the power

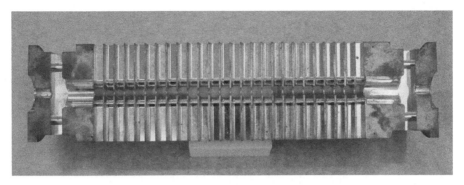

FIGURE 3.30 A sectioned CLIC accelerating structure operating in the $2\pi/3$ travelling mode; image courtesy of CERN.

only flows in a single direction and is absorbed in a load at the other end preventing reflections. The power is fed into the travelling-wave structure via an input coupler and any remaining power is removed at the other end via an output coupler. To avoid standing waves forming inside the travelling wave structure due to reflections at the couplers, each must be carefully matched individually to the structure so that there are no reflections inside the structure. A true travelling wave in vacuum would have a high group velocity requiring too high a power flow to be practical, and the phase velocity would be greater than the speed of light making synchronisation with a particle beam impossible. To avoid this, the waveguide must be 'loaded' to slow the wave down in both group and phase velocity. Whilst this can be done with a uniform dielectric loaded waveguide [31], it is more common to load the waveguide with aperture coupled disks, known as a disk-loaded waveguide [32]. A cutaway of a disk-loaded travelling-wave structure for CLIC is shown in Fig 3.30 [16].

As the disks are periodic, the wave will be reflected at each disk but will cancel every couple of cells due to the periodicity. As such they are not true travelling waves as each cell will have a longitudinal field variation, making it closer to a chain of standing-wave cells with a phase advance between them. However, the magnitude of the electric field will be identical in each cell and the structure will have a net power flow in one direction unlike a standing-wave structure. Using Floquet's theorem the field in each cell, E_{cz}, is identical in each cell other than a phase shift, as shown in Fig 3.31 for a $2\pi/3$ phase advance, and can hence be described using the field profile in a single cell E_z and the phase advance, ϕ_a as [33]

$$E_{cz} = E_z(z)[\exp(-i\phi_a) + \Gamma\exp(i\phi_a)], \tag{3.102}$$

where Γ is the reflected wave from the coupler, which is ideally zero for a matched structure. The travelling-wave structure for the AWAKE booster [34], is shown in Fig 3.32 showing the cell amplitude at a fixed point in time repeats every three cells, and is hence a $2\pi/3$ structure. The amplitude in each cell is constant but there is a phase difference between cells, so at any given point in time, the voltage in each cell will be different. The cell length and the phase advance is chosen such that the beam is always in the cell with the highest voltage. The length of each cell should satisfy

$$L_{cell} = \frac{\beta c \phi_a}{\omega} \tag{3.103}$$

in order for the beam and wave to be synchronous. For low-energy electrons where β varies in each cell, the cell length should be varied rather than the phase advance in order to minimise reflections. Given this, we can evaluate the phase advance and internal reflections inside a structure given the field in each cell. If we take the sum, Σ, and difference, Δ, of

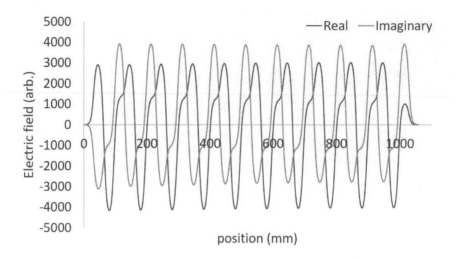

FIGURE 3.31   The real and imaginary components of the longitudinal electric field in a $2\pi/3$ travelling-wave structure.

FIGURE 3.32   The travelling-wave structure for the AWAKE booster.

the fields in the cells either side of a given cell

$$\Sigma = \frac{(E_z(z + L_{cell}) + E_z(z - L_{cell}))}{E_z(z)},$$

$$\Delta = \frac{(E_z(z + L_{cell}) - E_z(z - L_{cell}))}{E_z(z)} \tag{3.104}$$

then the phase advance can be found using

$$\cos(\phi_a) = \frac{\Sigma}{2} \tag{3.105}$$

and the reflected signal can be found from

$$\Gamma = \frac{2\sin(\phi_a) - i\Delta}{2\sin(\phi_a) + i\Delta}. \tag{3.106}$$

It should be noted that the internal reflection Γ is not the same as S_{11}, because if the input and output couplers are identical, the reflection from each coupler will cancel at the input giving $S_{11} = 0$ despite there being a reflected wave inside the cavity between the two couplers. For a matched travelling-wave structure, we hence require $S_{11} = \Gamma = 0$.

Each cell has a power flow into the cell, P_w, power loss in that cell due to ohmic losses or beam loading, and a power flow out of that cell. If the power flow is much larger than the other losses then the structure has a wider bandwidth, and the cavity behaves like a travelling wave with the filling time of each cell being short compared to the time for the power to flow through the structure. It is this increased bandwidth that makes travelling-wave structures insensitive to imperfections allowing longer structures to be used. They can be at least four times longer than a standing-wave structure. It is possible to load a short structure so that the group velocity, and hence the power flow, is much lower to increase efficiency. In such cases the individual cells fill slower, and reflections may occur during filling like a hybrid between a travelling- and standing-wave structure [35]. Due to ohmic losses the power flow decreases along the structure. The lower the group velocity, the higher the ohmic losses in the cell and hence the power flow will decrease faster along the length of the structure. If the structure is too long, the power will be too low in the end of the structure to achieve any usable gradient hence each structure has a maximum realistic length dependent on group velocity. If the structure is too short, then the power flow at the end of the structure will be large and will be absorbed in an RF load. However, having a lower group velocity also increases the stored energy per cell, and hence the gradient. Therefore for a given structure length, the group velocity should be chosen to maximise the average gradient. If the group velocity is chosen to be constant along the length, then the structure is said to be constant impedance.

We can calculate the accelerating voltage for a travelling-wave structure starting with the relation between power flow and group velocity in a cell,

$$P_w = v_g \frac{dU}{dz}. \tag{3.107}$$

The resistive power loss per unit length, P'_c is given as

$$P'_c = -\frac{dP_w}{dz}. \tag{3.108}$$

Considering the definition of Q factor and applying the power flow equation above we get

$$\frac{\mathrm{d}P_w}{\mathrm{d}z} = -\frac{\omega P_w}{Q v_g} \tag{3.109}$$

As the fields will decay exponentially along the structure, we can define an attenuation parameter, α_0, as

$$\alpha = \frac{\omega}{2Q v_g}, \tag{3.110}$$

such that

$$\frac{\mathrm{d}P_w}{\mathrm{d}z} = -2\alpha P_w. \tag{3.111}$$

Considering the definition of shunt impedance per unit length and applying the power flow equation above we get

$$\frac{\omega P_w}{Q v_g} = 2\alpha P_w = \frac{E_{acc}^2}{r_s}. \tag{3.112}$$

This can be rearranged to give the accelerating gradient in the a given cell cell, $E_a cc$,

$$E_{acc} = \sqrt{2 r_s \alpha P_w}. \tag{3.113}$$

Hence the power flow decays along the structure according to

$$P_w(z) = P_0 \exp(-2\alpha z). \tag{3.114}$$

Where P_0 is the power fed into the first cell via the input coupler. The exponent of the total decay along the structure, τ_0, is given by

$$\tau_0 = \alpha L = \frac{\omega L}{2Qv_g}. \tag{3.115}$$

As the power flow decays, so does the accelerating field

$$\frac{\mathrm{d}E_{acc}}{\mathrm{d}z} = -\alpha E_{acc}, \tag{3.116}$$

and hence the accelerating voltage is reduced, and can be found by using this equation to find the accelerating gradient along the structure length and integrating leading to

$$V_{acc} = E_0 L \frac{(1 - e^{-\tau_0})}{\tau_0} \cos\phi. \tag{3.117}$$

where E_0 is the field in the first cell. Hence, taking the equation for the accelerating gradient in the first cell into account we obtain

$$V_{acc} = \sqrt{2r_s P_0 L} \frac{(1 - e^{-\tau_0})}{\sqrt{\tau_0}} \cos\phi. \tag{3.118}$$

In the constant-impedance structure described above, all cells are identical. The gradient can be improved for a given input power and structure length by tapering the group velocity along the structure. For reasons of making power dissipation and breakdown uniform the optimum case is for the group velocity to be tapered such that the gradient is constant in each cell [36]; hence these are known as constant-gradient structures. Each subsequent cell will have a slightly lower group velocity than the one before it to account for RF losses in the previous cell. For a given input power, the group velocity in the first cell, and hence all subsequent cells, is chosen to be as low as possible to achieve the maximum gradient in each cell whilst still allowing a small amount of power to reach the final cell. The group velocity in each following cell has to be matched to the cell before it such that the gradient is the same. If the power flow is too low the structure will experience reflected power, lower bandwidths and more sensitivity to tolerances.

In both cases the group velocity is normally varied by changing the aperture radius but can also be varied by altering coupling slots placed in the disk near the cavity walls [35]. When increasing the aperture or coupling slot radii, the two adjacent cells are coupled via either electric or magnetic fields respectively, and the increased surface currents around the opening lead to higher peak electric and magnetic fields. Higher peak electric fields may result in breakdown and higher peak magnetic fields will increase the ohmic losses and hence decrease the shunt impedance. The input and output couplers must be matched to the structure so that the impedance of the coupler appears the same as an infinite structure, therefore ensuring there are no internal reflections. There will be a small reflection as the adjacent cell fills but this will be small for most structures due to the high power flow along the structure. The coupler should be designed to have an external Q factor, Q_e, given by

$$Q_e = \frac{c\phi_a}{v_g}. \tag{3.119}$$

For low-energy electron linacs, such as those used in radiotherapy, the electron velocity will change along the structure's length. In such cases the distance between the disks can be varied such that the phase velocity can be changed without changing the phase advance of the linac, and hence retaining the travelling wave without internal reflections.

3.8   Wakefields

When a relativistic electron beam travels down a conducting beam pipe, it generates an image current which travels with the beam. When the beam reaches a change in the cross-section of the beampipe, such as a cavity, the image current must take a longer conduction path hence it will slip behind the bunch. There will be a decelerating or deflecting force on the beam as it moves away from the image charge/current, and the beam will lose energy to the cavities' fields driven by these surface currents and charges. When the beam leaves the cavity. the energy transferred to the cavity modes will remain, and can interact with later bunches, or later particles within the same bunch. This effect is known as a wakefield, as it is the field left in the wake of the bunch, and can cause serious disruption in the beam.

Wakefields are discussed in Chapter 7 so here we limit ourselves to the RF effects. Wakefields will radiate over a wide range of frequencies, including the operating frequency of the cavity. This will cause a change in the cavity's operating modes fields known as *beam-loading*. Any radiation into other resonant modes of the cavity must be damped to avoid them growing to levels where they cause the beam quality to be degraded, which we will discuss later in this chapter.

If a cavity is driven by an external RF source then the wakefield will be superimposed on the driven fields in the cavity. If the beam current is synchronised with the peak of the RF voltage, known as on-crest, then only the amplitude of the operating mode will change, while if the beam is off-crest, there will be both a phase and amplitude change. The change in field will also change the matching conditions for the cavity, causing reflections at the coupler and hence requiring a change in external Q to re-match the cavity.

3.8.1   On-Crest Beam-Loading

For the case where the beam is on-crest, cavity behaviour can be described to the first order with some minor modifications to the equations without beam. In this case the beam-loading can be modelled as purely resistive, although it could have a negative resistance for the case of decelerating. The power transferred from the cavity to the beam in the cavity, ignoring the change in the cavity voltage due to the wake within a single bunch, is approximately given by

$$P_b = V_{acc} I_b \tag{3.120}$$

where V_{acc} is the accelerating voltage and I_b is the beam current, which must be replaced by the RF source, along with the power to replace ohmic losses in the walls, to maintain the cavity voltage. Looking at the cavity and beam from the RF source, cavity ohmic losses and on-crest beam-loading is indistinguishable, and hence we can define a new coupling factor

$$\beta_b = \frac{P_e}{P_c + P_b} \tag{3.121}$$

and hence the reflected power can be given as

$$P_r = \frac{\omega U}{Q_e} = P_f \left(\frac{1 - \beta_b}{1 + \beta_b} \right)^2 \tag{3.122}$$

and the stored energy becomes

$$U_0 = \frac{4 P_f \beta_b^2}{(1 + \beta_b)^2} \frac{Q_e}{\omega}. \tag{3.123}$$

3.8.2   Off-Crest Beam Loading

If the beam current is not in phase with the RF voltage, then the beam-loading gains a reactive component, either capacitive or inductive depending on the side of the crest on which the beam arrives. As such, the beam-loading will change the phase of the RF as well as the amplitude. Additional RF power will be required due to the reactance, as it will cause reflections at the input coupler. A similar effect will occur if the generator frequency and the cavity frequency are different, with the cavity presenting a reactance to the generator. It is useful to define a detuning angle, ψ, given by

$$\tan \psi = -2Q_L \frac{\Delta \omega}{\omega}. \tag{3.124}$$

Considering the power transferred to the beam as well as the reflections due to the change in reactance, or generator detuning, the required RF power, P_g, to keep the voltage constant is [30]

$$P_g = P_c \frac{(1+\beta)^2}{4\beta} \frac{1}{\cos^2 \psi} \left[\left(\cos \phi_s + \frac{V_b \cos \psi}{V_{acc}} \right)^2 + \left(\sin \phi_s + \frac{V_b \sin \psi}{V_{acc}} \right)^2 \right], \tag{3.125}$$

where ϕ_s is the phase shift between the cavity voltage and the beam current, P_c is the power required without the beam and V_b is the beam-induced voltage in the cavity given by

$$V_b = \frac{I_b r_s \cos \psi}{1 + \beta}. \tag{3.126}$$

The additional required power can be corrected by tuning the cavity to a different resonant frequency to cancel out the beam's reactance. In this case the cavity should be detuned by

$$\tan \psi = -2Q_L \frac{\Delta \omega}{\omega} = -\frac{I R_s \sin \phi_s}{V_{acc}(1 + \beta)}. \tag{3.127}$$

3.9   Superconducting RF

In 1911 Kammerlingh Onnes discovered that the resistance of some materials disappeared when cooled by liquid helium to a temperature of 4.2 K but it wasn't until 1956 that this effect was explained by Bardeen, Cooper and Schrieffer in what is known as BCS theory (after their initials), which won them a Nobel prize in 1972. Electrons are fermions and thus obey Fermi statistics; this means that the Pauli exclusion principle holds and only two electrons with opposite spins can occupy each energy level. However, in some materials there is a transition temperature, T_c, below which electrons with opposite spins experience an attraction via lattice vibrations and become weakly bound. The transition temperature for niobium, the most common SRF material, is 9.3 K. The bound electrons are known as Cooper pairs, which obey Bose-Einstein statistics; hence the Pauli exclusion principle no longer applies and the Cooper pairs can occupy the same energy state. The Cooper pairs all flow as one with the same velocity and the same direction and are not scattered by impurities, hence the material has zero resistance to DC currents. A key aspect of a superconductor (as opposed to an ordinary very good conductor such as gold) is that when cooled below its transition temperature, all magnetic fields will be expelled. This is known as the Meissner effect, and is caused by supercurrents flowing with no resistance to shield the magnetic field, which is energetically more favourable than allowing the field to enter inside the superconductor.

When cooled to absolute zero, all the electrons are bound in Cooper pairs, while at temperatures between zero and the transition temperature, some of the electrons remain unpaired and behave like normal electrons [18]. This can be considered as two conducting fluids in parallel, one with the normal conducting conductivity and the other with the superconducting state of conductivity. In a DC case, as the superconducting conductivity is so much higher, almost all the current is carried by the Cooper pairs which can flow without resistance. However, when an RF field is applied to a superconductor, the resistance is not zero, although it is very small. While the Cooper pairs move without friction they do have mass and inertia. Because of the inertia, the Cooper pairs do not screen applied time-varying fields perfectly as there is a delay between the current reversing direction and when the electric field is reversed. A time-varying electric field penetrates a small distance into the surface due to induction from the time-varying magnetic field inside the surface. This causes a small power dissipation as the fields at this depth can cause the normal electrons to carry some of the current. London derived two equations to describe the behaviour of a perfect conductor. London realised that the condition for the magnetic field expulsion is

$$\nabla \times \mathbf{j_s} + \frac{n_s e^2}{m} \mathbf{B} = 0, \tag{3.128}$$

where n_s is the number density of Cooper pairs and $\mathbf{j_s}$ is the current density induced in the superconductor's surface by an electric field E. This is known as the 2nd London equation. Using the London and Maxwell's equations we can show that the field will penetrate a short distance into a superconductor, known as the London penetration depth, λ_L, where the magnetic field parallel to the surface will decay though the superconductor as

$$H_z = H_o \exp\left(-\frac{x}{\lambda_L}\right), \tag{3.129}$$

where x is the distance from the surface. The London penetration depth for niobium, the most common superconducting (SRF) material, is 36 nm. London also postulated that the rate of change of the current in time was proportional to the applied electric field;

$$\frac{\partial \mathbf{j_s}}{\partial t} = \frac{n_s e^2}{m} \mathbf{E}, \tag{3.130}$$

known as the 1st London equation. This means that when an RF field is applied, the current will be out of phase with the voltage, and hence the surface impedance has both resistance and reactance. The surface reactance, X_s is given by

$$X_s = \omega \mu_0 \lambda_L. \tag{3.131}$$

The dissipated power can be given in terms of a surface resistance, which is much smaller than the reactance. The resistance of a superconductor at RF frequencies was derived from Bardeen, Cooper and Schrieffer and is hence known as the BCS resistance, R_{BCS} :

$$R_{BCS} = A\frac{\omega^2}{T} \exp\left(-\frac{\Delta}{k_B T}\right), \tag{3.132}$$

where Δ is the band gap of the superconductor, T is the temperature, and A is a material dependant constant. It is generally found that the surface resistance is proportional to the conductivity of the normal conducting state. From this equation it can be seen that the BCS surface resistance has the following dependence:

- R_{BCS} increases $\propto f^2$, shown in Fig 3.33;

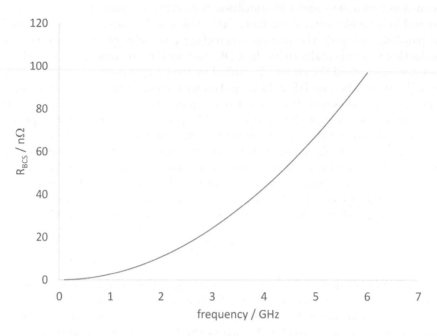

FIGURE 3.33   BCS surface resistance as a function of frequency for a niobium cavity at a temperature of 1.8 K.

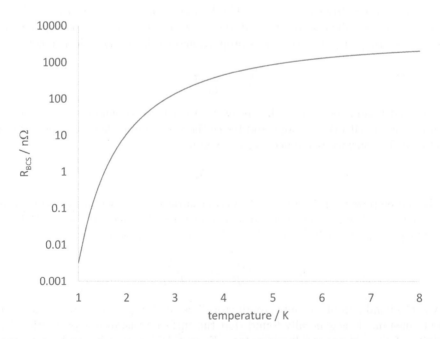

FIGURE 3.34   BCS surface resistance as a function of temperature for a niobium cavity at a frequency of 1.3 GHz.

- R_{BCS} increases exponentially with temperature, shown in Fig 3.34.

Superconducting RF (SRF) cavities have higher losses as they increase in frequency; for this reason there are few SRF cavities in accelerators with operating frequencies above 4 GHz. As SRF cavities have a very high ohmic Q factor, they also have a much larger shunt impedance (which relates voltage to power dissipated in the cavity) than a normal-conducting (NCRF) cavity. This means we can avoid nose cones and can have larger irises. As SRF cavities often operate at lower frequencies and have larger irises – and hence have lower geometric shunt impedance – SRF cavities have much lower wakefields. However in SRF cavities, any energy induced remains undamped in the cavity for long periods of time without special additional couplers. The most common SRF material is niobium (Nb), which has BCS resistance [18]

$$R_{BCS}[\Omega] = 2 \times 10^{-4} \frac{1}{T[K]} \left(\frac{f[\text{GHz}]}{1.5} \right)^2 \exp \left(-\frac{17.67}{T[\text{K}]} \right). \tag{3.133}$$

More recently studies have shown that doping or infusing Nitrogen into Niobium, by annealing in a partial pressure of Nitrogen followed by an electropolish, can provide a lower surface resistance that the predicted limit for bulk Niobium [37]

3.9.1 Residual Resistance

In addition to the BCS resistance there can be additional losses due to impurities or surface layers known as the residual resistance, R_{res}. The total resistance is given by

$$R_{total} = R_{BCS} + R_{res}. \tag{3.134}$$

Typically, a clean cavity operating at a frequency of around 400 MHz will have a residual resistance between 1 and 10 nΩ and the operating temperature is chosen such that the BCS resistance is less than the residual resistance. This means that low-frequency cavities (below 500 MHz) will usually operate at 4.2 K, which is the boiling point of He at atmospheric pressure. At higher frequencies (such as 1.3 GHz) the operating temperature is reduced to $\sim$2 K. As SRF cavities typically operate with resistances between 1 and 10 nΩ, this gives Q factors over 10^9 for elliptical cavities. The residual resistance is thought to increase with cavity frequency as well [38].

A major cause of residual resistance is flux pinning. If we look at what happens as we apply an external magnetic field to a normal conductor and cool it to a perfectly conducting state we can see that the flux lines become 'frozen in' where the conductor becomes magnetised. This causes a problem if a cavity with normal conducting impurities is cooled in the presence of an external magnetic field, H_{dc}, (such as the earth's magnetic field). Supercurrents flow around these trapped magnetic fields. When an RF magnetic field is applied to the cavity, the field lines will oscillate which leads to an increased surface resistance. The additional resistance is given by

$$R_{res} = \alpha_m H_{dc} \sqrt{f[GHz]}, \tag{3.135}$$

where α_m is around 0.2–0.3 nΩ/mG for Niobium. For this reason SRF cavities are normally shielded from all external magnetic fields. Cavities with a thin-film Niobium coating tend to be less sensitive to external magnetic fields. Residual resistance can also be caused by ohmic losses or dielectric losses in the impurity itself, such as a copper inclusion, where the generated heat is transferred to the superconductor, raising its temperature and hence the local BCS resistance. Recent studies have suggested that the creation of thermo-electric currents due to large temperature gradients when cooling the cavity down can cause magnetic fields which increase the residual resistance [39].

3.9.2   SRF Field Limitations

Due to the sharp increase in RF surface losses with temperature, the superconducting state is very delicate at high field. Heating caused by RF or electron phenomena can lead to a thermal runaway, although it should be noted that the thermal conductivity is also temperature dependant. There is also the possibility of phase transitions between the superconducting and normal conducting states at high field. SRF cavities are limited by these effects to just over 50 MV/m depending on the ratio of surface fields to gradient, however ∼30 MV/m is currently about the maximum gradient achievable repeatedly for accelerator applications. The European XFEL chose a design gradient of 23.6 MV/m to ensure a high manufacturing yield, i.e. so that most cavities either achieve or exceed the design gradient [40]; the International Linear Collider (ILC) has a global R&D program to demonstrate reliable performance at 31.5 MV/m [41].

Critical Magnetic Field

The superconducting state is more ordered than the normal conducting state, hence it has less Gibbs free energy, which is the maximum amount of reversible work that can be performed by a system at constant temperature. When an external magnetic field is applied, supercurrents flow in the penetration depth to cancel out the fields in the interior. This causes the Gibbs free energy to rise in the superconductor quadratically with field for the superconducting state. When the field is increased to a level where the free energy of the superconductor is equal to the free energy of the normal state the two phases are in equilibrium. This occurs at the thermodynamic critical magnetic field, H_c, above which all the flux enters the superconductor (although as we will see below, this is modified by surface energy barriers). At this point the cavity is no longer superconducting and the cavity is said to have quenched. The critical field varies with temperature as

$$H_c(T) = H_c(0) \left[1 - \left(\frac{T}{T_c} \right)^2 \right]. \tag{3.136}$$

The transition temperature T_c is the temperature where the superconductor changes between the normal and superconducting state.

There is also a surface energy barrier at the interface between the superconductor and a normal conducting region. This surface energy can be positive or negative. Type-I superconductors have a positive surface energy and all fields will enter the superconductor at H_c. In Type-II superconductors – such as niobium – the negative surface energy makes it energetically favourable for a normal conducting fluxoid to enter the superconductor at a lower magnetic field, known as H_{c1} creating small normal conducting flux tubes inside the superconductor which mostly remains superconducting. As the external magnetic field is increased, more fluxoids penetrate the superconductor in an ordered lattice. The fluxoids have a finite size given by the coherence length, which is the length scale of changes in the superconducting state, so eventually there will be so many fluxoids that they will touch each other and all flux will enter the cavity. This happens at a field higher than H_c, known as H_{c2}. The superconducting state can exist meta-stably in a superheated state higher than H_{c1} at RF frequencies up to the RF critical field, roughly equal to H_c for niobium at 2 K. For niobium at 0 K, we have H_{c1} =170 mT, H_c =200 mT, and H_{c2} =240 mT, and the coherence length is 64 nm.

Thermal Breakdown of Superconductivity

Thermal breakdown is when a superconductor abruptly becomes normal conducting, similar to a quench, caused by the surface temperature exceeding T_c. This is a runaway effect as

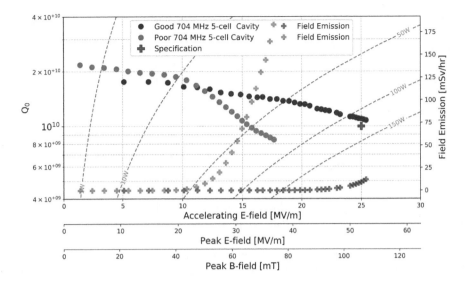

FIGURE 3.35 Q_0 plotted against E_{acc} for two cavities (one good, one poor) from CERN's bulk niobium high gradient SRF programme; image courtesy of CERN [42].

a superconductor's temperature raises its surface resistance, which hence increases power dissipation and temperature; this heats the surrounding area which in turn has its resistance increase. The main cause of a thermal breakdown is the heating of normal conducting impurities on the cavity surface (which heats the superconductor around it), or by heating due to field emission where the emitted electrons impact on the cavity surface depositing their energy. As mentioned previously, the temperature gradient between a cooled wall (in this case by liquid helium) and the RF surface where heat is applied is proportional to the wall thickness and a thermal boundary resistance known as the Kapitza resistance, hence this effect can be reduced by using thinner walls. The downside of this is making the cavities mechanically weak and prone to deformation. As a compromise, most SRF cavity walls are 3–4 mm thick. Another solution is to place a thin film of superconductor inside a copper cavity. As the copper has a high thermal conductivity, they can have thicker walls for a given temperature rise. However, coating a cavity with a superconductor is a developing field and cavity performance is still not comparable with bulk niobium cavities, but is suitable for low gradient applications, such as synchrotrons like the LHC.

Field Emission

In the presence of impurities or defects we also have an electric field limit. This is caused by field emission of electrons from regions of high electric field. These electrons will impact on the cavity surface which will locally increase the cavity temperature, leading to a higher surface resistance at that location. Impurities or defects can cause sharp points on the cavity surface, known as field emitters, which have a field enhancement on the edges causing higher local surface fields leading to field emission. The field emission is also usually accompanied by X-ray emission which can therefore be used to determine if field emission has occurred. It can be seen in Fig 3.35 for the case of the poor cavity that the Q factor drops off steeply in an SRF cavity when it starts to field emit as the heating leads to a higher surface resistance, while in the good cavity, field emission starts at a much higher field, likely due to a cleaner surface.

As previously mentioned, multipactor can also be a limiting factor in SRF due to the additional heat load. Pure niobium has a peak secondary emission yield of around 1.2, however, any oxide layers or contamination on the surface can increase the secondary emission yield above 2.0. Fortunately the electron impacts caused by multipactor act to clean the cavity surface, hence conditioning the surface and reducing the secondary emission yield. Multipactor that persists after a few hours of conditioning is known as a hard multipacting barrier, while multipactor that dissipates after conditioning is known as a soft multipacting barrier.

3.9.3   Cavity Cleaning

In order to achieve high gradients, the cavities must be specially cleaned to remove any particulates which could lead to field emission or increased surface heating. They must be washed with ultra-pure water (in clean rooms), rinsed using high-pressure water jets and have the walls smoothed using acid. Two methods of surface preparation have been developed for surface preparation of SRF Nb cavities. The first is buffered chemical polishing (BCP). This method uses a mixture of three acids: hydrofluoric acid, nitric acid and phosphoric acid. Nitric acid reacts with Nb to form niobium pentoxide, Nb_2O_5. Hydrofluoric acid reacts with the pentoxide to form niobium fluoride, NbF_5, which is soluble, creating a polished Nb surface. The phosphoric acid serves as a buffer to help keep the reaction rate constant. The other surface preparation method is electro-polishing (EP). Here an acid mixture of mostly sulphuric acid with some hydrofluoric acid is used; the cavity acts as an anode and a cathode electrode is placed inside the cavity, with a potential difference of 10–20 V applied, which activates the polishing process. The enhanced electric field at any protrusion will cause the Nb surface to oxidise there first, thereby smoothing the surface [18].

3.9.4   Microphonics and Tuners

All mechanical structures have mechanical resonances, where the transfer of mechanical vibrations from the source (such as a vacuum pump) to the structure is enhanced. In SRF these effects are called microphonics. As mentioned previously, SRF cavities have very high Q factors giving very small bandwidths, usually less than 1 Hz. Mechanical vibrations coming from ground motion, vacuum pumps and other environmental noise will cause the resonant frequency to vary in time by up to 1 kHz, which is three orders of magnitude more than the cavity's bandwidth. When testing a cavity without beam, the LLRF system can rapidly vary the drive frequency to follow the cavity frequency, but in an operating accelerator the drive frequency is fixed. This requires the cavity bandwidth to be increased by decreasing the external Q of the fundamental power coupler.

The klystron frequency, which is normally fixed to a stable reference clock, must be within the cavity bandwidth, hence the cavity frequency must be tuned within that range by squashing its shape using a mechanical tuner. As the bandwidth is so small, any small perturbation must be accounted for. In order to fast tune for microphonics, the mechanical tuner can be fitted with a piezoelectric crystal which expands or contracts depending on a voltage applied to it, allowing a smaller cavity bandwidth to be used.

3.9.5   Cryogenics

The ohmic losses may be much lower for SC cavities than normal conducting cavities by a factor of around 10^5, however significant additional electrical power is required in the system to remove the heat and/or to re-condense the helium as cryogenic refrigerators are very inefficient. The cryogenic system requirements reduce the efficiency of superconducting structures although they are still more efficient overall than normal conducting structures. All refrigerators have a technical efficiency, η_T, of 20–30 %. In addition, we are limited by the Carnot efficiency, η_c, which is the maximum theoretical efficiency any heat engine working between a hot, T_{hot}, and cold, T_{cold}, temperature reservoir can operate at, given by

$$\eta_c = \frac{T_{cold}}{T_{hot} - T_{cold}}. \tag{3.137}$$

The dynamic heat load, P_c, is the RF power dissipated in the cavity walls by the RF fields. A static heat load, P_s, adds additional heating (the static heat load is the power dissipated with no RF in the cavity due to supports and other connections).

Liquid helium transfer lines are another static heat load that typically requires $\sim$0.1 W per metre of cooling (although some flexible connections may have higher losses), so total loss is length L multiplied by loss per metre, i.e. 0.1 L. It is standard to fill to an overcapacity, O, in case extra cooling is required. Hence we can calculate the total electrical power needed for cooling each cryostat, P_{cryo}, as

$$P_{cryo} = \frac{O(P_c + P_s + 0.1L)}{\eta_T \eta_c}. \tag{3.138}$$

Apart from the power required to extract the heat, SRF cavities have very few problems operating with long pulses at their maximum gradient; hence SRF cavities are currently favoured for CW (continuous) applications.

3.10   RF Couplers

The RF is coupled into the cavity from the waveguide via a fundamental power coupler (FPC). The interface between the cavity and the coupler can couple via the electric fields, magnetic fields or both. The coupler can come in rectangular waveguide or coaxial configurations. In the case of waveguide couplers, the field in the waveguide mode (normally the TE_{10} mode) should be matched to the fields in the cavity, with electric and/or magnetic fields aligned in the same direction on either side of the interface. The coupling between the electric fields can be found by matching the cavity field at the interface $\mathbf{E_{cav}}$ to an expansion in terms of the modes inside the coupler

$$\mathbf{E_{cav}} = \sum_{n=1} a_n \mathbf{E_{n,coup}}, \tag{3.139}$$

where $\mathbf{E_{n,coup}}$ is the electric field of the nth waveguide mode at the same interface and a_n is the amplitude of that waveguide mode. Similarly, the magnetic field at the cavity interface, $\mathbf{B_{cav}}$, is expanded as

$$\mathbf{B_{cav}} = \sum_{n=1} b_n \mathbf{B_{n,coup}} \tag{3.140}$$

where $\mathbf{B_{n,coup}}$ is the magnetic field of the nth waveguide mode at the interface and b_n is the amplitude of that waveguide mode. This equation can be solved for each waveguide mode to find the coupling to each mode.

For coaxial couplers we have a choice in the geometry at the end of the coupler where the cavity and coupler meet, that we can optimise to ensure the cavity is critically coupled. If we leave the inner conductor un-terminated with no connection to the outer conductor (known as probe termination), as shown in Fig 3.36, then the electric field of the cavity can create a charge difference between the inner and outer conductor which varies with time,

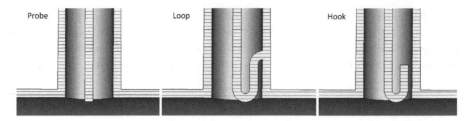

FIGURE 3.36   The three types of termination for a coaxial coupler: probe, loop and hook.

hence acting as a current source in parallel with the capacitance between the inner and outer conductor. The current, I, is given by

$$I = -\frac{dQ}{dt} = -\epsilon_0 \frac{d \int_{tip} \mathbf{E}.d\mathbf{S}}{dt}, \tag{3.141}$$

where E is the electric field on the tip of the inner conductor and S is the surface area of the tip of the inner conductor. If we connect the inner conductor to the outer conductor via a loop then the magnetic field can create a voltage across the hook loop via magnetic induction. This has an equivalent circuit diagram of a voltage source in series with an inductor. The voltage is given by

$$V = -\frac{d\Phi}{dt} = -\frac{d \int_{loop} \mathbf{B}.d\mathbf{S}}{dt}, \tag{3.142}$$

where Φ is the magnetic flux through the loop. A magnetic loop has difficulties in assembly as the inner and outer conductors need to be joined. It is also possible to instead have an inductive hook at the end of the coaxial lines inner conductor that has a small capacitive gap between itself and the outer conductor, also shown in Fig 3.36. Such a termination can be excited by both electric and magnetic fields, however each has a slightly different equivalent circuit. For the hook, the inductor and capacitor are in series with each other, however for magnetic field coupling, this series LC circuit is also in series with the voltage source and for electric coupling, the series LC circuit is instead in parallel with the current source. As the capacitor, C_{gap}, and inductor, L_{loop}, are in series, they form a resonant circuit which acts as a bandstop filter for electric fields and a bandpass filter for magnetic fields, with a resonant frequency

$$\omega_f = \frac{1}{L_{loop}C_{gap}}. \tag{3.143}$$

The equivalent circuit for each type of coupling is shown in Fig 3.37. The choice between types will depend on the chosen coupler location, the cavity fields at that location, and the RF heating on the coupler tip.

3.10.1   Fundamental Power Couplers

The RF is fed into the cavity via a fundamental power coupler (FPC), which is designed to handle high power flow. By varying the geometry of the coupler, hence altering their capacitance and/or inductance, we can vary the external Q of the coupler, in order to match the RF systems. For high-frequency normal conducting cavities, the FPC is almost always waveguide for power handling reasons, while for low-frequency cavities (below 400 MHz) coaxial coupling is preferred to reduce the size. The couplers can be placed in the cavity equator, known as on-cell couplers, or beside the cavity to couple via the beam pipe. SRF cavities normally prefer coaxial couplers, even at higher frequencies, to reduce

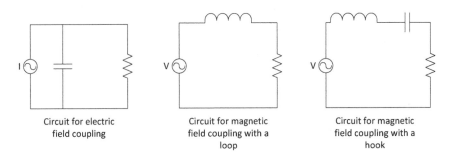

Circuit for electric field coupling

Circuit for magnetic field coupling with a loop

Circuit for magnetic field coupling with a hook

FIGURE 3.37 Equivalent circuits for electric and magnetic coupling.

the heat transport through large waveguides, although for synchrotrons where high power is required, rectangular waveguide couplers are sometimes used as it avoids the problem of cooling the inner conductor. The presence of a coupling hole near the cavity equator enhances the magnetic field and may cause premature thermal breakdown in the case of superconducting RF cavities; hence, SRF couplers are normally placed in the beampipe away from the cavity, although some low-field SRF cavities use on-cell couplers.

In normal-conducting cavities, couplers in the beam pipe can either be placed next to the cavity so that the waveguide couples via the iris, or separated from the cavity via a longer, larger diameter circular waveguide in the beam pipe (such that the beam pipe is not cut-off) known as a mode launcher. The advantage of a mode launcher is that the structure that couples the rectangular waveguide to the circular waveguide can be manufactured separately and connected to the cavity via a flange, although it takes up more space longitudinally. In many linacs there is a requirement to make the fields as symmetric as possible to avoid a transverse electric or magnetic field on the beam axis which may disrupt the beam. To avoid this two transversely opposing waveguide feeds are often used so that power is fed from both sides.

For SRF couplers the design is complicated by the requirement to minimise the heat conduction between the room-temperature interface and the liquid helium vessel. To minimise the thermal conduction, couplers are often made from steel with a thin coating of copper to minimise ohmic losses on the RF surfaces. For a given coupler length, it is inefficient to simply have a temperature gradient between the cold and warm parts; typically there are several stages held at fixed temperatures by cooling with liquid helium at the lowest temperature stage, then helium gas or liquid nitrogen at an intermediate stage in order to minimise the heat deposited at the lowest temperature. Due to the temperature gradient, bellows must be used to allow the coupler to thermally contract when cooling down.

In addition, to keeping the cavity clean, the coupler will have one or two RF windows which are transparent to RF but which are vacuum tight. The windows will be made from a high-resistivity ceramic – such as alumina (aluminium oxide) or beryllia (beryllium oxide) – meaning that the windows have the problem of charging up if they are struck by electrons; hence, care is taken to avoid any line of sight from the beam to the window. However, the window can still be impacted by electrons due to field emission causing them to charge up. This leads to the possibility of multipactor, vacuum arcs, or flashover – the latter where electrons are attracted to the charged ceramic, which on impact produces more secondary electrons, leaving a net positive charge, which in turn are also attracted to the ceramic by the positive charge to give an avalanche. These phenomena can lead to coupler damage, and eventually window metallisation or detuning of the coupler. Multipactor can be avoided in coaxial couplers by providing a DC bias between the inner and outer conductors. Another major cause of window failure is mechanical stress caused by thermal gradients along the window.

Many coaxial couplers for SRF cavities operating at frequencies above 0.4 GHz will connect to a rectangular waveguide, and hence a special coupler known as a doorknob is used to transition between the coaxial line and the waveguide. All of the features of an FPC need matching to the RF at the resonant frequency which results in the coupler having a narrow bandwidth.

Common causes of failure in superconducting cavity FPCs include:

- Vacuum leaks/cracked window;

- Overheating;

- Arcing/breakdown;

- Window metalisation;

- Multipactor;

- Band-pass detuning.

3.10.2   HOM Couplers

As was discussed earlier, a bunch of charged particles will decelerate and deposit RF energy into undesired modes in the cavity, in a process known as wakefields. These wakefields excite the higher-order modes (HOMs) of the cavity (i.e. modes of higher order than the fundamental TM_{110} mode of the cavity), which can then have unwanted effects on the beam. In order to reduce the effects of these wakefields it is necessary to damp (reduce the energy in) these modes using special couplers to remove this power. HOM couplers are designed to couple power from cavity HOMs out of the cavity to a resistive load. However, they must not take power out of the cavity at the fundamental frequency. To avoid this, the coupler must use a high-pass or band-stop filter. This can be implemented in two ways:

- use a waveguide with the cut-off frequency above the fundamental frequency (high-pass);

- use a band-stop filter in a coaxial line using inductive stubs (metal cylinders connecting the inner and outer conductor) on the inner conductor with a small capacitive gap in the stub.

Waveguide couplers are often larger than coaxial couplers but can handle higher powers. They are very simple, but their size can often be a problem in SRF applications as it provides a thermally-conducting path between the cold and hot parts of the cryomodule. Waveguide couplers often have stronger coupling to the HOMs, and can handle higher HOM powers with less RF loss, are simpler to cool, and have less chance of electron activity such as multipactor and are hence favoured for high-current applications.

Coaxial HOM couplers are very complicated and include many inductive stubs and capacitive gaps in order to minimise coupling at the fundamental frequency and maximise coupling at the most problematic HOMs. If the inner conductor is large enough, it may be possible to have water or helium flow inside of it for cooling. The complicated geometry can also cause problems with multipactor or arcing.

If the beam current is high enough where wakefields are an issue in normal conducting cavities waveguide couplers are mounted on every cell, with an RF load composed of a Silicon Carbide (SiC) wedge installed in each waveguide. For very high-current applications, RF absorbers can be placed in the cavity beam pipe allowing frequencies above the waveguide cut-off to be strongly damped. Modes with frequencies less than the beam pipe cut-off will decay exponentially in the beam pipe, with the decay sharper at lower frequencies, hence at frequencies close to the cut-off the mode can still be damped if the fields haven't decayed before the absorbers.

3.10.3 Coaxial HOM Couplers

The complex pass-band structure of coaxial couplers are often modelled using equivalent circuits. Like FPCs they can have capacitive or inductive coupling. The reactive coupling element will reduce the power deposited in the load, but at a single frequency the reactive element can be compensated for by using another reactive element with the opposite sign. Capacitive coupling can be compensated with a parallel inductor, taking the form of a stub, and an inductive coupler can be compensated with a series capacitance (a gap). The compensation frequency, ω_{comp}, for capacitive coupling (with capacitance C) compensated with a stub is given as

$$\omega_{comp} = \frac{1}{L_c C} \tag{3.144}$$

where L_c is the inductance of the compensating stub. The reactance of an element, of impedance Z_s, can be varied by using a transmission line, of impedance Z_c and length L, between the element and the measurement point. The impedance, Z, at the measurement point is given by

$$Z = Z_c \frac{Z_s + i Z_c \tan k_z L}{Z_c - i Z_s \tan k_z L}. \tag{3.145}$$

As it is easier to implement a stub than a gap – since stubs also provide mechanical support and cooling while gaps need additional support structures – any coupling element can be compensated by a stub and a length of transmission line [17].

Compensating at one frequency to get higher transmission, by cancelling the antenna's reactance with a component with the opposite reactance at that frequency (normally the frequency of the highest shunt impedance HOM), will cause the reactance to be higher at other frequencies, as a capacitor's reactance will decrease with frequency while an inductor's reactance decreases with frequency. This can also result in stopbands, where no power is transported in a finite frequency band, due to resonances between two reactances separated by a distance at high frequencies where the gap is comparable to a quarter or half of the wavelength depending on the exact components.

In order to filter the fundamental mode frequency we can place a gap between the stub and the outer conductor, giving a capacitance, C_f i series with the inductance, as shown in Fig 3.38, with the filter centre frequency, ω_f given by

$$\omega_f = \frac{1}{L_c C_f} \tag{3.146}$$

where C_f is the filter capacitance. The addition of this capacitance will slightly alter the compensation frequency as well.

A real HOM coupler for the LHC crab cavities [43] is shown in Fig 3.39. Here a hook coupler is used as the coupler is placed in a region of high magnetic field but low electric field, with a hook chosen over a loop in order to have the couplers be demountable. A cylindrical electrode is placed between the inner and outer conductor, supported by a stub on the inner conductor, with the capacitance and inductance chosen to reject any coupling at 400 MHz (the cavity's operational frequency). The inner conductor has a large radius and is attached to the top of the can to provide strong cooling. The coupler bends by 90° at the top, before having a capacitive gap between the inner conductor and the pick-up. By altering the distances between elements, we can create a high-pass filter and provide additional damping at the frequencies of the most dangerous HOMs.

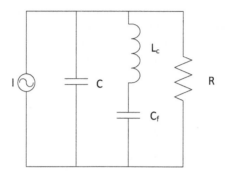

FIGURE 3.38   Equivalent circuit of a coaxial HOM coupler.

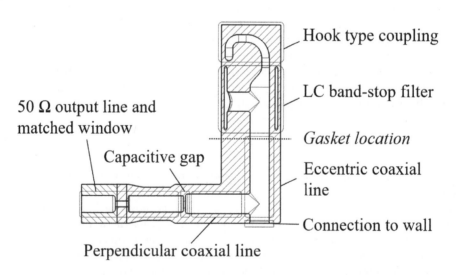

FIGURE 3.39   Coaxial HOM coupler for the LHC double quarter-wave crab cavity.

TABLE 3.2 Comparison of CLIC and ILC parameters.

Parameter	Units	CLIC	ILC
structure type		$2\pi/3$ TWS	Coupled-cavity SW
frequency	GHz	11.9942	1.3
gradient	MV/m	100	31.5
E_{peak}	MV/m	250	63
Q_0		7245	$> 5 \times 10^9$
shunt impedance	MΩ/m	95.4	2590000
input power	MW	62.4	0.311
cavity length	m	0.233	1
filling time	μs	0.066	565
min aperture	mm	2.35	70

3.11   Cavity Geometries

There are several different types of cavity geometry depending on the velocity and species of the particles to be accelerated. Some have higher shunt impedance for low particle velocities and small gaps, while others are more suited to particles travelling at virtually the speed of light. At low particle velocity, RF defocusing is an issue, as will be discussed in Chapter 5, requiring low-frequency cavities. Low frequency cavities typically require special cavity shapes to keep the cavity size to practical limits. In proton and ion synchrotrons it is necessary to change the cavity frequency as the beam is accelerated which requires cavities that can quickly and repeatably alter their frequency.

We have already discussed disk-loaded structures and side-coupled standing-wave cavities. As these operate in the π and bi-periodic $\pi/2$ modes respectively, they are best suited to high particle velocities. As the gap length is reduced to maintain synchronism with lower-velocity particles the distances between the disks is reduced, as shown by Equation 3.100. This reduces the shunt impedance since more of the length is now taken up with the disks which have a finite thickness reducing the gradient, and the increased RF losses on the disks increases the power losses. As a result, other cavity shapes may be more effective for use with low-velocity protons and ions. Typically disk-loaded cavities or side-coupled cavities are used for particle velocities above $0.5c$; however, structures have been realised at lower particle velocities [44]. Since we use the symbol $\beta = v/c$, cavities designed for low, medium and high particle velocities are referred to as low-beta, medium-beta and high-beta cavities.

For lower frequencies ($<$200 MHz) the cavity size can be difficult to realise practically for disk-loaded cavities, with diameters 0.76–1 times the wavelength depending on the cell length and aperture size. TEM- and TE-mode cavities – which can have smaller diameters with respect to wavelength – may be more practical.

The choice of a superconducting or a normal conducting cavity changes the cavity parameters greatly. In Table 3.2 we see a comparison between the two proposed designs for the next big linear lepton collider, CLIC [16] and ILC [15, 45]. As can be seen the NCRF CLIC cavity has a gradient reach three times higher than the SRF ILC cavity and the cavity fills 4 orders of magnitude faster, due to the higher frequency and high group velocity. However the ILC cavity needs 100 times less RF power and has an aperture 30 times larger, reducing the wakefields considerably.

3.11.1   Elliptical Cavities

Disk-loaded cavities have issues with multipactor at high gradient with electrons performing cyclotron orbits every half RF period in the magnetic field at the equator. For normal-conducting cavities this is not a major issue, but for superconducting cavities the heat

FIGURE 3.40   Five-cell elliptical cavity for LEP; image courtesy of CERN.

generated can severely limit cavity operation. Initial SRF cavities were limited in this way but later cavities avoided this by using elliptical geometries as mentioned previously [26]. Electrons strike the surface at different angles depending on the orbit radii, causing the secondaries to move in successive orbits towards the centre where the electric field is zero and electrons cannot be accelerated to sufficient energy to create new secondary electrons.

Initially, the equator ellipse size was limited to ensure a sloped wall angle [15], to allow acid and water to drain more effectively from the cavities during cleaning, but this requirement is no longer felt to be necessary [46]. By varying the wall angle we can change the ratio between the peak surface electric and magnetic fields. Early elliptical SRF cavities were limited by field emission and hence a large positive slope was used to minimise the peak electric field. Modern cleaning methods have reduced field emission such that the cavities are now limited by magnetic field effects such as heating, and hence smaller slopes, or even negative slopes can be used. The elliptical cavities for LEP are shown in Fig 3.40.

3.11.2   RF Electron Guns

RF guns are electron sources with a photocathode installed inside an RF cavity. The electrons will leave the cathode at an energy of a few eV, and should be accelerated as quickly as possible to avoid the beam being blown up by its own self-fields (so-called space-charge forces). Electrons will become relativistic in a single 3 GHz cell at gradients above 30 MV/m, hence only the first cell needs modification, normally being around a half-cell long. At higher frequencies further cells may need their length modified as the smaller gap means the beam will not be fully relativistic at the exit of the first cell.

As the beam is at low energy it can be very sensitive to dipole or quadrupole components of the field caused by coupler asymmetry; these may cause the electrical centre of the cavity to shift off the beam axis (dipole) or cause the field to vary with radius differently in the horizontal and vertical planes (quadrupole). This is avoided in two ways, either by using two couplers to make the field symmetric to avoid dipole components (while using an elliptical cross section to reduce the quadrupole component), or by using a coaxial coupler inside the

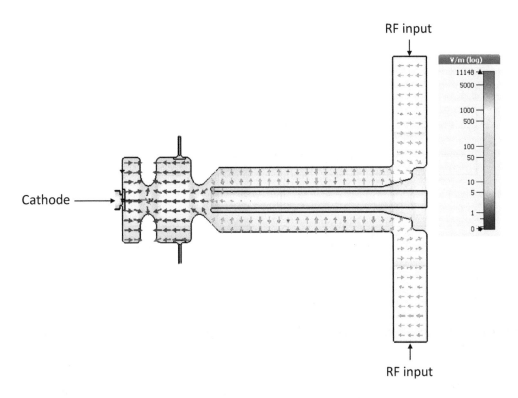

FIGURE 3.41   RF gun for the CLARA accelerator with coaxial coupling.

beam pipe to maintain azimuthal symmetry as shown in Fig 3.41.

If a coaxial coupler is used it will have a door-knob transition to a waveguide away from the cavity. Normally this is fed from a single side, with a short circuit on the other side at a fixed distance to cancel out any reflections. This will excite an additional dipole component, but if the dipole mode is cut-off in the coaxial line, this will not be transmitted to the cavity. The cut-off frequency of a coaxial line, of inner conductor radius a and outer conductor radius b, is given by

$$\omega_c = \frac{2c}{a+b}. \tag{3.147}$$

The size of the inner conductor must be large enough to allow the laser beam to be brought in to the cathode, and hence in many cases the dipole mode will not be cut off. In such cases a dual feed door-knob can be utilised as shown in Fig 3.41 [47].

3.11.3   Half-Wave Resonators and Spoke Cavities

For low-energy proton and ion beams the gap must be reduced to keep the transit-time factor high for single cells and to maintain synchronism for multi-cell structures, as the particle velocity is lower. As the gap gets smaller elliptical cavities become less mechanically stiff, and microphonics becomes a limiting issue. A common geometry for low-beta cavities is the coaxial resonator, made up of a length section of coaxial line, with conducting walls at both ends, shown in Fig 3.42. As the electric field component parallel to the walls must be zero on the conducting walls at the ends, the resonator will have resonant frequencies such that there is an integer number of half wavelengths between the two ends in the TEM mode, making it smaller than an elliptical cavity at the same frequency, as it only needs to be long in one axis. As the resonator is operated in the TEM mode, there are no longitudinal field

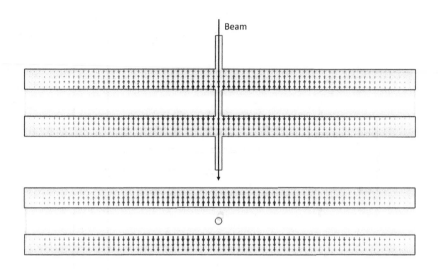

FIGURE 3.42   Electric field pattern inside a half-wave resonator; (top) the beam is travelling down the page; (bottom) the beam is travelling out of the page.

components and hence the structure should be oriented such that the beam will travel in the cavities radial direction, between the outer and inner conductor, to be accelerated by the radial electric field. The cylindrical shape and the inner conductor provide additional mechanical stiffness reducing the sensitivity to microphonics. For medium-beta ($\beta \sim 0.4$) the half-wave cavity starts to become less mechanically stable. This can be remedied by varying the outer conductor orientation, in a structure known as a spoke resonator [48]. In these cavities a half-wavelength rod is placed radially across a cylinder, as shown in Fig 3.43. These structures work well at intermediate particle velocities, $0.15 < \beta < 0.62$. Spokes can be sensitive to multipactor; however, altering the shape of the outer conductor can mitigate this [49]. If multiple cells are required to maximise the voltage that can be obtained in a finite length, several rods can be placed inside one cylinder along the length creating multi-cell cavities [50]. The rods are strongly coupled to each other as there are no walls between them to prevent the field from one rod reach the next rod. Spoke resonators have also been proposed for accelerating relativistic electrons as they are smaller radially than elliptical cavities [51].

3.11.4   Quarter-Wave Cavities

For even lower frequencies, even half-wave resonators become too large due to the need to be a half-wavelength long in one axis. The resonator size can be reduced by a factor of two in the long axis by using a quarter-wave cavity instead [53]. Quarter-wave cavities are again coaxial resonators; however, while one side has a conducting wall at the end, the other side has a gap between the inner conductor and the wall creating a capacitive loading of the resonator. At the capacitive gap, the potential on the inner conductor produces a longitudinal electric field across the gap, allowing the electric field to be maximised at one end in the gap and zero at the other end at the conducting wall, making the resonator approximately a quarter wavelength long and so making it half the transverse size of a half-wave resonator. If we consider that the admittance at the end of the inner conductor,

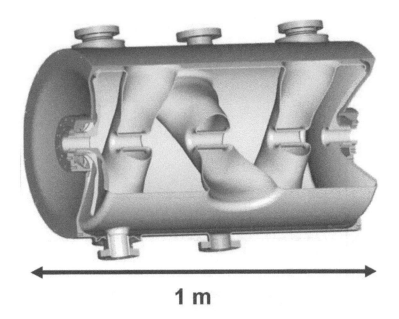

1 m

FIGURE 3.43 Sectioned view of a 345 MHz triple-spoke-loaded cavity for $\beta = 0.5$ from [52].

at the capacitive gap, should be zero, we can state that the admittance of the capacitor should be equal and opposite the admittance of the line. The impedance of the line varies along the line, l, as, $Z(l) = Z_c \tan(k_z l)$, hence the resonant frequency for a line of length L can be given as

$$-\frac{i}{\omega C} = Z_c \tan k_z L \tag{3.148}$$

where $k_z = \omega/c$ and C is the capacitance of the gap at the end.

As there are electric fields in the gap between the inner conductor and the end plate as well as between the inner and outer conductor, a quarter-wave resonator can be oriented to accelerate electrons travelling either radially or longitudinally depending on if it is better suited to make it compact longitudinally or transversely, respectively. Where the electron beam travels between the inner and outer conductor, there is typically a small beam tube cut radially through the inner conductor near the tip of the inner conductor, creating an accelerating gap on either side of the inner conductor; this is shown in Fig 3.44. There will still be some magnetic field at the beam tube and hence care must be taken to ensure the beam's trajectory isn't disrupted due to this. The quarter wavelength is in the transverse direction, hence these cavities are transversely large. Quarter-wave cavities can also have electron beams travel longitudinally parallel to the coaxial line rather than radially, but in these cavities the ratio of the cavity length (around $1/4$ of the wavelength) to the accelerating gap is large but the radius can be very small. There is a beam pipe cut into the inner conductor and the beam only experiences acceleration in the small gap between the inner conductor and the end plate, as shown in Fig 3.45. Such geometries have been proposed as low-frequency RF electron guns, and for very low-frequency cavities such as the 56 MHz cavities in RHIC [54].

Another example of a quarter-wave resonator is the RF system in many cyclotrons. In a cyclotron the acceleration takes place in the capacitive gap between two electrodes, known as Dees as the original designs had 'D'-shaped electrodes. However the Dees themselves are not resonant structures at the low frequencies required in cyclotron RF systems; hence in order to have a resonant structure, the Dees are each connected to a quarter-wave resonator

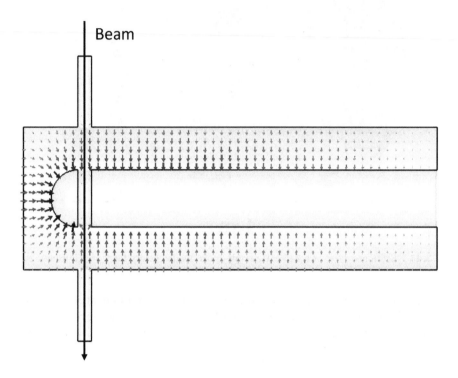

FIGURE 3.44   Electric field pattern inside a vertically-oriented quarter-wave resonator.

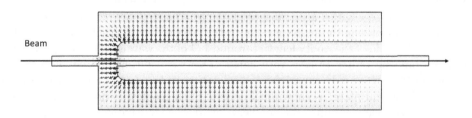

FIGURE 3.45   Electric field pattern inside a longitudinally-oriented quarter-wave resonator.

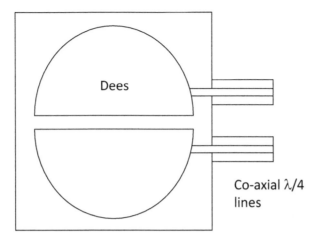

FIGURE 3.46 Geometry of the RF electrodes used in early cyclotrons, showing the Dees and the quarter-wave lines. The entire RF system and liner is inserted between the two cyclotron poles from the side.

called a stem, made of a coaxial line with a short at the end and the Dees forming the capacitive ends. These stems can be seen in Fig 3.46 and a more modern vertical stem in Fig 3.47. Often these shorts are movable to provide frequency variation. In very large fixed-frequency cyclotrons – such as the PSI 590 MeV cyclotron [55] – the quarter-wave resonators are sometimes replaced with large waveguide resonators. It is also possible to use a double-gap system such that the central electrode can support a TEM mode, and the electrodes are made to be a half-wavelength long [56].

In proton and ion synchrotrons the RF frequency, and hence the cavity frequency, has to change as the beam is accelerated and as the revolution time changes. One method of doing so is to load the cavity with a ferrite, which is a ferromagnetic material with lower RF losses [57]. A ferrite has a permeability that varies as a function of applied magnetic field. An electromagnet can be used to bias the ferrite, changing the permeability and hence the resonant frequency. These are typically longitudinally-oriented quarter-wave cavities with rings of ferrite placed in the base of the cavity where the magnetic field is strongest. Amorphous and nano-crystalline magnetic alloy materials can also be used that have much higher permeability and a much lower Q. The low Q gives a wide frequency range such that tuning may not be required.

A comparison of quarter-wave, half-wave, spoke and elliptical cavities/resonators is shown in Fig 3.48 showing the particle velocity and frequency range where each is most effective. Generally, higher frequency cavities are preferred as they are smaller, however one would not use a low-beta cavity at high frequency due to transverse defocusing as discussed in Chapter 5. A high cavity frequency may also limit the beam pipe aperture creating stronger wakefields.

3.11.5 Drift-Tube Linacs (DTLs)

A fraction of the RF losses in a disk-loaded waveguide occurs on the disks. The distance between the centres of any two disks is proportional to the beam velocity for a fixed phase advance, and so the number of disks per metre, and hence the RF losses on the disks, increases as the beam energy decreases. At a certain point disk-loaded waveguides start to become very inefficient and hence accelerating cavities without disks are required. In

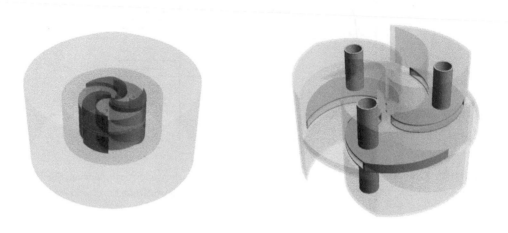

FIGURE 3.47 Illustration of the Dees and RF liner in a modern AVF cyclotron that includes pole hills and valleys. On the left is shown the entire yoke, pole and RF system. On the right can be seen the 3 Dees with vertical stems, surrounded by liners situated in the valleys.

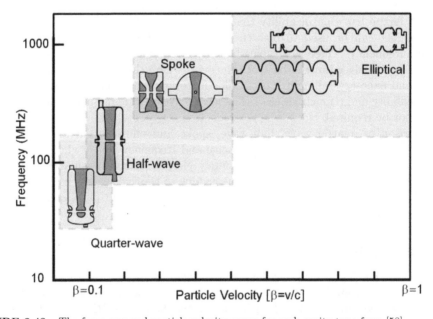

FIGURE 3.48 The frequency and particle velocity range for each cavity type from [58].

drift-tube linacs the distance between gaps is longer than the distance the beam travels in a half RF period and hence the beam would normally be in the gap during the decelerating phases. To avoid the beam being decelerated, the particles need to be shielded from the RF fields during phases that would decelerate the beam by having the beam pass through small aperture beam pipes, referred to as drift-tubes, as previously mentioned. Hence these devices are known as drift-tube linacs (DTLs).

There are two standard drift-tube types:

- Wideröe linacs operate at very low frequencies in a TEM mode, with every drift-tube held at the opposite potential from the drift-tube at either side, which alternates in time such that the RF is accelerating when the beam arrives at the gap between drift tube;

- Alvarez linacs have a TM_{010} mode in a long (compared to the distance the beam will travel in an RF period) cylindrical tank with several drift tubes separated by an integer number of wavelengths, shown in Fig 3.49.

In both cases as the beam is shielded from the RF most of the time, the gradient and shunt impedance is low, as well as the gradient which is typically 5–10 MV/m, but at low beam velocity they perform well compared to other cavities. An Alvarez DTL typically has a shunt impedance of about 50 MΩ/m at 20 MeV, which drops to around 20 MΩ/m at 200 MeV [59] whilst a disk-loaded cavity typically has a higher shunt impedance at higher energies. The drift tubes are supported by stalks, which in the case of Alvarez linacs, are a quarter wavelength long making them resonant, which makes the fields less sensitive to variations in dimensions. In order to focus the beam, quadrupole magnets can be placed inside some of the drift tubes. As the beam velocity increases, it is possible to use a hybrid geometry, where a coupled cavity linac can have one or two drift tubes inserted inside each cell, giving a higher shunt impedance at intermediate beam energies. Such cavities are known as coupled-cavity, drift-tube linacs (CC-DTLs).

Typically, Wideröe linacs are used for very-low-velocity particles like heavy ions, where we need a very low frequency to reduce RF defocusing (see Chapter 7). Alvarez DTLs are commonly used for proton machines at intermediate particle velocities – like Linac4 at CERN – for particle velocities in the range $0.05 < \beta < 0.5$. The Alvarez DTL in CERN's Linac4 operates at a frequency of 352 MHz and this requires a tank diameter of 500 mm. It is subdivided into three tanks and is 19 m long in total, with about 110 drift tubes to accelerate protons from 3 MeV to 50 MeV (β varies from 0.08 to 0.31), taking the cell length from 68 mm at the entrance to 264 mm at the exit [60].

3.11.6 TE Mode Linacs

In order to accelerate particles, we require a longitudinal electric field, hence TE modes (known as H modes in some countries) in constant cross-section cavities cannot be used to accelerate charged particles. However, by inserting crossbars, shown in Fig 3.50, inside a resonant tank we may perturb the fields to give them a local longitudinal electric field near the crossbar. As it only extends a short distance from the crossbar these structures, known as CH ('crossbar H-mode') structures, are only useful at very low velocities. A similar device uses interdigital stalks, shown in Fig 3.51, to achieve the same effect; this is known as an IH ('interdigital H-mode') structure. These structures are smaller than Alvarez DTLs for a given frequency, which is useful when using very-low-frequency systems where the wavelength can be several metres in size.

TE modes typically have lower surface magnetic fields than the TM_{110} mode and hence ohmic losses are reduced allowing very high shunt impedance, close to 100 MΩ/m at low

FIGURE 3.49 The Alvarez drift-tube linac for Linac4 at CERN; image courtesy of CERN.

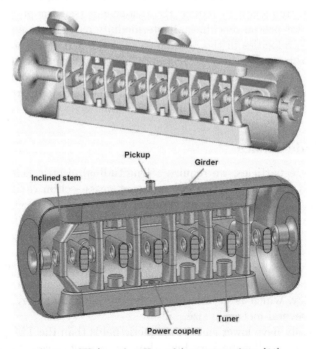

FIGURE 3.50 Superconducting CH (crossbar H-mode) resonator from [61].

FIGURE 3.51 IH (interdigital H-mode) resonator for ISAC radioactive ion beam facility at TRIUMF; [62]

particle velocities; however, this drops sharply with increasing particle velocity. For this reason CH structures are only used for $\beta < 0.3$ [63].

3.11.7   Radio-Frequency Quadrupoles (RFQs)

For very-low-energy hadrons (such as protons) the space charge in the beam at low energy can blow the beam apart; hence, for intense beams we need to focus, bunch, and accelerate the particles at the same time. As we will see in Chapter 5, RF focusing and bunching occur at opposite phases, such that focusing phases are debunching and defocusing phases are bunching. Focusing of the beam can be achieved, while simultaneously accelerating the beam, by using electrostatic quadrupoles; four electrodes are used, with each electrode having the opposite potential to the electrodes on either side of it. This creates a focusing electric field in one plane (say, x), and defocusing in the other plane. The focusing plane is alternated by the longitudinal oscillation of the wave on the line to achieve a net focusing effect in both planes. If we have a corrugation on the surface of each electrode but with a longitudinal separation in the peaks, we can create an additional longitudinal field component, which can be used to bunch and accelerate. Such a device is called a radio-frequency quadrupole (RFQ). There are two types of RFQ: vane and rod. A four-vane RFQ operates in a TE mode, with azimuthal index $m = 2$ giving a quadrupole mode. These are simpler to manufacture but are only of feasible size at higher frequencies above 200 MHz. The four-vane RFQ for Linac4 at CERN is shown in Fig 3.52. A 4-rod RFQ has a corrugated longitudinal rod as each electrode, allowing the structure to operate in a TEM mode making the transverse size independent of the operating frequency; hence these structures are mostly used where lower frequencies are required [30].

FIGURE 3.52   Four-vane RFQ for Linac4 at CERN; image courtesy of CERN.

3.12   RF Sources

RF structures require input powers between tens of kW to tens of MW to reach gradients of tens of MV/m. Typically, for large accelerators, the RF source is always an amplifier where the output signal is a higher-power copy of the input signal. Unlike low-power RF oscillators, high-power oscillators are typically not stable enough to use when two or more sources are required to be combined or synchronised, although phase-locked oscillators are possible for long pulses but are rarely used. This is often due to electron loading or thermal effects at high-power. For small industrial accelerators – where there is no need to synchronise – oscillators can be used. Typically an RF system will comprise a high power RF (HPRF) amplifier and a low-level RF system (LLRF), which will take feedback from the cavity and send the correct drive signal to the amplifier to keep the cavity voltage at the setpoint voltage and phase.

RF amplifiers are typically characterised by a few key parameters:

- Saturated output power: This is the maximum RF power an RF source can produce when overdriven. No accelerators operate at this power level as the control system will need to increase and decrease power to keep the cavity voltage constant in the presence of disturbances, so typically the operating power is 1 to 3 dB less than the saturated output power.

- Gain: This is the ratio of RF output power to RF input power, typically expressed in dB. In many devices this is constant at low to intermediate power but decreases with increasing output power. The 1 dB compression point is the output power where the gain is reduced by 1 dB from the gain at intermediate powers. Gain is given by $G = 10 \log P_{out}/P_{in}$. Typically an amplifier will not have enough gain to go from the LLRF power to the operating power in a single stage, hence a series of lower power amplifiers are often required.

- RF efficiency: This is the ratio of the saturated RF output power to the DC input power. High-power RF sources typically operate at efficiencies at or below 50%, significantly increasing the electricity costs of the facility. Any remaining energy in the electron beam must be dissipated in a load known as a collector. Proposed future high-energy lepton colliders have a total RF power usage of 100–180 MW so the difference between a 40% efficient and an 80% efficient amplifier has a major impact on running costs.

- Harmonic content: This is the ratio of the output power at the design frequency to the output power at the harmonics of the drive frequency. This is measured in dBc (decibels relative to the carrier).

High-power RF sources for accelerators come in many varieties, and different types depending on the frequency and power required [64].

3.12.1 Gridded Tubes

In these devices a biased metal wire grid is placed close to the cathode in a vacuum diode. As the bias grid is closer to the cathode than the anode, it can create the same electric field with a lower voltage and can hence control the space-charge limited emission current in time by varying the bias voltage, with emitted electrons being accelerated to the full anode-cathode voltage. The wires in the grid are thin to avoid intercepting the electrons. Typical devices are triodes and tetrodes (which include a 4th screening grid), shown in Fig 3.53. These devices are typically low gain, around 13 dB, and have issues at higher frequencies as the electron must pass the cathode-grid gap in a half RF period, and hence tend to be used below 500 MHz.

A more efficient coupling from the beam to the RF can be obtained by replacing the anode with a resonant cavity with a high shunt impedance. This type of device is known as an inductive output tube (IOT) and is shown in 3.54. These devices have more gain than a tetrode (20–30 dB) and operate up to ~3 GHz but tend to have relatively low output powers of under 100 kW.

In all gridded tubes the grid can be DC biased to change the current waveform, known as different amplifier classes. If the DC and RF voltages are equal, known as class A, the device conducts at all phases providing perfect sinusoidal current profiles, but at the cost of efficiency due to the large DC component. If there is no DC bias, known as class B, the current waveform is a half-wave rectified sinusoid, and hence has higher harmonics of the RF frequency. In class C amplifiers a negative DC bias is used so the device only conducts for a small fraction of the RF period, giving even more harmonics but highest efficiencies.

3.12.2 Klystrons

To obtain higher powers and/or frequencies we cannot utilise a grid. The grid can be avoided by utilising velocity bunching where a DC beam traverses an input RF cavity which accelerates some electrons and decelerates others. As the electron beam travels down the beampipe the faster particles catch up with the slower ones forming discrete bunches. This effect can be enhanced by using several additional intermediate bunching cavities, which are excited by the bunched beam but phased to provide further bunching, as shown in Fig 3.55. When fully bunched the beam passes through an output cavity, tuned to maximise the power output, and is then dumped [64]. Klystrons can provide powers of tens of MW up to frequencies of tens of GHz, and have very high gain (~ 50 dB). However, velocity bunching isn't perfect, and due to the requirement to operate below maximum power to allow overhead for RF control, klystrons do not operate at very high efficiencies ($\sim$30–40% at operation). More recently developments have investigated high efficiency klystrons providing maximum efficiencies of 80% [65]. The lower the beam current for a given voltage, the higher the

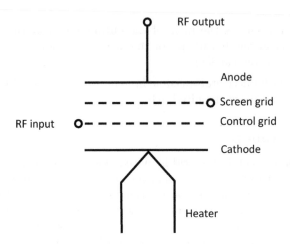

FIGURE 3.53   Layout of a tetrode gridded tube.

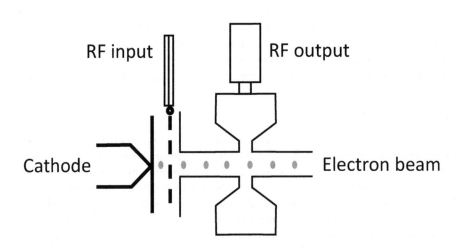

FIGURE 3.54   Diagram of an IOT.

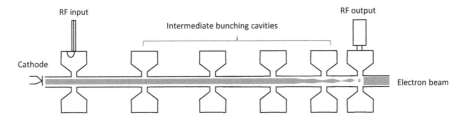

FIGURE 3.55 Geometry of a klystron.

efficiency so splitting the beam current into several lower current beamlets that traverse the same cavities, known as a Multi-Beam Klystron (MBK), can provide higher efficiencies. As klystrons can be quite long a solenoid magnet is required to confine the beam. High power klystrons can require very high cathode voltages up to 500 kV, requiring the high voltage end to be operated inside an oil tank to prevent arcing.

3.12.3   Solid-State Power Amplifiers

For other lower-power applications, semiconductor transistor amplifiers are commonly used, but these are limited to around 100 W for laterally diffused metal oxide semiconductors (LDMOS). For high average power applications, thousands of these LDMOS transistors can be combined to achieve hundreds of kW of RF power. At higher frequencies, above 1 GHz, GaN transistors are preferred for their higher efficiency. The big advantage of solid-state amplifiers is that the transistors fail gradually and with regular maintenance the amplifier can be made to run without downtime. This is particularly important for 3rd generation light sources. They also operate at much lower voltages and can be air cooled. As the size and cost is dominated by the peak power, such devices are large and not cost effective for short pulse, high peak power applications.

3.12.4   Magnetrons

Most particle accelerators require several RF cavities to be individually powered, but each cavity should be synchronised with the beams arrival time, meaning that only amplifiers where the phase can be tighly controlled can be used. For applications requiring only a single RF structure with a DC electron beam, oscillators can be used instead as there is no requirement for synchronising. The magnetron is the most common high-power RF oscillator, due to its compact size, low cost, and high efficiency, and is commonly utilised in industrial and medical linacs. In a magnetron, electrons are launched from the cathode at the inner conductor of a coaxial line. The electrons are made to follow circular orbits due to an external axial magnetic field. The magnetic field is chosen so the electrons fall just short of the anode/outer conductor and return to the cathode. Interaction with an RF field means that electrons that are decelerated in the first half cycle (hence giving energy to the RF field) will lose energy and will have a larger cyclotron radius and will hence hit the anode, rather than returning to the cathode and being accelerated on the 2nd half cycle. To enhance the process the anode is formed into a series of resonant cavities, with the use of vanes. The electrons that gain energy from the RF form a cloud around the cathode known as the sub-synchronous zone while the electrons losing energy to the RF form spokes as can be seen in Fig 3.56. Magnetrons for accelerators operate up to 9.3 GHz and provide a few MW of RF power but their oscillation frequency can vary by 0.1 % due to thermal expansion, reflected power, magnetic field changes and power supply ripple. For this reason they are typically only used to drive single cavity industrial and medical linacs where the drift in drive frequency is not an issue.

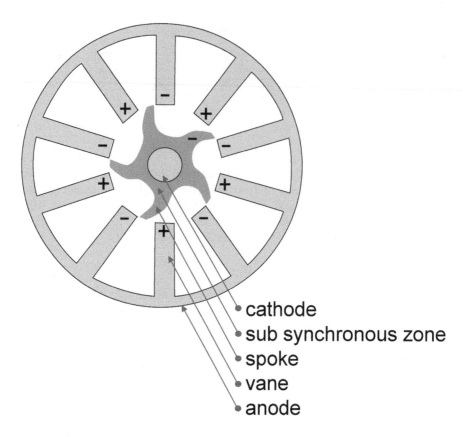

FIGURE 3.56   A vaned magnetron showing the electron beam spokes.

It is possible to seed a magnetron such that the oscillations will phase-lock to an externally injected RF signal, however due to the frequency variations mentioned above, it takes a significant amount of RF power to lock a magnetron. More recently there has been research into providing feedback to reduce the frequency variation by altering the power supply or the magnetic field [66]. The feedback means that the magnetron will phase lock at reduced input powers, opening the door to cheaper RF sources for high average power applications.

3.12.5   Dielectric Laser Accelerators

To achieve higher gradient we can use higher-frequency RF, moving to millimetre waves, THz frequencies or even higher. The breakdown rate is known to scale with frequency and pulse length, both of which allow higher field strengths at higher frequencies. However, as the frequency increases, the wavelength decreases, making the structures much smaller and harder to manufacture. At smaller wavelengths, a particle bunch will cover a wider range of RF phases making capture of electron beams more difficult. There are several methods of interacting with high-frequency accelerators:

- diffraction gratings [67],
- inverse free-electron laser [68],
- scaled-down RF cavities [69],
- photonic bandgap structures [70],

- waveguide loaded with dielectric or corrugations [31] [71].

Significant gradients, above 300 MV/m, have already been achieved at optical and infrared wavelengths [67], but as the beam is often longer than the wavelength, the beam obtains a large energy spread. To avoid this we need to go to longer wavelengths, with the ideal wavelength being around 0.6 mm (i.e. a frequency of around 0.5 THz) [31]. At this frequency, high-power radiation sources are typically of wide bandwidth and hence the structures also need to be wideband to utilise all the THz pulse energy. The bandwidth is limited by the fact that not all frequencies will travel at the beam velocity, and hence the pulse will slip out of synchronism with the beam at higher and lower frequencies. Making cavity-like structures is difficult at higher frequencies, so other ways are required to maintain synchronism; these include dielectric-loaded waveguide, corrugated waveguide, and all-dielectric accelerators made from photonic bandgap structures. The sources to drive the fields include lasers and Čerenkov generation in non-linear crystals, but they can also be excited by the beam in a dielectric wakefield accelerator [72]. Here, either the head of a bunch is decelerated and the tail accelerated or one drive bunch drives a wake to accelerate a separate witness bunch.

3.12.6   Plasma Accelerators

Another method to achieve gradients beyond the breakdown limit of copper at microwave frequencies is to use plasma to accelerate particles. A strong electromagnetic field coming from either an intense laser [73] or a charged particle beam [74] creates a channel in the plasma where the electrons are repelled, or at lower intensities, a displacement of electrons occurs. As the electrons return to the channel, attracted by the positive charge generated by the newly created ions, they develop a large travelling electric field which can be used to accelerate a short bunch of electrons. Such an accelerator can generate very high gradients in the GV/m range; however, issues remain in trying to achieve beams of sufficient quality to be utilised for most applications. Technical issues also need to be solved around laser efficiency, the ability to use more than one acceleration stage, stability, and increasing average beam power. The concept was originally devised by Tajima and Dawson in 1979 and was experimentally verified by Joshi in 1984. The current record generates 7.8 GeV in 20 cm providing a gradient of 39 GV/m using a petawatt laser [75]. The energy gain is inversely proportional to the plasma density, n_0, as the gradient scales with $\sqrt{n_0}$ and the laser depletion length scales as $n_0^{-3/2}$. More recently, laser heating techniques have sought to circumvent this, in which case the interaction length is limited by dephasing between the wake and the accelerated electron beam. To increase the acceleration length in a plasma, the AWAKE collaboration has demonstrated the use of energetic proton beams from the super proton synchrotron (SPS) at CERN to drive a wake in a 10 m plasma, which then accelerated an injected electron beam from 19 MeV to 2 GeV. Such a concept could be extrapolated to use the 13 TeV proton beam from the LHC to create a TeV-scale electron collider [74].

Exercises

1. A cathode operating in temperature-limited thermonic emission has a diameter of 3 cm and a work function of 2 eV. Calculate the temperature of the cathode to have an emission current of 200 mA, and the minimum voltage required to ensure temperature-limited emission.

2. Derive the shunt impedance per unit length for a pillbox cavity at 12 GHz, where the cavity is designed to accelerate electrons travelling at $\beta=1$.

3. If a cavity has a shunt impedance of 100 MΩ and has an accelerating voltage of 100 MV, calculate the power required to accelerate a beam current of 10 mA.

4. A 1.3 GHz cavity has an ohmic Q factor of 10^{10}, an external Q factor of 10^6 and is driven by a 10 kW RF source. What is the stored energy in the cavity, and the reflected power for a steady-state situation?

5. A $\pi/2$ side-coupled structure at 3 GHz has 15 accelerating cells and 14 coupling cells. If there is a 1 MHz difference between the accelerating and side-coupled cell frequencies calculate the coupling required to have the accelerating fields in each cell within 1% of each other.

6. A 60 cell, 12 GHz, $2\pi/3$ constant-impedance travelling-wave structure is fed with a 10 MW amplifier and each cell has a Q_0 of 10,000. Calculate the group velocity required to maximise the accelerating voltage, and the external Q of the couplers.

7. Calculate the maximum surface magnetic field due to pulsed RF heating on a 12 GHz copper cavity, if the RF pulse duration is 1 μs.

8. A 500 MHz niobium SRF cavity is cooled to a temperature of 4.2 K. If it has a geometry factor, G, of 100 Ω and a residual resistance of 5 nΩ, what is the ohmic Q of the cavity?

9. A probe-type HOM coupler has a coupling capacitance of 5 pF. If it is mounted on a 1.3 GHz cavity, with the highest-impedance HOM at 2.5 GHz, design a circuit to filter the operating mode's frequency and compensate at the HOM frequency; calculate the values of any capacitors or inductors used.

References

1. E Skaria and BM Varghese. DC-DC booster with cascaded connected multilevel voltage multiplier applied to transformer less converter for high power applications. *J. Electr & Electron. Eng*, 9(5):73–78, 2014.

2. JD Cockcroft and ETS Walton. Experiments with high velocity positive ions.¯(i) further developments in the method of obtaining high velocity positive ions. *Proceedings of the Royal Society of London. Series A, Containing Papers of a Mathematical and Physical Character*, 136(830):619–630, 1932.

3. C. Young, M. Chen, T. Chang, C. Ko, and K. Jen. Cascade cockcroft˘walton voltage multiplier applied to transformerless high step-up dc˘dc converter. *IEEE Transactions on Industrial Electronics*, 60(2):523–537, Feb 2013.

4. MA Kemp. *Solid-state Marx Modulators for Emerging Applications.* 2012.

5. YY Lau, Youfan Liu, and RK Parker. Electron emission: From the fowler–nordheim relation to the child–langmuir law. *Physics of Plasmas*, 1(6):2082–2085, 1994.

6. T Siggins, C Sinclair, C Bohn, D Bullard, David Douglas, A Grippo, J Gubeli, GA Krafft, and B Yunn. Performance of a DC GaAs photocathode gun for the Jefferson lab FEL. *Nuclear Instruments and Methods in Physics Research Section A: Accelerators, Spectrometers, Detectors and Associated Equipment*, 475(1-3):549–553, 2001.

7. Richard G Forbes and Jonathan HB Deane. Reformulation of the standard theory of fowler–nordheim tunnelling and cold field electron emission. *Proceedings of the Royal Society A: Mathematical, Physical and Engineering Sciences*, 463(2087):2907–2927, 2007.

8. K.L Jensen, Y.Y Lau, D.W Feldman, and P.G O'Shea. Electron emission contributions to dark current and its relation to microscopic field enhancement and heating in accelerator structures. *Physical Review Special Topics: Accelerators and Beams*, 11(8):081001, 2008.

9. D Faircloth. Ion sources for high-power hadron accelerators. *arXiv preprint arXiv:1302.3745*, 2013.

10. WD Kilpatrick. Criterion for vacuum sparking designed to include both RF and DC. *Review of Scientific Instruments*, 28(10):824–826, 1957.

11. PF Dahl. Rolf widere: Progenitor of particle accelerators. *SSCL-SR-ll86*, 1992.

12. LW Alvarez, H Bradner, JV Franck, H Gordon, JD Gow, LC Marshall, F Oppenheimer, WKH Panofsky, C Richman, and JR Woodyard. Berkeley proton linear accelerator. *Review of Scientific Instruments*, 26(2):111–133, 1955.

13. DW Fry, RBR-S Harvie, LB Mullett, and W Walkinshaw. A travelling-wave linear accelerator for 4-MeV electrons. *Nature*, 162(4126):859, 1948.

14. EL Chu and W.W Hansen. Disk-loaded wave guides. *Journal of Applied Physics*, 20(3):280–285, 1949.

15. Bernard Aune, R Bandelmann, D Bloess, B Bonin, A Bosotti, M Champion, C Crawford, G Deppe, B Dwersteg, DA Edwards, et al. Superconducting TESLA cavities. *Physical Review Special Topics: Accelerators and Beams*, 3(9):092001, 2000.

16. M. Aicheler, P. Burrows, M. Draper, T. Garvey, P. Lebrun, K. Peach, N. Phinney, H. Schmickler, D. Schulte, and N. Toge, editors. *A multi-TeV linear collider based on CLIC technology: CLIC Conceptual Design Report*, volume CERN-2012-007. 2012.

17. D.M Pozar. *Microwave Engineering; 3rd ed.* Wiley, Hoboken, NJ, 2005.

18. H. Padamsee, J Knobloch, T Hays, et al. *RF Superconductivity for Accelerators*, volume 2011. Wiley Online Library, 2008.

19. TJ Boyd Jr. Kilpatrick's criterion. *Los Alamos Group AT-1 Report*, 82:28, 1982.

20. Alberto Degiovanni, Walter Wuensch, and Jorge Giner Navarro. Comparison of the conditioning of high gradient accelerating structures. *Phys. Rev. Accel. Beams*, 19:032001, Mar 2016.

21. A. Grudiev, S. Calatroni, and W. Wuensch. New local field quantity describing the high gradient limit of accelerating structures. *Phys. Rev. ST Accel. Beams*, 12:102001, Oct 2009.

22. F. Djurabekova, S Parviainen, A Pohjonen, and K Nordlund. Atomistic modeling of metal surfaces under electric fields: Direct coupling of electric fields to a molecular dynamics algorithm. *Physical Review E*, 83(2):026704, 2011.

23. K. Nordlund and F. Djurabekova. Defect model for the dependence of breakdown rate on external electric fields. *Physical Review Special Topics: Accelerators and Beams*, 15(7):071002, 2012.

24. M. A. Furman and M. T. F. Pivi. Probabilistic model for the simulation of secondary electron emission. *Physical Review Special Topics: Accelerators and Beams*, 5:124404, Dec 2002.

25. G. Burt and A. C. Dexter. Prediction of multipactor in the iris region of rf deflecting mode cavities. *Physical Review Special Topics: Accelerators and Beams*, 14:122002, Dec 2011.

26. Pasi Yla-Oijala. Electron multipacting in TESLA cavities and input couplers. *Particle Accelerators*, 63:105–137, 1999.

27. D.P Pritzkau and R.H Siemann. Experimental study of RF pulsed heating on oxygen free electronic copper. *Physical Review Special Topics: Accelerators and Beams*, 5(11):112002, 2002.

28. M. Jenkins, G Burt, AV Praveen Kumar, Y. Saveliev, P. Corlett, T. Hartnett, R Smith, A Wheelhouse, P McIntosh, and K Middleman. Prototype 1 MeV X-band linac for aviation cargo inspection. *Physical Review Accelerators and Beams*, 22(2):020101, 2019.

29. S. Pitman. *Optimisation studies for a high gradient proton Linac for application in proton imaging: ProBE: Proton Boosting Linac for imaging and therapy.* PhD thesis, Lancaster University, 2019.

30. T. P. Wangler. *RF Linear Accelerators, Second Edition.* Wiley, 2008.

31. MT Hibberd, AL Healy, DS Lake, V Georgiadis, EJH Smith, OJ Finlay, TH Pacey, JK Jones, Y Saveliev, DA Walsh, et al. Terahertz-driven acceleration of a relativistic 35 MeV electron beam. In *2019 44th International Conference on Infrared, Millimeter, and Terahertz Waves (IRMMW-THz)*, pages 1–2. IEEE, 2019.

32. C. Nantista, S. Tantawi, and V. Dolgashev. Low-field accelerator structure couplers and design techniques. *Physical Review Special Topics: Accelerators and Beams*, 7(7):072001, 2004.

33. N. M. Kroll, C. K. Ng, and D. C. Vier. Applications of time domain simulation to coupler design for periodic structures. In *Proceedings of 20th International Linac Conference, Linac 2000, Monterey, USA*, pages 614–617.

34. K Pepitone, S Doebert, G Burt, E Chevallay, N Chritin, C Delory, V Fedosseev, Ch Hessler, G McMonagle, Oznur Mete, et al. The electron accelerator for the AWAKE experiment at CERN. *Nuclear Instruments and Methods in Physics Research Section A: Accelerators, Spectrometers, Detectors and Associated Equipment*, 829:73–75, 2016.

35. S Benedetti, A Grudiev, and A Latina. High gradient linac for proton therapy. *Physical Review Accelerators and Beams*, 20(4):040101, 2017.

36. WJ Gallagher. Design of travelling wave electron linear accelerators. In *IEEE Transactions on Nuclear Science*, number 3, page 282, 1967.

37. A Grassellino, A Romanenko, D Sergatskov, O Melnychuk, Y Trenikhina, A Crawford, A Rowe, M Wong, T Khabiboulline, and F Barkov. Nitrogen and argon doping of niobium for superconducting radio frequency cavities: a pathway to highly efficient accelerating structures. *Superconductor Science and Technology*, 26(10):102001, aug 2013.

38. T. Junginger. EuCARD-BOO-2012-004, 2012.

39. J-M Vogt, O Kugeler, and J Knobloch. High-Q operation of superconducting RF cavities: Potential impact of thermocurrents on the RF surface resistance. *Physical Review Special Topics-Accelerators and Beams*, 18(4):042001, 2015.

40. D Reschke et al. Challenges in SRF module production for the European XFEL. In *Proceedings of the 15th International Workshop on RF Superconductivity, Chicago, Ill., USA*, 2011.

41. D. Broemmelsiek, B. Chase, D. Edstrom, E. Harms, J. Leibfritz, S. Nagaitsev, Y. Pischalnikov, A. Romanov, J. Ruan, W. Schappert, et al. Record high-gradient SRF beam acceleration at Fermilab. *New Journal of Physics*, 20(11):113018, 2018.

42. A. Macpherson, K. Hernndez-Chahn, C. Jarrige, P. Maesen, F. Pillon, K. Schirm, R. Torres-Sanchez, and N. Valverde Alonso. CERN's bulk niobium high gradient SRF programme: developments and recent cold test results. page MOPB074. 5 p, 2015.

43. J. Mitchell. *DQW Crab Cavity HOMs and Dampers for the HL-LHC.* Lancaster University, 2019.

44. U. Amaldi, P Berra, K Crandall, D Toet, M Weiss, R Zennaro, E Rosso, B Szeless, M Vretenar, C Cicardi, et al. LIBO ¯ a linac-booster for protontherapy: Construction and tests of a prototype. *Nuclear Instruments and Methods in Physics Research Section A: Accelerators, Spectrometers, Detectors and Associated Equipment*, 521(2-3):512–529, 2004.

45. et al Adolphsen C. *The International Linear Collider Technical Design Report: Volume 3. II: Accelerator Baseline Design.* 2013.

46. J. Sekutowicz, K Ko, L Ge, L Lee, Zenghai Li, C Ng, G Schussman, Liling Xiao, I Gonin, T Khabibouline, et al. Design of a low loss SRF cavity for the ILC. In *Proceedings of the 2005 Particle Accelerator Conference*, pages 3342–3344. IEEE, 2005.

47. B. Militsyn, L. Cowie, P. Goudket, J. McKenzie, and A. Wheelhouse. Design of the high repetition rate photocathode gun for the clara project. In *Proceedings of Linac2014, Geneva, Switzerland*, 2014.

48. G Olry, JL Biarrotte, S Blivet, S Bousson, F Chatelet, T Junquera, A Le Goff, J Lesrel, C Milot, AC Mueller, et al. Development of SRF spoke cavities for low and intermediate energy ion linacs. In *Proceedings of the 9th International Workshop on RF Superconductivity*, volume 3, page 76, 2003.

49. Z Yao, RE Laxdal, B Matheson, BS Waraich, and V Zvyagintsev. Design and fabrication of balloon single spoke resonator. In *Proceedings of the 18th International Workshop on RF Superconductivity*, 2017.

50. G Apollinari, I Gonin, T Khabiboulline, G Lanfranco, F McConologue, G Romanov, and R Wagner. Design of 325 MHz single and triple spoke resonators at FNAL. *IEEE Transactions on Applied Superconductivity*, 17(2):1322–1325, 2007.

51. C.S Hopper and J.R Delayen. Superconducting spoke cavities for high-velocity applications. *Physical Review Special Topics: Accelerators and Beams*, 16(10):102001, 2013.

52. Michael Kelly. Superconducting spoke cavities. 2006.

53. J. Delayen. Low and intermediate beta cavity design-a tutorial. Technical report, 2003.

54. D Naik and I Ben-Zvi. Suppressing multipacting in a 56 mhz quarter wave resonator. *Physical Review Special Topics: Accelerators and Beams*, 13(5):052001, 2010.

55. W Wagner, M Seidel, E Morenzoni, F Groeschel, M Wohlmuther, and M Daum. Psi status 2008: Developments at the 590 mev proton accelerator facility. *Nuclear Instruments and Methods in Physics Research Section A: Accelerators, Spectrometers, Detectors and Associated Equipment*, 600(1):5–7, 2009.

56. *CERN Accelerator School: Cyclotrons, Linacs and their Applications. Proceedings, editor=Turner, S, year=1996, institution=European Organization for Nuclear Research.*

57. ISK Gardner. CERN Accelerator School on RF engineering for particle accelerators. 350, 1992.

58. M Kelly. Superconducting radio-frequency cavities for low-beta particle accelerators. *Reviews of Accelerator Science and Technology*, 5:185–203, 2012.

59. F. Gerigk. Cavity types. *arXiv preprint arXiv:1111.4897*, 2011.

60. S Ramberger, N Alharbi, P Bourquin, Y Cuvet, F Gerigk, AM Lombardi, E Sargsyan, M Vretenar, and A Pisent. Drift tube linac design and prototyping for the cern linac4. *proc. Linac08*, 2008.

61. Chuan Zhang, Michael Busch, Florian Dziuba, Horst Klein, Holger Podlech, and Ulrich Ratzinger. Recent studies on a 3-17mev dtl for eurotrans with respect to rf structures and beam dynamics. *IPAC 2010 - 1st International Particle Accelerator Conference*, 01 2010.

62. A.K. Mitra, Pierre Bricault, Ken Fong, R.E. Laxdal, Raymond Poirier, and A. Vasyuchenko. Rf measurement summary of isac dlt tanks and dtl bunchers. pages 951 – 953 vol.2, 02 2001.

63. G. Clemente, U. Ratzinger, H. Podlech, L. Groening, R. Brodhage, and W. Barth. Development of room temperature crossbar-h-mode cavities for proton and ion acceleration in the low to medium beta range. *Physical Review Special Topics: Accelerators and Beams*, 14(11):110101, 2011.

64. R.G Carter. *Microwave and RF Vacuum Electronic Power Sources*. Cambridge University Press, 2018.

65. D. Constable, A. Baikov, G. Burt, I. Guzilov, V. Hill, A. Jensen, R. Kowalczyk, C. Lingwood, R. Marchesin, and C. et al Marrelli. High efficiency klystron development for particle accelerators. In *58th ICFA Advanced Beam Dynamics Workshop on High Luminosity Circular e+ e− Colliders (eeFACT'16), Daresbury, UK, October 24-27, 2016*, pages 185–187, 2017.

66. A. C. Dexter, G. Burt, R. G. Carter, I. Tahir, H. Wang, K. Davis, and R. Rimmer. First demonstration and performance of an injection locked continuous wave magnetron to phase control a superconducting cavity. *Physical Review Special Topics: Accelerators and Beams*, 14:032001, Mar 2011.

67. EA Peralta, K Soong, RJ England, ER Colby, Z Wu, B Montazeri, C McGuinness, J McNeur, KJ Leedle, D Walz, et al. Demonstration of electron acceleration in a laser-driven dielectric microstructure. *Nature*, 503(7474):91, 2013.

68. E. Curry, S Fabbri, J Maxson, P Musumeci, and A Gover. Meter-scale terahertz-driven acceleration of a relativistic beam. *Physical Review Letters*, 120(9):094801, 2018.

69. M. Fakhari, A. Fallahi, and F.X Kartner. THz cavities and injectors for compact electron acceleration using laser-driven THz sources. *Physical Review Accelerators and Beams*, 20(4):041302, 2017.

70. J.R. England, R.J Noble, K. Bane, David H Dowell, C. Ng, J.E Spencer, S. Tantawi, Z. Wu, R.L Byer, and E. et al Peralta. Dielectric laser accelerators. *Reviews of Modern Physics*, 86(4):1337, 2014.

71. A. L. Lake D. S. Georgiadis V Smith E. J. H. Finlay O. J. Pacey T. H. Jones J. K. Saveliev Y. Walsh D. A. Snedden E. W. Appleby R. B. Burt G. Graham D. M. Jamison S. P. AU Hibberd, M. T. Healy. Acceleration of relativistic beams using laser-generated terahertz pulses. *Nature Photonics*, Aug 2020.

72. BD O'Shea, G Andonian, SK Barber, KL Fitzmorris, S Hakimi, J Harrison, PD Hoang, MJ Hogan, B Naranjo, OB Williams, et al. Observation of acceleration and deceleration in gigaelectron-volt-per-metre gradient dielectric wakefield accelerators. *Nature Communications*, 7:12763, 2016.

73. S.M Hooker. Developments in laser-driven plasma accelerators. *Nature Photonics*, 7(10):775, 2013.

74. E. Adli, A Ahuja, O Apsimon, R Apsimon, A-M Bachmann, D Barrientos, F Batsch, J Bauche, VK Berglyd Olsen, M Bernardini, et al. Acceleration of electrons in the plasma wakefield of a proton bunch. *Nature*, 561(7723):363–367, 2018.

75. A. J. Gonsalves, K. Nakamura, J. Daniels, C. Benedetti, C. Pieronek, T. C. H. de Raadt, S. Steinke, J. H. Bin, S. S. Bulanov, J. van Tilborg, C. G. R. Geddes, C. B. Schroeder, Cs. Tth, E. Esarey, K. Swanson, L. Fan-Chiang, G. Bagdasarov, N. Bobrova, V. Gasilov, G. Korn, P. Sasorov, and W. P. Leemans. Petawatt laser guiding and electron beam acceleration to 8 GeV in a laser-heated capillary discharge waveguide. *Phys. Rev. Lett.*, 122:084801, Feb 2019.

4

Magnets for Beam Control and Manipulation

An enormous strength associated with particle accelerators is the ability we have to steer, focus, and otherwise manipulate the charged particle beams. This enables us to create accelerators with a circular geometry so the particles continuously and stably pass around the machine time and again or to generate very tightly focused beams down to the nanometre level, for example. Our ability to steer and focus particles has some similarities to using mirrors and lenses in conventional optics. One limitation of optics which is often overlooked however, is that they rely on the material properties of the item itself, the consequence of this being that a lens, for example, will only properly function over a restricted part of the electromagnetic spectrum. So, you can't focus X-rays with a lens that focuses visible light. Since we manipulate charged particles with magnetic fields rather than relying upon specific materials, we do not have this limitation – any charged particle of any energy is effected in an entirely predictable and repeatable way. There are no particle energies which are 'off-limits' because Nature hasn't provided a material or coating with the right properties! This chapter will explain how the standard magnetic field distributions of dipole, quadrupole, and so on can be generated with high quality in the real world using coils and steel poles. It will also consider many of the practicalities involved in designing and manufacturing highly reliable magnets, either static or time-varying. Finally, the application of the alternative magnet technologies of permanent magnets and superconducting magnets will also be covered.

4.1 The Family of Standard Magnetic Field Profiles

The majority of magnetic field distributions that are used in particle accelerators are quite simple. The first one is the uniform, constant field which provides a bending force to the beam, making it take a circular path. The second one is where the field on the horizontal axis increases linearly with horizontal position x and passes through the origin. This applies a force to the beam that depends on its distance from the axis. If the beam is on-axis then it sees no field and passes straight through, but if it is off-axis, then it feels a force which bends it towards the axis proportional to x, much like an optical lens. So, this linear field variation with x applies focusing to the beam. A third popular field shape is one where the field increases with x^2 on the horizontal axis, which is used to correct focusing aberrations due to the beam of particles not all having exactly the same momentum. It turns out, as we shall see, that to make a pure constant field requires a two-pole magnet, called a dipole. One which varies with x requires four poles, and so is called a quadrupole, and the one which varies with x^2 requires six poles, and so is called a sextupole. Clearly there is a very simple pattern emerging here for these pure, ideal, field shapes in terms of the power of the field variation with x and the number of poles required to generate such a field. Hopefully, it is now clear why the term 'multipoles' is used in the accelerator community when discussing magnetic fields and their impact on the beams. Each multipole (dipole, quadrupole, sextupole, etc.) actually represents an independent term on the infinite polynomial series $B_n x^n$ as we shall discuss in more detail later.

A nice feature of magnets is that these different, pure, field shapes can be added together to make a more complex field pattern, if the beam requires it, with the ideal pole arrangement being readily determined. An example of this is when a combined focusing and bending field is required. In this case the ideal field varies linearly with x but is non-zero at the origin so even the beam which passes through the centre of the field feels an overall bending force. This field shape, called a gradient dipole or combined function dipole, along with the others mentioned above, are sketched out schematically in Fig 4.1. How the pole shape and number of poles is determined by the field shape required will now be explored, closely following the approach described by Tanabe [1], which provides more detail if required.

We start from the two Maxwell equations which are relevant to static (i.e. do not vary with time) magnetic fields and also make the further assumption that there are no current sources. Since the charged particle beams pass through the gap between the magnet poles, well away from current-carrying conductors, this is a good assumption in general:

$$\nabla \cdot \mathbf{B} = 0, \tag{4.1}$$

$$\nabla \times \mathbf{B} = 0. \tag{4.2}$$

Next we introduce the vector potential, $\mathbf{A}$, and scalar potential, V. These two potentials are commonly used in vector calculus to develop an understanding of the field being analysed. In our case it turns out that the vector potential maps out the lines of flux and the scalar potential maps out the family of ideal pole shapes required for a particular magnetic field. Either of these potentials can be used to determine the magnetic field since, due to standard results from vector calculus, we can also write

$$\mathbf{B} = \nabla \times \mathbf{A}, \tag{4.3}$$

$$\mathbf{B} = -\nabla V. \tag{4.4}$$

Both $\mathbf{A}$ and V satisfy the Laplace equation

$$\nabla^2 \mathbf{A} = \nabla^2 V = 0,$$

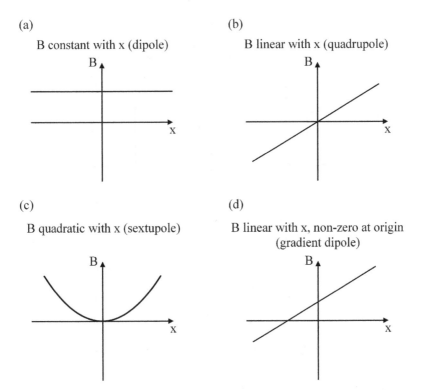

(a)

B constant with x (dipole)

(b)

B linear with x (quadrupole)

(c)

B quadratic with x (sextupole)

(d)

B linear with x, non-zero at origin
(gradient dipole)

FIGURE 4.1 Common magnetic field distributions used by accelerators and their associated name or pole configuration.

which means that the complex function, $F = \mathbf{A} + iV$ also satisfies the Laplace equation. If we now constrain ourselves to working in two dimensions, then we can find the potentials which satisfy the Maxwell equations above. First, we note that any analytic function of the complex variable $z = x + iy$ also satisfies the Laplace equation and so we can use a convenient function $C_n z^n = \mathbf{A} + iV$ to help us find the vector and scalar equipotentials (i.e. contours of a particular constant value) which will map out the lines of flux and possible pole shapes for some standard magnet types. The potentials will also enable us to calculate the magnetic field according to the equations

$$B_x = \frac{\partial A}{\partial y} = -\frac{\partial V}{\partial x}, \tag{4.5}$$

$$B_y = -\frac{\partial A}{\partial x} = -\frac{\partial V}{\partial y}. \tag{4.6}$$

Note that the magnetic field is given by the gradient of the potential. This tallies with our understanding that when flux lines are densely packed together the fields are highest.

4.1.1 Case $n = 1$: Dipole

In the general case we can write

$$\sum_{n=1}^{\infty} C_n z^n = \mathbf{A} + iV,$$

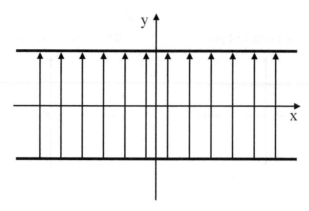

FIGURE 4.2 The scalar equipotential for a uniform magnetic field is a horizontal steel pole surface. The vector equipotentials map out the flux lines; as they are equally spaced, the magnetic field must be perfectly uniform.

but for now we will restrict ourselves to the simplest case of $n = 1$ only. In this case then

$$C_1 z = C_1(x + iy) = A + iV. \tag{4.7}$$

If C_1 is real we can gather the real and imaginary terms and see that the potentials are given by

$$A = C_1 x, \tag{4.8}$$
$$V = C_1 y. \tag{4.9}$$

Differentiating, according to the equations above, to find what value of B_x and B_y these potentials represent gives us

$$B_x = \frac{\partial A}{\partial y} = 0, \tag{4.10}$$

$$B_y = -\frac{\partial A}{\partial x} = -C_1. \tag{4.11}$$

And so the case $n = 1$ gives us a constant magnetic field in the vertical plane. The equipotentials for this case are plotted in Fig 4.2 and, as expected for a perfect vertical field, the vector potential maps out the equally-spaced vertical flux lines, which are orthogonal to the lines of scalar potential (this is always true in fact). These scalar potential lines define the perfect steel pole surface that will generate these magnetic fields. In this case a pair of horizontal, parallel, steel poles (a dipole) equally spaced about the horizontal axis, extending out to infinity in $\pm x$ are required. One pole is determined by V and the opposite one by $-V$. Note that each and every scalar equipotential line represents a possible pole surface; there is not just one unique position for the poles, there is a whole family of poles which will create this ideal field. The magnet designer can choose the optimum pair of poles which meet the physical and magnetic requirements for that particular application.

4.1.2 Case $n = 2$: Quadrupole

For this example we have that

$$C_2 z^2 = C_2(x + iy)^2 = A + iV,$$
$$C_2(x^2 + 2ixy - y^2) = A + iV.$$

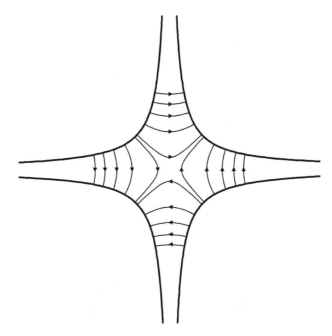

FIGURE 4.3 The equipotentials for a quadrupole which generates a field that is linear with position from the origin (see Fig 4.1 (b)). The vector equipotentials (shown with arrows) map out the flux lines. The field is zero at the centre of the magnet.

Gathering the real and imaginary terms gives us

$$A = C_2(x^2 - y^2), \tag{4.12}$$
$$V = 2C_2xy. \tag{4.13}$$

Differentiating the vector potential, as before, then gives us the magnetic fields

$$B_x = -2C_2y, \tag{4.14}$$
$$B_y = -2C_2x. \tag{4.15}$$

So, the vertical magnetic field on the horizontal axis is linear with x, and zero at the origin, as required for a focusing magnet. The field along the vertical axis is horizontal and linear with y (with the same coefficient as B_y) and so is also providing a focusing effect. Unfortunately, due to B_x and B_y having the same sign in the equations above, one axis will focus the beam towards the origin whilst the other axis will defocus the beam away from the origin. This well-known concept that a quadrupole focuses in one plane and defocuses in the other is fundamental, as we can now see. The fields must obey Maxwell's equations and this is a direct consequence of that requirement.

The equipotentials for this case are plotted in Fig 4.3 with both A and V mapping out rectangular hyperbolas (which means the asymptotes are perpendicular to each other). Again, we can see from the lines of vector equipotential that they become more densely packed away from the origin, indicating the field strength increase. The scalar equipotentials map out ideal steel pole surfaces, which in this case extend to infinity along both the x and y axes. There must be four poles – one per quadrant – and hence this is called a quadrupole.

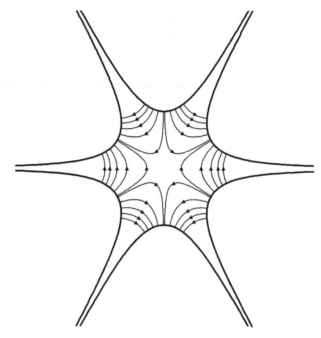

FIGURE 4.4 The equipotentials for a sextupole which generates a field that is quadratic with position from the origin (see Fig 4.1 (c)). The vector equipotentials, shown with arrows, map out the flux lines. The field is zero at the centre of the magnet.

4.1.3   Case $n = 3$: Sextupole

For this example we have that

$$C_3 z^3 = C_3(x + iy)^3 \quad = A + iV,$$
$$C_3(x^3 - 3xy^2 + 3ix^2y - iy^3) \quad = A + iV.$$

Following the same procedure as before gives us

$$A \quad = C_3(x^3 - 3xy^2), \tag{4.16}$$
$$V \quad = C_3(3x^2y - y^3). \tag{4.17}$$

Differentiating the vector potential, as before, then gives us the magnetic fields

$$B_x \quad = \quad -6C_3xy, \tag{4.18}$$
$$B_y \quad = \quad -3C_3x^2 + 3C_3y^2. \tag{4.19}$$

Now we can see that the vertical field on the x axis is quadratic in x and zero at the origin. The equipotentials for this case are plotted in Fig 4.4. The scalar equipotentials have asymptotes at $0°$, $60°$, $120°$, ... and so there are six poles required for this field shape – hence the term sextupole.

4.1.4   Case $n = 1, 2$: Gradient Dipole

As mentioned earlier, a common magnet which combines two multipole types is the gradient (or combined-function) dipole, which has a non-zero field on axis and has a field varying linearly with x; see Fig 4.1 (d). As we know that we want a combination of dipole and quadrupole, we follow the same procedure as before but this time include the terms for

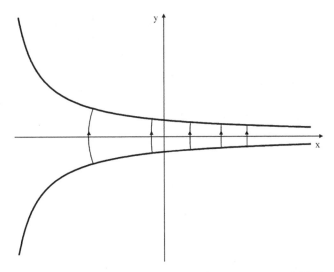

FIGURE 4.5   The equipotentials for a gradient dipole which generates a field that is non-zero at the origin and linear with position (see Fig 4.1 (d)). The vector equipotentials, shown with arrows, map out the flux lines.

both $n = 1$ and $n = 2$:

$$C_1 z + C_2 z^2 = C_1(x + iy) + C_2(x + iy)^2 \quad = A + iV,$$
$$C_1 x + iC_1 y + C_2(x^2 + 2ixy - y^2) \quad = A + iV.$$

Gathering the real and imaginary terms gives us

$$A \quad = \quad C_1 x + C_2 x^2 - C_2 y^2, \tag{4.20}$$
$$V \quad = \quad C_1 y + 2C_2 xy. \tag{4.21}$$

Differentiating the vector potential then gives us the magnetic field

$$B_y = -\frac{\partial A}{\partial x} = -C_1 - 2C_2 x. \tag{4.22}$$

So, the field varies as required and the ideal pole shape is found by plotting lines of constant V for the required values of C_1 and C_2, see Fig 4.5. This particular magnet can also be considered to be a simple quadrupole, but with the beam axis offset from the physical centre of the quadrupole so that the field at the origin is non-zero. If one applies a geometric shift of the origin along the x-axis to a normal quadrupole description then exactly the same pole shape equation is found.

4.2   Generating an Arbitrary Magnetic Field Shape

For the standard magnet types the pole shapes are well known, with several examples being given in the previous section. So, when asked to design a quadrupole, for example, the ideal pole shape is already known to be a hyperbola and the designer must optimise the magnet for maximum efficiency, which normally means minimizing the magnet aperture. They must also choose how to approximate the pole shape to the ideal, which extends to infinity, given a field quality specification over a particular physical region which is required by the accelerator. Such choices are important, to make sure a magnet performs as expected and will be covered later in this chapter, but not particularly challenging from a physics perspective. A more challenging problem for a magnet designer is to be given a magnetic field profile (i.e. B_y as a function of x), that does not correspond to a well-known type, and to design a magnet that will generate the required fields. The first step in this case is to fit the field profile to a polynomial of the form:

$$B_y = B_1 + B_2' x + B_3'' x^2 + ..., \tag{4.23}$$

where B_2' represents the dB/dx (quadrupole) term and B_3'' represents the $d^2 B/dx^2$ (sextupole) term and so on. Then we can write

$$
\begin{aligned}
B_y = -\frac{\partial V}{\partial y} &= B_1 + B_2' x + B_3'' x^2 + ... \\
V &= -\int (B_1 + B_2' x + B_3'' x^2 + ...) dy \\
V &= -(B_1 + B_2' x + B_3'' x^2 + ...) y \\
y &= -\frac{V}{B_1 + B_2' x + B_3'' x^2 + ...}.
\end{aligned}
\tag{4.24}
$$

A line of constant scalar potential will then define the ideal pole shape that will generate the required field. The optimal value of V is normally the one which minimises the magnet aperture, and hence the required Ampere-turns in the coils, within the physical boundary conditions set by the other factors at play, such as the beam aperture requirements or achieving a particular vacuum level. Later in this chapter we will look at how these ideal magnetic fields and pole shapes, which extend to infinity, are dealt with in real-life situations. The skill of the engineer or magnet designer is to generate the magnetic field of the correct shape and of sufficient quality in the region where it is required by the beam in as efficient a manner as possible. Here, efficiency normally equates to cost to build and cost to operate.

4.3 Magnet Multipoles

We have already noted that we describe different magnetic field distributions in terms of 'multipoles', with examples being pure dipole, pure quadrupole, and so on. In this section we will define multipoles more formally and explain how we use them to specify and judge the quality of a magnetic field. In general, all physical distributions of magnetic field in two dimensions in a region free from steel and coils can be described by an infinite sum of all multipoles [2].

$$
\begin{aligned}
B_y + iB_x &= \sum_{n=1}^{\infty} C_n z^{n-1} \\
&= \sum_{n=1}^{\infty} C_n (x + iy)^{n-1}.
\end{aligned}
\tag{4.25}
$$

A pure multipole has $C_n \neq 0$ for just one term in the series ($n = 1$ is a dipole, $n = 2$ is a quadrupole, etc). We also note that C_n is a complex constant so

$$B_y + iB_x = \sum_{n=1}^{\infty} (J_n + iK_n)(x + iy)^{n-1}. \tag{4.26}$$

The coefficients J_n and K_n characterise the strength and orientation of each multipole component. The units of these coefficients are different for every value of n (e.g. J_1 is in

T, J_2 is in T/m, J_3 is in T/m^2) which can get cumbersome. A common approach is to *normalise* the coefficients so they become dimensionless. This is achieved by multiplying the expression by a reference field B_{ref} and dividing by a reference radius R_{ref} raised to the power n. Note that B_{ref} is the actual magnitude (in T) of the main field (i.e. the actual multipole we are interested in) measured at the position R_{ref}. The actual choice of what R_{ref} to select is arbitrary but should be stated, with a typical value being 2/3 of the magnet inner radius since this is often a good approximation to the full extent of the beam within the magnet. We can now write

$$B_y + iB_x = B_{ref} \sum_{n=1}^{\infty} (j_n + ik_n) \left(\frac{x + iy}{R_{ref}} \right)^{n-1}. \tag{4.27}$$

So, for the pure vertical dipole case, where $B_{ref} = J_1$ (in T) we can see that $j_1 = 1$ (dimensionless). Similarly, for the quadrupole, with $B_{ref} = J_2 R_{ref}$, we can see that $j_2 = 1$. This clearly demonstrates that we have normalised the multipole expansion. For a good-quality magnet, the other coefficients would be expected to be <0.01% of the main component and so to make discussion and comparison of different magnets a little easier, it is also common to multiply the expansion again by the constant 10^{-4} so that the main component has the value 10,000 and the other components have values of around unity. Magnet designers will (confusingly!) talk about how many 'units' of a particular multipole are present in their magnet, and it is this further normalised case that they are referring to.

In the accelerator environment there are two orientations of multipole field that are utilised. The first is called *normal* and the second is called *skew*. The normal cases are those where the magnetic field is vertical on the horizontal axis and, in fact, all of the cases considered earlier were of this type (see Figs. 4.2 to 4.4). The skew cases are those where the magnetic field is horizontal on the horizontal axis. We can see from the figures that skew magnets are simply normal magnets that are rotated by $\pi/2n$ about the axis. More formally, the normal cases are characterised by C_n being real and the skew cases when C_n is imaginary. In other words, the J_n terms represent the normal multipoles and the K_n terms the skew multipoles. It is easy to see that if we repeat the $n = 1$ example from earlier (Section 4.1), but this time assuming that C_1 is imaginary, we will find that the magnetic field is still a perfect dipole but that it is now oriented in the horizontal plane (i.e. it is a skew dipole).

Field Errors

Of course, a perfect multipole only contains one term in the multipole expansion and as such contains no field errors. Unfortunately, such magnets require infinitely wide steel poles or equally as unrealistic current density distributions (if we choose not to use any steel, as we shall see later in Section 4.6). In practice, we must design a magnet which is an excellent *approximation* to the ideal, which in this case means the steel poles have a finite extent. This unavoidably introduces systematic field errors even if we then build our design with no physical imperfections. This type of error is sometimes called an *allowed* error. The possible multipole terms which can generate these allowed errors are limited by symmetry and polarity [1] to those which generate a field in the same direction if they are rotated by π/n and have their polarity reversed. So in the $n = 1$ case, if we rotate the dipole by 180° and reverse the polarity, the field direction is the same, but if we rotate a quadrupole by 180° and reverse the polarity, then the field direction is misaligned. If we try the same rotate-and-reverse process on the sextupole, then it is aligned and so a dipole magnet will contain a sextupole error term. The formal generalisation of this is that the allowed multipoles are those that satisfy

$$n_{allowed} = n(2m + 1), \tag{4.28}$$

where $m = 0, 1, 2, \ldots$. Clearly $m = 0$ corresponds to the multipole we are trying to generate but the subsequent ones are all error terms. So a dipole magnet will include multipole errors of sextupole, decapole, 14-pole, etc. A quadrupole will contain errors due to 12-pole, 20-pole, and so on.

Manufacturing tolerances mean that the magnets are not perfect and that symmetries will be broken. This implies that a real magnet can and will have non-zero values for all the J_n and K_n coefficients. In fact, certain multipole errors can point to particular fabrication errors in terms of what symmetry has been broken [3].

A common method of specifying the field quality of any particular magnet is to put absolute limits on each of the multipole terms up to a sufficiently high order. The limits should be determined by thorough beam dynamics simulations as should the determination for which orders are critical. An alternative approach is to define the field quality in terms of how the absolute field level is allowed to vary within the good field region. For example, a dipole magnet may be specified to have a maximum field variation locally (i.e. at some specific longitudinal position within the magnet) of up to 0.01% of the main field, or a quadrupole might be specified to have a maximum gradient deviation locally of up to 0.01%. This alternative method puts absolute limits on the magnetic field performance but makes no comment on the actual multipole content. It is also important that *integrated* field levels and quality are specified in both cases. This means that the field quality should be judged through the length of the magnet (the fields are integrated in the beam direction) since this is what the beam will do! A high-quality dipole in a storage ring would be expected to have an integrated field variation of better than 0.01%, but in a single pass accelerator a level of 0.1% may well be sufficient. Similar numbers apply to integrated quadrupole gradient errors.

4.4   Electromagnets

The magnet type of choice for most accelerator applications is one based upon the use of current-carrying resistive (or normal conducting) coils. The alternative technologies which are based upon superconducting coils or permanent magnets have very important applications in accelerators, and will be discussed later, but they are generally employed when the standard electromagnet is unable to meet the required needs of the accelerator. The electromagnet is popular because they are well understood, relatively straightforward to design and build, are extremely reliable, available from industry, and easily adjusted by simply changing the current in a coil. In this section we will look at three different types of electromagnet; DC, AC, and pulsed. The first is DC (direct current) which means that the current is held constant and so the field is static. Of course, this does not mean that the field cannot be changed by altering the current, just that a static field is required by the accelerator. This type is used, for example, in a storage ring or transfer line which operates at a fixed beam energy day after day. The second type is AC (alternating current) which means the current has a time-varying, periodic, waveform. This is used to generate time-varying, periodic, magnetic fields as is required in a synchrotron, for example. Strictly speaking, AC implies that the current reverses direction in the circuit but this is often not the case in accelerator magnets where AC is used as shorthand to indicate that the magnetic field is periodically varying with time between a minimum and a maximum value, often with the same polarity. Also, the current waveform is typically not sinusoidal. The waveform is determined by the needs of the accelerator within the limitations of the magnet and power supply circuit. The third type of electromagnet is the pulsed magnet. These are magnets which are energised by a current pulse as and when required, and are off the remainder of the time. This type of magnet might be used to capture a beam injected into a storage ring

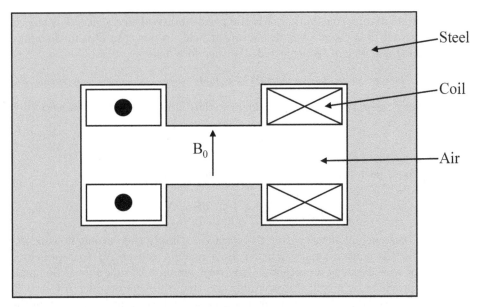

FIGURE 4.6 Cross section through an H-dipole, which is used to generate a uniform magnetic field. The electron beam is travelling into the page at the centre of the magnet, in the region between the two poles. The upper and lower poles each have a coil encircling the pole. The cross and dot within the coil cross section denotes the current flowing into and out of the page.

or to rapidly discard (or dump) a beam as part of a machine protection system, for example. In this section we will consider each of these three electromagnet types in practical terms and discuss the issues which should be taken into account when designing and building them.

4.4.1 Practicalities of DC Magnets

From the earlier section looking at the families of standard accelerator magnets we now have a theoretical understanding of the ideal pole shapes for each type. We will next look at how these theoretical pole shape curves are turned into real devices, starting with the example of a DC dipole.

Dipoles

For a perfect uniform magnetic field we found in Section 4.1.1 that we need a pair of infinitely wide horizontal, parallel, steel poles. We approximate these with a pair of finite-width, parallel poles which are energised by a coil wrapped around each pole. To complete the magnetic circuit efficiently, we need to connect the two poles with steel, away from the region of interest where the particle beam will travel, since steel has a much higher relative permeability than air. A popular dipole design is the so-called H-type, illustrated in Fig 4.6, because it is symmetric, supported mechanically on both sides, and has coils which are a simple shape to wind. The name simply comes from the H shape that the steel parts define in the central air region.

To calculate how the current flowing in the coils relates to the magnetic field at the centre of the dipole we need to refer back to another of Maxwell's equations

$$\nabla \times \mathbf{B} = \mu_r \mu_0 \mathbf{J} + \mu_r \mu_0 \frac{\partial \mathbf{D}}{\partial t}, \tag{4.29}$$

where μ_r is the relative permeability, μ_0 is the permeability of free space, $\mathbf{J}$ is the electric current density, and $\mathbf{D}$ is the electric displacement field. In our case $\mathbf{D}$ is unchanging with time (and is zero) and so can be neglected. We therefore have

$$\nabla \times \mathbf{B} = \mu_r \mu_0 \mathbf{J}. \tag{4.30}$$

Applying Stokes's theorem from vector calculus, which states that if $\mathbf{F}$ is a smooth vector field, then

$$\int_S \nabla \times \mathbf{F} \cdot d\mathbf{S} = \oint_P \mathbf{F} \cdot d\mathbf{l}, \tag{4.31}$$

where P is a closed path that is the boundary of the surface S, we obtain

$$\oint_P \frac{\mathbf{B}}{\mu_r \mu_0} \cdot d\mathbf{l} = \int_S \mathbf{J} \cdot d\mathbf{S} = NI. \tag{4.32}$$

The integral of the current density over the surface is simply the current flowing through the surface, which is conventionally written for a magnet coil as NI to represent a coil with N turns of wire carrying a current I. It is very common to talk about the number of Ampere-turns provided by a coil; this is simply shorthand for the product NI.

So, now we can see that if we know what magnetic field we want in our system then by integrating this field along a closed path – of our choosing – we can determine what current will be needed to develop this field. Returning to the H dipole of Fig 4.7, the closed path has been chosen to be made up of three parts. Path 1 is in the air region ($\mu_r = 1$), starting from the centre to the pole tip (a distance $g/2$, where g is the full magnet gap), and the field is uniform with value B_0, path 2 is in the steel where the magnetic field will be of similar magnitude to B_0 but where μ_r will be very large, and path 3 completes the loop through both steel and air but at the midplane of the dipole where B is always orthogonal to the horizontal axis and so the dot product will be zero. So, in the limit where the relative permeability of the steel tends towards infinity, we have that

$$
\begin{aligned}
NI &= \int_{P1} \frac{\mathbf{B}}{\mu_r \mu_0} \cdot d\mathbf{l} + \int_{P2} \frac{\mathbf{B}}{\mu_r \mu_0} \cdot d\mathbf{l} + \int_{P3} \frac{\mathbf{B}}{\mu_r \mu_0} \cdot d\mathbf{l}, \\
NI &= \frac{g B_0}{2 \mu_0}.
\end{aligned}
\tag{4.33}
$$

Remember that this value of NI is for the top coil (enclosed by our selected path); there will also need to be the same number of Ampere-turns in the bottom coil to generate B_0 across the full magnet gap. An interesting point to note is that, in the H dipole configuration, winding the coils around the return yoke, or back leg (as shown in Fig 4.8) is not an efficient solution. One might assume that using 4 coils of NI Ampere-turns instead of two would double the field at the centre of the dipole, but in fact it just generates the same field as before since the line integral of B bounds the same NI as the case where the coil is wound around the pole. In fact, the coils around the back leg act to generate field outside and away from the dipole, in the region where it is not required! A second point to note is that in this idealised case, where the relative permeability of the steel tends towards infinity, the pole width does not appear in the equation. So, no matter how wide the pole is, the same magnetic field will be generated in the air gap between the poles, at least in the central region. Also, the integral of B_y/μ_o in the air region along any vertical path parallel to path 1 is a constant. In the central region of the magnet, the field is uniform in y but towards the sides where the pole terminates, the vertical field in the plane of the magnet (the x axis) begins to fall away. We can conclude from this that, since the integral is constant, the vertical field, B_y, must therefore increase with y, as we get closer to the steel surface to compensate. This is indeed the case; B_y increases in the region near to the pole corner.

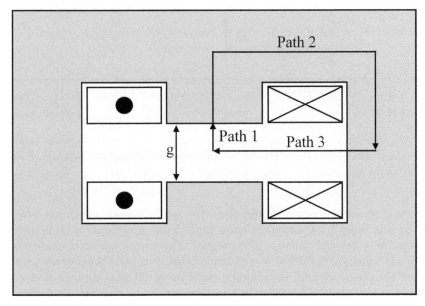

FIGURE 4.7   Cross section through an H-dipole showing the closed integration path chosen to calculate the current needed to generate a particular field value.

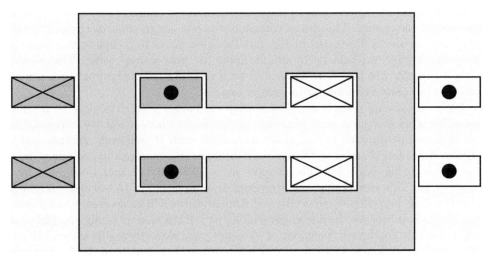

FIGURE 4.8   Cross section through an H-dipole where four coils are wound around the back leg instead of two around the pole. This generates the same field at the magnet centre as the two-coil version and not double the field as one might intuitively expect.

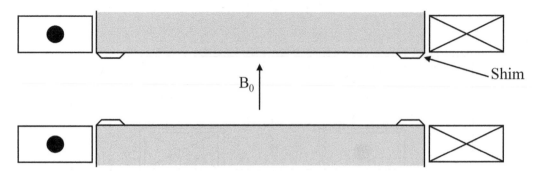

FIGURE 4.9 Close-up of dipole pole region illustrating the use of small pole shape adjustments, or shims, to counteract the B field decay towards the pole edges. The shims can be made of separate steel pieces fastened to the pole or, more typically, be integral to the pole itself.

Returning to our practical DC dipole, the accelerator designer will define a region where the magnetic field must meet some particular field quality specification. This region is often referred to as the *good field region* of the magnet. The size, shape, and quality of the field in this region is usually determined by extensive beam dynamics simulations to understand what part of the magnetic field the charged particles could pass through under a number of scenarios and how they will behave in an imperfect (i.e. a realistic) field. The magnetic field outside of the good field region is of no relevance to the beam since it should never encounter this part of the field. So, the magnet designer will optimise the magnet to achieve the required specification in the good field region and no more. In a dipole a particular absolute field level and uniformity is typically specified over a certain region. The magnet designer will choose a pole width which *just* achieves this. In our simple H dipole design, the vertical magnetic field is constant in the central region between the poles but towards the edges, where the pole is terminated to allow space for the coil, the field starts to decrease. By making some small reduction to the pole gap in this part of the magnet, this intrinsic field decay can be counterbalanced, and so the useful field of the dipole is extended horizontally with a very simple change. This minor change to the pole shape towards the pole corner is called *shimming* and is illustrated in Fig 4.9. The alternative to including this extra steel at the pole extremities would be to simply make the pole a little wider. This would be perfectly acceptable but just more expensive. As a general rule, the cost of a magnet scales with its mass so more material means greater expense.

So far we have only considered ideal steel performance with extremely large relative permeability. This is a good approximation at low fields where a relative permeability of several thousand is common, but μ_r is not a constant with B and tends towards 1 at very high fields. An example graph of relative permeability for a good-quality, common, magnet steel is shown in Fig 4.10. For this example $\mu_r > 1000$ until around 1.4 T, and by 2 T it is around 50. This means that the approximation used earlier to calculate NI at such high fields will no longer hold and additional Ampere-turns will be necessary to achieve the required field in air because the line integral along path 2 is no longer negligible. Calculating the impact of this non-linear behaviour of μ_r is not possible analytically; instead there are several magnet modelling software tools which can be used to calculate the fields numerically, either in two or three dimensions.

An alternative to the H dipole is the C dipole, which is illustrated in Fig 4.11. The name here reflects the C shape of the steel yoke. It is effectively one half of an H-type magnet. This design is less rigid mechanically because it is only supported on one side, but access from the side is now possible, which makes magnetic measurements easier and can also be important for certain applications; these include synchrotron light sources where beams of

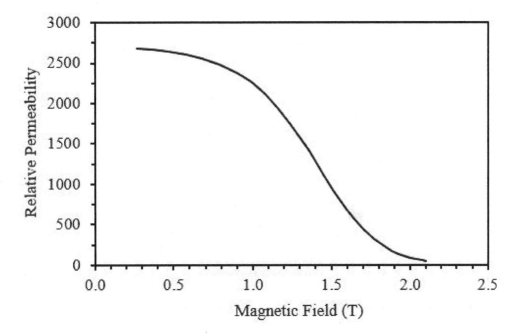

FIGURE 4.10   Graph showing the relative permeability as a function of magnetic field for a good-quality, low-carbon, commonly used magnet steel (XC10/AISI 1010).

X-rays from upstream undulators, or the dipole itself, would otherwise intercept the steel back leg as they exit from the accelerator.

Quadrupoles

Next we will consider the DC quadrupole, starting with the pole shape determined in Section 4.1.2 and visualised in Fig 4.3. The ideal steel pole surface extends out to infinity horizontally and vertically and adjacent poles approach each other. The consequence of this is that, at face value, there seems little prospect of finding the space to wrap a coil around each pole. Fortunately, since we are only interested in generating high-quality magnetic fields in the good field region, we can choose to terminate the pole asymptote at an appropriate position and then shape the steel away from this region so that space for coils is created along with a return path for the magnetic flux within the steel. An example of a standard quadrupole cross section is given in Fig 4.12. To determine the Ampere-turns required to achieve a particular quadrupole field gradient, we must follow a similar integral along a closed path calculation to that used earlier for the dipole. Path 1 is in the air region where the field at the centre of the magnet is zero and it increases linearly, reaching B_0 at the pole tip, which is at a radius r_0 from the magnet centre (so the field integral along path 1 is $r_0 B_0 / 2$). Note that the field direction is radial only (i.e. co-linear with Path 1). Path 2 is within the steel region, which has very large relative permeability, and path 3 completes the loop through both steel and air, but at the midplane of the quadrupole where B is always orthogonal to the horizontal axis and so the dot product will be zero.

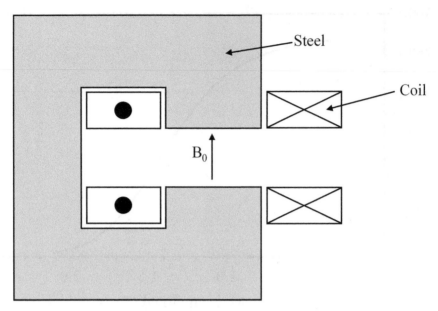

FIGURE 4.11   Cross section through a C-dipole, which is used to generate a uniform magnetic field. The electron beam is travelling into the page in the region between the two poles. The upper and lower poles each have a coil encircling the pole. The cross and dot within the coil cross section denote the current flowing into and out of the page.

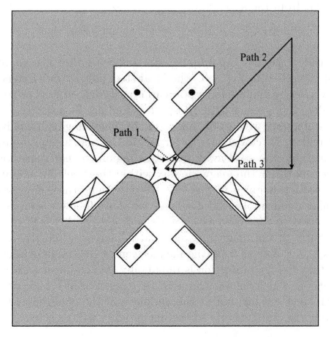

FIGURE 4.12   Cross section through a quadrupole. The electron beam is travelling into the page in the central region between the four poles. The cross and dot within the coil cross section denote the current flowing into and out of the page.

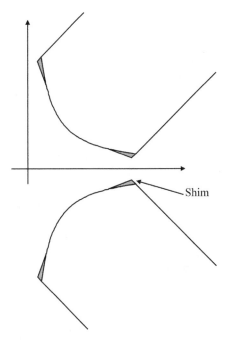

FIGURE 4.13   Close-up of quadrupole pole region illustrating the use of small pole shape adjustments, or shims, to counteract the B field decay towards the pole corners. The shims are typically integral to the pole itself and are tangent to the pole asymptotic shape but this is not essential.

$$NI = \int_{P1} \frac{\mathbf{B}}{\mu_r \mu_0} \cdot d\mathbf{l} + \int_{P2} \frac{\mathbf{B}}{\mu_r \mu_0} \cdot d\mathbf{l} + \int_{P3} \frac{\mathbf{B}}{\mu_r \mu_0} \cdot d\mathbf{l},$$

$$NI = \frac{r_0 B_0}{2\mu_0}. \tag{4.34}$$

The field gradient generated by the quadrupole, with NI Ampere-turns per coil, is therefore

$$\frac{\mathrm{d}B_y}{\mathrm{d}x} = \frac{B_0}{r_0} = \frac{2\mu_0 NI}{r_0^2}. \tag{4.35}$$

This analysis is effectively the same as for the dipole and it can similarly be further extended for higher-order magnets as required. For example, in the case of a sextupole, which is parameterised by the second field derivative, we find that

$$\frac{\mathrm{d}^2 B_y}{\mathrm{d}x^2} = \frac{6\mu_0 NI}{r_0^3}. \tag{4.36}$$

As for the dipole case, the magnet designer has the option of making adjustments to the quadrupole steel pole shape at the extremities to counteract the natural field roll off due to the pole corner. A small amount of extra steel is included at the pole corner, to bring the field back up to the required level over a short distance. This is a cost-effective change that can readily be made to any design, maximising the good field region for a particular steel pole width. Fig 4.13 illustrates how this shim is often included as a simple tangent to the hyperbolic pole shape.

Coils

The conductor which is wound to form the coils is typically made of copper, although aluminium is also sometimes used. Even though copper is a very good electrical conductor, each coil will have a finite resistance and so ohmic heating is an important consideration for electromagnets. To achieve a particular field level, we need to provide sufficient Ampere-turns, NI, via the coils. The choice of how many turns, N, to have in each coil obviously dictates the current that the conductor has to carry. A very large number of turns means that a relatively low current is needed, whereas very few turns in the coil means that a relatively large current must be supplied. The designer must select N and I so that the coil is practical to build and operate but should bear in mind that there are a range of solutions possible, to some extent this is somewhat of an arbitrary choice. A practical coil is one which can operate at full current continuously without breaking down or overheating. To prevent overheating, even of a well-designed coil, some form of cooling is generally required. For coils which consume low power, a heat sink might be adequate but more often than not active water cooling is required. For moderate power this can be indirect cooling, which generally means a water-cooled surface (e.g. a copper plate) is thermally attached to the coil, but for higher powers, direct water cooling is required to extract the heat. Direct cooling means that water flows through the copper conductor itself by using a cross section of conductor with a central hole for the water. Effectively the conductor is a thick-walled tube wound to make the coil. The key feature of this direct cooling is that the water flows directly through every turn in the coil and so the inner windings of the coil – which are the hottest and so most vulnerable to thermal failure as they are fully surrounded by other turns and so have no surface exposed to air – also receive the direct benefit of the cooling water. For this reason directly-cooled coils can tolerate higher currents than those cooled passively or indirectly. Strictly speaking, it is a higher current density within the conductor which can be tolerated by directly-cooled coils since the electrical resistance of the conductor is proportional to the conductor cross-sectional area. As a rule of thumb, if the current density is below 1 A/mm^2 then air convection cooling is sufficient; indirect water cooling can be used up to 2 or 3 A/mm^2, and direct water cooling used above this value. There is no clear limit on current density for direct-cooled coils although staying below 10 to 15 A/mm^2 is often quoted as good practice. However, all of these numbers are just provided for guidance and examples can be found where low current densities do need cooling, and vice versa. The magnet designer must consider the thermal performance of every coil to satisfy themselves that it will operate safely and reliably and to define the water-cooling requirements, such as inlet temperature, outlet temperature, flow rate, and pressure. Again, there is no absolute maximum value for the water temperature rise which can be tolerated, but water temperature increases between the inlet and the outlet of between 10 and 20°C are typical.

Since the cooling water flows directly through the current-carrying conductor, it must have low conductivity to prevent unwanted breakdowns. Demineralised (sometimes called de-ionised) water, which has had almost all of its mineral ions removed, is usually employed with a resistivity of around 5 MΩ cm. This water can be rather corrosive, leaching out material from the coils and piping, and so care must be taken with the correct use of compatible materials and avoiding particular combinations of materials at joints, for instance [4].

Steel Yoke

The steel structure which forms the main body of the magnet is called the yoke. The steel is shaped to form a continuous, low magnetic reluctance, high permeability flux path except in the air gap region of the beam. Steel that is close to the beam air gap, deliberately shaped to create the required magnetic field shape, is called a pole and steel that joins the poles to form this continuous path, away from the air gap, is called a back leg. In reality it is normal for the complete steel yoke to not be made from a single piece for practical engineering reasons. These reasons could be the feasibility of physically mounting the coils around a pole, or allowing the whole magnet to be split into two halves so that the beam vacuum chamber can be installed. Such joints in the steel yoke have a negligible impact on the field levels, assuming any potential air gap at the joint is minimised, but care must be taken to ensure that the mechanical quality of the yoke is maintained by making sure such joints can be made repeatedly without changing the physical shape of the yoke. This repeatability is often guaranteed, to within tight tolerances (e.g. to within $\sim$20 to 30 μm), by the use of pins or dowels.

The steel pieces that form the yoke can be machined from solid steel or built up by stacking thin ($\sim$1 mm) steel sheets or laminations. When time-varying magnetic fields are required then laminations must be used because of eddy current effects, as we shall see later. However, for static magnetic fields, both solid and laminated options are feasible. Laminations are shaped, using a stamping tool, to form the transverse cross section of the magnet and then stacked up in the longitudinal beam direction before being permanently joined together (often using gluing but also involving welding or mechanical fixtures sometimes) to form a single unit. The use of a stamping tool means that lamination-to-lamination shape repeatability is very good and also is a quick process so can be cost-effective. The tool or die, which has to be specially made for each magnet, is quite expensive though, so when small numbers of magnets are required, it is often more cost-effective to use solid steel yokes. The accuracy of laminations due to the stamping process also means that the required mechanical tolerances ($\sim$20 to 30 μm for the pole surface) for long magnets can be achieved more readily since machining over long pieces of steel reduces the precision achievable. Another advantage of using laminations is that any variation in the magnetic properties from batch to batch from the steel manufacturer can be considered more readily by mixing up – or *shuffling* – the laminations from the various batches within each single yoke. Such steel property variation can otherwise lead to small but significant differences in performance from magnet to magnet.

Longitudinal Issues

This section has so far concentrated on designing transverse cross sections for dipoles and quadrupoles and then considerations for the coil requirements to achieve the required field levels. Establishing these transverse designs is the first crucial step for any magnet design. The next step is to consider the longitudinal design (i.e. along the length of the magnet, in the direction of the beam). The length requirement is set by the needs of the particle beam, what angle the dipole must bend over for example. In general, the cross section is held constant through the majority of the magnet with some modification at both the entrance and exit. As for the transverse pole shape case, where we avoid abrupt steps in the steel shape since sharp 'corners' can readily become highly saturated, the same is true longitudinally. It is standard for a dipole to have a smooth change or roll-off in magnet gap at the beam entrance and exit to lower this saturation effect. The transverse shims may well need to be adjusted in size to compensate. Overall the integrated dipole field (or specific integrated multipole terms) through the magnet must be kept within predetermined limits and the end terminations are an important contributor to these integrals that must be carefully assessed and minimised with an appropriate end design. A similar approach is also taken in quadrupoles, although in general terms, the impact and need for the roll-off becomes less critical at higher orders and straight angular cut-offs or chamfers are sometimes implemented as an acceptable approximation to a smoother profile.

Forces and Stored Energy

By considering the inductance of, and so the stored energy in, a solenoid, it is relatively easy to show that the magnetic energy density in a system (energy per unit volume, dE/dV) is

$$\frac{dE}{dV} = \frac{d^3 E}{dxdyds} = \frac{B^2}{2\mu_0 \mu_r}, \tag{4.37}$$

so to calculate the total stored energy in any magnet, we need to integrate this equation over the full volume, including all the yoke and coils. This is not a trivial integral and accurate solutions are only possible by using numerical codes. In a simple approximation for a uniform dipole field, B_0, where the steel is not saturated and so μ_r is very large in the yoke, we can choose to ignore the stored energy within the steel and just calculate the stored energy in the air gap. In this case the approximate stored energy is

$$E = \frac{B_0^2 g A}{2\mu_0}, \tag{4.38}$$

where g is the gap between the poles and A is the area under a pole, and the product gA is simply the volume between the poles down the full length of the magnet. This can be useful to estimate the inductance L of a dipole, remembering that the energy stored by an inductor is $LI^2/2$;

$$L = \frac{B_0^2 g A}{\mu_0 I^2}. \tag{4.39}$$

Substituting the approximate formula for the full Ampere-turns of a dipole required to achieve the field B_0 we get

$$L = \frac{\mu_0 N^2 A}{g}. \tag{4.40}$$

So, the inductance of a dipole depends upon the number of turns in the coils as well as the physical extent of the field and the magnet gap, it does not depend upon the peak field or the current in the coil.

If we use the standard result that the work done (energy) is force × distance, i.e. $dE = \mathbf{F} \cdot d\mathbf{y}$, we can express the force exerted between the two poles as

$$\mathbf{F} = \iint \frac{dE}{dV} dxds = \iint \frac{B^2}{2\mu_0} dxds. \tag{4.41}$$

So, in a region of constant magnetic field, the force would be

$$\mathbf{F} = \frac{B_0^2 A}{2\mu_0}. \tag{4.42}$$

Again, in the real world, to calculate forces accurately we must use a numerical code, but the above equation is useful to check that the code is giving realistic values. We also need to consider the force on the coils. We know that magnetic fields exert forces on charges, since this is why we use magnets in accelerators in the first place! We know that the force $\mathbf{F}$ on a charge q moving with velocity $\mathbf{v}$ within the presence of a magnetic field $\mathbf{B}$ is

$$\mathbf{F} = q\mathbf{v} \times \mathbf{B}. \tag{4.43}$$

If there are N charges per unit volume, the number in a small volume dV of the coil is NdV. The total magnetic force on the volume dV is simply the sum of the forces on the individual charges, so that

$$d\mathbf{F} = NdV(q\mathbf{v} \times \mathbf{B}). \tag{4.44}$$

If we remember that the current density, $\mathbf{J}$, is the flow of current per unit area through the wire, then we can see that $\mathbf{J} = Nq\mathbf{v}$ and so

$$d\mathbf{F} = (\mathbf{J} \times \mathbf{B})dV, \tag{4.45}$$

and the force on a coil is given by the integral

$$\mathbf{F} = \int \mathbf{J} \times \mathbf{B}dV. \tag{4.46}$$

It is clear that to calculate the force acting on a coil is not trivial, requiring an understanding of the magnetic field distribution within the coil, and so we rely on numerical codes once again. To understand the *direction* that the force is acting on the coil, we just need to apply the well-known right-hand rule. Consideration of the forces on conventional DC magnet coils is generally quite superficial, but this is far more important when working with superconducting magnets, where both $\mathbf{J}$ and $\mathbf{B}$ can be very high, as we shall see later in this chapter.

Reliability Issues

DC electromagnets can be extremely reliable components in a particle accelerator and sometimes failures are so infrequent that we can become complacent. We have worked on several accelerators which have suffered no major magnet failures at all during their lifetime. Such high reliability is built on making good design and material choices, having thorough testing and acceptance criteria before installation, and routine maintenance schedules. Often when a magnet system causes accelerator downtime it does not mean there has been complete magnet failure, and instead it is likely to be a water leak, water blockage, or a poor electrical connection. It is relatively rare for a coil to fail and need replacing and very rare for a steel yoke to fail. A survey of accelerator magnet reliability [5] found that water leaks from hoses or fittings were the most common cause of problems. Non-conductive hoses must be used when hollow direct water-cooled coils are employed and these hoses are always made from organic molecules (thermoplastics or elastomers) which can be damaged with ionizing radiation, obviously a problem for high-energy particle accelerators. The hoses degrade with time and eventually will become brittle and crack. Regular replacement of hoses before they fail, at a rate determined by the level of radiation they are subject to, is highly recommended and easy to overlook. Water fittings are a rather mundane item for a particle accelerator and so it is possible to not pay enough attention to them. Our experience, born out by this survey, is that water fittings do fail or leak and it is worth buying the more expensive fittings (which are still cheap when compared to the magnets themselves!) to increase reliability. The second highest cause of failure found was water leakage at brazed joints. Brazed joints are very difficult to avoid completely (e.g. they are used to connect water fittings to the hollow conductor ends) but when we procure the magnet we can certainly insist that we don't want them hidden away *inside* of the magnet coils. The best way to avoid this failure is to have a thorough coil and water fitting testing regime before the magnets are assembled. Should a braze fail, then the best fix is likely to be to change the coil, which means having a spare one available. So, when procuring magnets we always buy at least one spare coil of each type. Having said that, we have not had to exchange a coil for many many years, which probably reflects the high manufacturing standards routinely offered now by magnet suppliers, as well as the factory tests mentioned earlier. We continue to buy spare coils though just in case! The survey also noted that almost half of the accelerators surveyed suffered from water blockages in the cooling system *every year*. Such a blockage can be relatively simple to fix by flushing water through the system in the opposite flow direction. However, it is far better to carry out routine maintenance to prevent the buildup of deposits which are causing the blockages. Of course, if water flow levels are dropping day by day then this is a warning that it is time to intervene before the coil becomes completely blocked.

Other issues which were noted by the survey were that the electrical connections, which are typically bolted joints for the higher current magnets, can work loose over time or not be tightened sufficiently on installation. This is an issue which regular routine maintenance and inspection should find and resolve easily. Another issue to note, that we certainly have encountered in the past, is that the polarity of a magnet is easy to get wrong by incorrect electrical wiring to the magnet coils or at the power convertor. We carry out polarity checks on all magnets, using a simple hand-held Hall probe, whenever any electrical connections are touched. The time required to carry out these checks is much smaller than the time wasted in the control room later if a magnet is acting in the opposite mode to that expected!

Specifying and Procuring of Magnets

One of the most daunting things to face a newcomer to the field is being asked to procure a batch of magnets for a particular accelerator project. Designing a magnet with bespoke magnet software is one thing, but then converting your design into a real working magnet is another step up. The good news is that it *appears* far more daunting than it really is. The best place to start, if possible, is to talk to someone more experienced with magnet procurement, someone who has been through the process before. They can help by sharing experience, sharing specifications from previous procurement exercises, and by passing on relevant contacts in industry.

There are really two alternative approaches that can be adopted for procurement with the difference being a question of who takes responsibility for the overall magnet performance; the procurer or the supplier. Both approaches can work perfectly well but it should be clear from the start which approach is being followed. Some of the larger accelerator laboratories will choose to take full responsibility for the magnet design, including the complete mechanical and electrical designs. They effectively produce a pack of drawings, lay out a strict production process, and define all materials to be used, and they then pass this fabrication pack to a magnet manufacturer and ask them to build exactly as drawn. This approach, sometimes called 'build to print', means that the procurer takes full responsibility for the complete magnet design. The manufacturer is responsible for fabricating to the drawings and standards, but if the magnet does not perform as expected then, so long as they did what was asked, they will not be held responsible. The second approach is where the procurer specifies the magnet performance required and the other constraints, such as the space available and beam apertures, but offers no design at all to the supplier. The magnet supplier then takes full responsibility for all aspects of the magnet design and if the magnet does not perform to specification, it is the supplier's responsibility to resolve. This second approach is more expensive at face value, because the supplier has to design the magnet as well as build it, but it does save the procurer significant design effort. We have used this second approach successfully for numerous DC magnet procurements for at least twenty years, even though we are perfectly capable of carrying out the full magnet design ourselves, because we prefer to pass responsibility onto the supplier and we appreciate that magnet companies are more expert than us in the mechanical and electrical design of magnets since they do this every day and we only need to do it occasionally. Prior to procurement of standard DC dipoles, quadrupoles, and so on, we carry out some simple magnet design simulations, often only in 2D, to confirm that we are requesting a feasible magnet and also to gain an appreciation of any particular challenges associated with the magnet. Then, we generate a detailed specification explaining exactly what magnetic performance is required; this specification will typically be about ten to twenty pages in length.

The specification covers all aspects of the magnet but crucially does not provide any design at all. It is quite detailed so that the manufacturer has all of the required information to design and build a magnet that is exactly fit for purpose. Of course, once the contract is in place, we have regular communication with the company to ensure the flow of information is two-way and that no wrong assumptions have been made by the manufacturer. We also insist on a full design review with the manufacturer prior to them starting to actually cut metal or wind coils.

The typical contents of our procurement specification will provide a brief introduction to the project, a clear statement that the manufacturer is responsible for the complete magnetic, mechanical, electrical, and thermal design as well as the construction, testing, and all magnetic measurements. We accept magnets primarily on the basis of the magnetic measurements provided as well as mechanical, electrical and thermal checks. Within the specification some magnet parameters will be *mandatory*, such as dipole field level, field quality, and bend angle, whereas others will be *nominal* or even undefined to provide the manufacturer with scope to optimise the magnet in an efficient manner. Examples of nominal parameters might be the physical dimensions of the magnet, the number of turns in a coil, and the conductor cross section. The specification will also describe the mechanical interface to the accelerator, such as how the magnet is expected to be mounted onto whatever girder or stand is planned, and including the need for survey monuments and lifting brackets. We also specify on what side of the magnet we want the power and water connections to be placed, since this is important for the installed accelerator infrastructure.

It is very often required that accelerator magnets can be physically split in half so that a vacuum vessel can be installed, and it is important that this need is requested and that the magnet performance is unaffected by it being split and reassembled. This means that mating and alignment features must be built into the steel yokes to ensure they can repeatably be reassembled without affecting their physical shape to quite tight tolerances. We specify the magnetic measurement facility performance that is required in order to ensure the manufacturer is capable of carrying out adequate measurements after manufacture of the magnets.

For every magnet we provide a table of all of the essential parameters. For a dipole this would, as a minimum, include the type of dipole required (e.g. H or C, sector or parallel ended), the magnetic field, integrated magnetic field through the magnet, bend radius, the field uniformity locally and integrated through the magnet, the horizontal and vertical dimensions over which this field uniformity is required (the *good field region*), the minimum pole gap, and the physical space constraints (maximum width, height and length of the magnet).

For a quadrupole this would, as a minimum, include the integrated gradient strength through the magnet, the allowed integrated gradient variation within the good field region, the size and shape of the good field region, and the physical constraints. We also specify some level of thermal performance, such as maximum temperature rise allowed in the cooling water (typically 10 to 20°C), but do not specify water flow rates or water channel dimensions, for example. With regards to the steel yokes, we allow the manufacturers to propose either solid or laminated yokes and also they can choose any suitable magnet steel above some minimum level of acceptability.

For the coils, it is crucial that the insulation is adequate to prevent any electrical breakdown between turns or between the coil and the yoke. It is common to request that the coils be insulated using fibre glass tape wound around the conductor as the coil is fabricated and for the full coil to then be mechanically consolidated with a radiation-resistant epoxy resin under vacuum impregnation to ensure full penetration within the coil. We do not allow any joints in the conductor within a single coil as this is a possible source of unreliability or failure, as mentioned earlier. We insist on a set of electrical and thermal tests for every coil prior to magnet assembly. All the coils being thermally cycled several times to confirm the epoxy consolidation is mechanically robust. For the electrical tests we check the inter-turn insulation and insulation to ground.

After magnet assembly, we insist on further electrical checks at high voltage between the coil terminals and the yoke to ensure there is no breakdown and also thermal tests at full operating current to ensure the maximum temperature rise is below the limit specified. In general, we insist on a full set of prescribed magnetic measurements for every magnet built, although in some circumstances we have accepted a detailed set of measurements for the first few magnets and a reduced set of measurements for subsequent ones where we are confident that the overall risk to the project is small. Our experience of magnet procurement using this approach, from several different European manufacturers, has been excellent. We have always received magnets that have performed as required and there have been no serious failures at all.

4.4.2   Practicalities of AC Magnets

We use the term AC magnets as shorthand to refer to periodically time-varying magnets, sometimes called *cycling* magnets. We understand that alternating current strictly refers to current that periodically reverses direction and is often associated with a sinusoidal waveform. However, to be clear, that is not what we mean when we use the term AC as many magnets are designed to have periodically time-varying fields, such as in a synchrotron, where the field does not reverse polarity and nor does it follow a sinusoidal waveform. The magnetic field waveform required can take many forms and to first order the overall shape is not as critical to the magnet designer as the peak rate of change of the field and the repetition rate.

When dealing with time-varying fields, we have to take account of two extra effects, *eddy currents* and *hysteresis*. Both of these lead to additional power losses in the magnet on top of the resistive (ohmic) losses which DC magnet coils also suffer from. Eddy currents can also generate unwanted magnetic fields that can perturb the beam. The power supply for an AC magnet also has significant extra challenges which can limit how rapidly the fields can be changed.

Eddy Currents

Eddy currents are loops of electrical current induced in a conductor by a changing magnetic field. The name comes from the analogy to water forming eddies or whirlpools in areas of turbulence. The induced current is due to Faraday's Law which states that the voltage, V, induced in a loop of conductor in a region of varying magnetic field is given directly by the rate of change of the magnetic flux, Φ, as

$$V = -\frac{d\Phi}{dt}. \qquad (4.47)$$

Eddy voltages are induced equally in all conductors which experience the same time-varying fields, irrespective of the material, and so the magnitude of the eddy current depends inversely on the resistivity of the conductor. This means that non-magnetic materials, such as copper and aluminium, and low relative permeability materials such as stainless steel that is often used for vacuum chambers, will all experience eddy currents to an extent that depends upon how conductive the material is. Since currents are flowing, there is an associated heating which can be very significant. The currents flow in the plane perpendicular to the magnetic field direction.

To calculate the power deposited in a conducting material due to eddy currents we start from the simple conceptual layout of Fig 4.14. The magnetic field, B, is perpendicular to

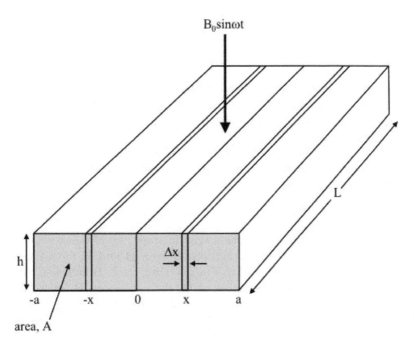

FIGURE 4.14 Conceptual sketch for calculating the average power loss per unit length in a conductor.

the block with time-varying strength $B_0 \sin \omega t$. The voltage induced in a loop made from two thin strips of width Δx at position $\pm x$ is given by the rate of change of flux. We assume that the two strips are joined at either end to form a complete loop and that the width of this joint is much less than the length of the strips, $2x \ll L$. Since the field is normal to the surface, the flux is simply the product of the area of the loop and the rate of change of the field itself.

$$V = -2xL\omega B_0 \cos \omega t.$$

The maximum, or peak voltage, induced in the loop is therefore

$$V_p = 2xL\omega B_0.$$

The resistance of the loop, R, depends upon the resistivity of the material, ρ, the cross-sectional area of the strip, $h\Delta x$, and the length of the strip, L;

$$R = \frac{2L\rho}{h\Delta x},$$

and so the peak current in the loop will be

$$I_p = \frac{x\omega B_0 h \Delta x}{\rho}.$$

Now we can calculate the peak power, $P_p = I_p^2 R$, as

$$P_p = \frac{2x^2 \omega^2 B_0^2 h \Delta x L}{\rho}.$$

By integrating the peak power in the loop with respect to x, from 0 to a, we can determine the peak power in the full block;

$$P_{block} = \frac{2a^3 \omega^2 B_0^2 hL}{3\rho}.$$

Since the average value of $\sin^2$ is $1/2$, the *average* power loss per unit length in the block is

$$P_{loss} = \frac{a^2 A \omega^2 B_0^2}{6\rho}, \qquad (4.48)$$

where A is the cross-sectional area of the block, $2ah$. So, a copper conductor, with resistivity of 1.7×10^{-8} Ωm, of cross section 10 mm x 10 mm, in a 1 T peak field oscillating at 50 Hz, will be absorbing a power loss due to eddy currents of 2.4 kW/m. This power loss scaling also explains why AC magnet yokes must be laminated in the xy plane that is parallel to the magnetic field, with an insulating coating between laminations. The thickness of the laminations, $2a$, is selected to be small enough to ensure the power loss is manageable yet not so small that the number of laminations per yoke becomes unwieldy.

Hysteresis

Hysteresis describes the dependence of the magnetic properties of the steel yoke on its past history. Fig 4.15 shows the typical relationship between **B** and **H** for a nominal magnet steel. When the material is first exposed to a magnetizing force, **B** increases with increasing **H** along path a. At sufficiently large values of **H**, the increase in **B** levels off and we say that the material is *saturated*. Now, if the magnetizing force is reversed, **B** follows path b. When the magnetizing force reaches zero, there is still a magnetic field in the material, and in the air gap of any magnet built using this material. If we continue to reverse the magnetizing force, then the material will again reach saturation at the opposite polarity. If we then increase **H** back towards zero, then the material follows path c and at zero the material is again magnetised but with opposite polarity to previously. The field in the air gap at zero excitation will be different depending upon which path the material has been taken through. This is why we must degauss magnets to ensure that the magnetic fields are repeatable for the same current in the coils. To be precise, degaussing implies that the remanent field in a magnet is reduced to zero but, in general in an accelerator, this is less important than repeatability from day to day, which can be achieved by following the same magnet excitation cycle and not necessarily requiring the reversing of the magnet polarity or the field at zero excitation being exactly zero. If the remanent field in the steel is required to be zero, then a comprehensive degauss process should be followed whereby the material is taken around the hysteresis path repeatedly whilst the magnet excitation levels are progressively decreased towards zero. Eventually the loops shrink in area until they are negligibly close to the origin and the remanent field is then zero.

The *hysteresis loop* described by taking the material into saturation at both extremes defines the boundary of other possible loops that the material would follow if the material is not excited so strongly. The power loss in the steel due to hysteresis is proportional to the area enclosed by the loop for the particular excitation regime that it is subject to. The greater the extremes of the excitation, the greater the losses. The losses are also proportional to the volume of the steel yoke and the frequency with which it is excited. The losses also depend upon the material choice; not all steel alloys are the same. Steels with high silicon content (a few %) have significantly lower AC losses than the usual low-carbon steels that are utilised in DC magnets (such as XC06/AISI 1006 and XC10/AISI 1010). Steel manufacturers will provide measurements of AC losses for different grades of steel, in W/kg, usually at the transformer frequencies of 50 and 60 Hz. Note that these losses combine both hysteresis and eddy losses.

Pulsed Kicker Magnets

There are some accelerators which require very fast pulsed dipole magnets or *kickers*, such as for injection or extraction of beams. In these cases the magnets are off for most of the

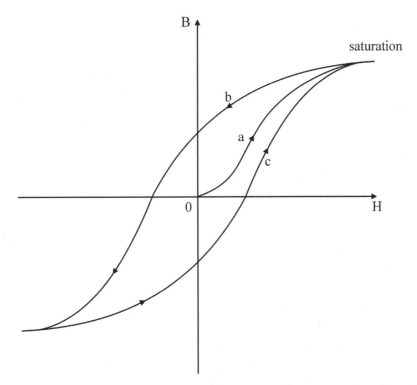

FIGURE 4.15 Typical variation of magnetic field as a function of magnetizing force for steel.

time and only fire as and when needed. Strictly speaking, they are in a different class than the AC magnets considered above since they do not necessarily follow a periodic waveform, but as they are certainly time varying, they have many issues in common with AC magnets and so are covered in this section. Kickers are characterised by very short pulse durations, μs or even ns, and consequently can only have relatively low field strength (tens of mT). There are two common classes of injection and extraction kickers. The first one is purely inductive, with one or a few turns of conductor and with the power supply as close as possible to the magnet to minimise stray inductance. The second, which is capable of faster rise times, is a transmission line or delay line system which matches the impedance of the kicker (capacitance and inductance) to the line, and the rise-time depends on the propagation time of the pulse through the magnet. Matching the impedance is not trivial and can lead to complex and expensive kicker and line designs. Purely inductive kicker magnets follow the same design principles as for other dipole magnets, except that steel laminations are no longer practical and ferrites must be used instead. Ferrites are ceramics that include a large proportion of iron oxide and are ferrimagnetic so they are used to enhance magnetic fields in a similar manner to steel although they do have much lower saturation levels (hundreds of mT). Ferrites have very high resistivity, meaning eddy currents can be neglected and the hysteresis losses are small even at very high frequency. Kickers are often mounted outside of the vacuum to avoid increasing the impedance encountered by the beam and so in this case conductive vacuum chambers cannot be used as the eddy currents would be too severe and impact on the field level in the beam region, and so ceramic vacuum vessels are employed which have a thin conductive coating on the inside to provide a conducting path for the beam image current.

Pulsed Septum Magnets

Septum magnets are typically combined with kickers to form complete injection or extraction systems. Kickers can differentiate beams in *time* by pulsing so fast that only the beam that should encounter the field does encounter the field. The very fast pulses required from a kicker limit the magnetic fields that they can generate. Septum magnets differentiate between beams *spatially*; they generate a strong dipole field to steer the injected or extracted beam but have zero field a very short distance away (~5 to 10 mm) so the stored beam is unaffected. There are DC and pulsed versions (the pulses are typically tens of μs); this section will briefly discuss two pulsed types. The first, with cross section illustrated in Fig 4.16 (a), has a simple C-shaped yoke, fabricated using thin steel laminations, and a single turn coil; the return leg of the coil separates the high-field region from the (very close to) zero field region. To achieve the required field levels (~1 T, say) the current in the single turn must be very large and so cooling of the coil is required but challenging, especially in the return leg which is desired to be as thin as possible so the physical separation between the two beams is minimised. The alternative design, with cross section illustrated in Fig 4.16 (b), has a simpler coil arrangement wound around the back leg of the yoke, making cooling more straightforward, and then a passive conducting screen separates the two magnetic field regions. When the coil is pulsed, eddy currents are induced within this screen which then act to shield the field. The eddy current screen is very effective with field levels less than 1% of the main field being achieved [6]. If lower leakage fields are required, then the eddy current screen arrangement can be extended to create a full return box around the magnet and a thin steel magnetic screen also added on the outside of the eddy current screen in the critical region, in which case field levels less than 0.1% of the main field are achieved [7]. The eddy currents themselves do not disappear as soon as the current pulse has reached zero current; they decay on a timescale set by the resistivity of the material within which they are flowing.

4.5   Permanent Magnets

Electromagnets are the standard solution employed for the vast majority of particle accelerators with permanent magnets (PMs) being used in some significant but still niche applications, such as for undulators in accelerator-driven light sources. There is currently an increasing interest in the application of PMs to more mainstream solutions for common magnets such as dipoles and quadrupoles. Clearly, PM-based solutions are only useful in generating static magnetic fields (equivalent to DC), not time-varying ones, although it should be recognised that the field is not necessarily *fixed*, and many designs exist which enable adjustable dipole and quadrupole fields using PMs. One reason for this increasing interest is that PMs do not consume any electricity in coils or associated cooling water infrastructure and so the electrical power demand, and hence the operating costs, of a facility can be significantly reduced. A second reason is that PMs are very powerful and they can generate very strong magnetic fields, competitive with normal conducting electromagnets, and when physical space for the magnet is tight they can often exceed the capabilities of electromagnets. Other advantages are that since no cooling water is required, then a potential cause for magnet vibrations can be eliminated, no high-precision power supply is required, and they are extremely stable and reliable from day to day as there is very little that can fail. The Fermilab Recycler ring is a 3.3 km antiproton storage ring which was and remains the first large-scale accelerator project built where all of the main magnets are based upon PMs and not electromagnets [8]. The ring has operated for many years now and the experience has been very positive [9].

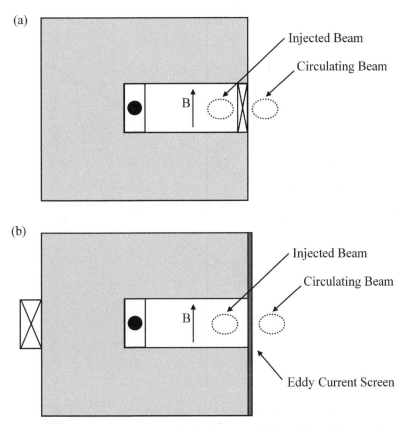

FIGURE 4.16 Cross section of two types of pulsed septum: (a) is a direct drive version and (b) is an eddy current version.

Permanent Magnet Materials

A material is said to be a permanent magnet or magnetically *hard* if it can independently support a useful flux in an air gap of a device. PMs are ferromagnetic and as such they have a characteristic hysteresis loop (or **BH** curve); the loop for an ideal PM is shown in Fig 4.17. As the magnetizing force increases, the magnetic field increases with gradient μ_0. Remember that for magnetically *soft* iron alloys, the gradient is $\mu_r\mu_0$ where μ_r is the relative permeability of the material, which is non-linear with **H** and much larger than 1, in general, until the steel is completely saturated. As the magnetizing force decreases to zero, the PM material remains fully magnetised and exhibits a strong remanent field, B_r. The material remains magnetised and resists any negative **H** until large values are reached and then the PM effectively flips polarity. The value of **H** which is required to reduce **B** to zero is called the coercivity, H_c. The value of **H** where the flip occurs is called the intrinsic coercivity, H_i, and this is a very useful number for comparing different grades of material since it effectively describes just how *permanent* the material is, which is naturally an important requirement! It should be noted that the ideal PM is linear in the second quadrant; this is important since this is the zone where the PM will be operating to deliver flux (**+B**) into an air gap (**-H**). All PMs are affected by temperature changes, with their values for B_r and H_c decreasing as the temperature increases. Since accelerator magnets operate at around room temperature (ignoring superconducting cryogenic magnets for now) there is little danger of the environmental temperature causing irreversible changes to the PM. A more important consideration is the temperature drifts which the PMs might

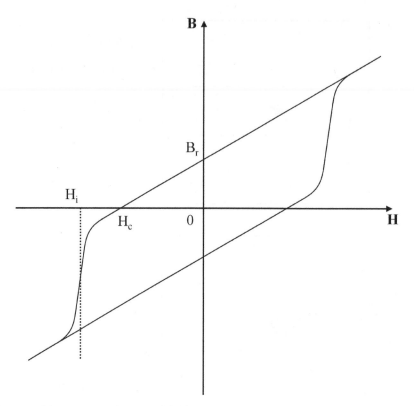

FIGURE 4.17   The variation of magnetic field with magnetizing force for an ideal PM.

encounter within the accelerator facility since these can cause small magnetic field changes which could be important. Undulator magnets may be mounted in temperature-stabilised enclosures to counter this effect, although it is more common now for the full accelerator to be temperature stabilised and not just the undulators. A counter to the loss in field strength as the temperature increases is that if the temperature decreases the field will increase. This intrinsic effect is made use of now by undulators which are cryogenically cooled (often to 77 K) in order to benefit from this increased magnetic field performance from the PMs.

There are two common types of PM that are used for accelerator magnet and undulator applications, with a third type now starting to be used for cryogenic applications. All three types exhibit behaviours close to the ideal described above. The two standard types are samarium-cobalt ($SmCo_5$ or Sm_2Co_{17}) and neodymium-iron-boron ($Nd_2Fe_{14}B$). The third, new type is praseodymium-iron-boron ($Pr_2Fe_{14}B$), which is favoured now when wanting to take advantage of enhanced magnetic properties by working at cryogenic temperatures. The other two types do benefit from increasingly enhanced magnetic properties as they are cooled but only down to an intermediate temperature of around 150 K, below which the magnetic properties degrade due to a spin reorientation effect [10]. $Pr_2Fe_{14}B$ does not suffer from this effect and so it can be operated at temperatures that are able to be maintained easily, such as the boiling point of nitrogen (77 K). Table 4.1 summarises the key features of the two main types of PM employed. The characteristics of praseodymium-based magnets are not so well established yet although the remanent field and the coercivity have been measured to be ~ 1.3 T and ~ 1500 kA/m respectively at room temperature and ~ 1.6 T and ~ 6000 kA/m at 77 K [11, 12]. Note that the relative permeabilities of these materials is very close to unity and so they behave magnetically in a similar manner to a coil in air, with a good approximation being that the fields from a group of PM blocks can be

TABLE 4.1 Comparison of the typical characteristics of the two main PM materials commonly used for accelerator magnets.

Characteristic	SmCo	NdFeB
Remanent Field at 300 K (T)	$0.9 - 1.1$	$1.1 - 1.4$
Coercivity, H_c (kA/m)	$600 - 800$	$800 - 1000$
Intrinsic Coercivity, H_i (kA/m)	$1000 - 2000$	$1000 - 2500$
Maximum Energy Density (kJ/m^3)	$150 - 250$	$200 - 400$
Temperature Coefficient of B_r (%/°C)	-0.035	-0.10
Maximum Operating Temperature (°C)	$250 - 350$	$50 - 200$
Relative Permeability, μ_r	~ 1.03	~ 1.1

added linearly to calculate the total field within a volume. In detail, the permeability is in fact anisotropic, it is marginally different in one plane of the material to another, and this might need to be taken into account when carrying out detailed magnetic calculations for a particular system.

The parameter ranges associated with the magnetic characteristics listed within Table 4.1 give an indication of the broad spread available by selecting different *grades* of the same basic alloy, using alternative *additives*, that the manufacturers are able to create to match the needs of a particular application. The manufacturers can in fact predict and control the parameters very precisely to meet the needs of the customer. The PM manufacturers publish catalogues which are available online, listing the various PM grades available and their precise characteristics.

As well as the temperature variation of the material characteristics, which can potentially lead to unwanted magnetic field variations over time, there are two other issues which need to be considered. The first is the concept of *aging* and whether or not the characteristics of a material are likely to degrade over months or years and the second is that of *radiation damage* and whether or not a PM can and should be employed in an accelerator environment. The aging of PM blocks is difficult to quantify since it is affected by the grade, coercivity, the shape of the block, the magnetic circuit, the regular cycling or variation of the fields within the circuit, the temperature variation, and so on. However, in reality, since the accelerator environment is maintained at room temperature, and large coercivity materials are employed, aging over time is not a serious concern. When we procure PM blocks we request that they are all heated to well above room temperature, perhaps 50°C or so, for a few hours to ensure that any irreversible aging due to moderate temperature variation has already been *built into* the blocks. This temperature aging has a minor impact on the material characteristics and subsequently we have not observed any measurable variation of PM performance over a timescale of more than twenty years.

Radiation damage to PM materials is a significant issue that must be considered when working within an accelerator environment. There are certainly several examples of PM-based accelerator magnets that have been directly affected by ionizing radiation or direct impact from high-energy particle beams. Most often these are undulators which have PM blocks very close to the beam axis (often only a few mm). The effect of the radiation is to degrade the magnetic properties of the PM material, locally leading to loss of magnetic field and poorer field quality. This is especially important for undulators which have very strict field quality requirements to optimise the synchrotron radiation output that they emit. However, the vast majority of undulators installed in accelerators have suffered little or no damage. If beam losses in the vicinity of an undulator are carefully controlled then undulators can operate extremely well for tens of years with apparently no loss of performance. We have measured the magnetic field in an undulator that was removed from a 2 GeV electron storage ring after more than twenty years of continuous service and could measure no discernible degradation in the magnetic field levels at all.

There have been a wide variety of experiments exposing PMs to radiation in a number of different scenarios. An excellent literature review summarising the data has been written recently [13]. The results of the experiments show certain trends, but it is not possible to establish clear quantifiable predictions based on the results. The general trends that have been established are that temperature is very important, experiments above room temperature show that damage is worse and cooling magnets to well below room temperature confers increased resistance to damage. The radiation resistance of SmCo is consistently shown to be better than NdFeB. Both of these effects are likely to be due to the higher relative intrinsic coercivity. It has also been demonstrated that the block aspect ratio (length to diameter ratio) makes a significant difference, as does the direction of the magnetization within the block (the easy axis) in relation to the direction of the beam generating the radiation. The broad conclusions supported by the studies show that radiation resistance is improved by using a grade with a higher coercivity, choosing SmCo over NdFeB (Sm_2Co_{17} being more resistant than $SmCo_5$), altering the shape of the magnet or the geometry to select a more optimal working point for the material, decreasing the temperature of the PM, pre-baking the PMs to thermally stabilise them, and positioning the PMs as far from the beam axis as feasible to reduce the dose and to allow the possible addition of radiation shielding.

PM-Based Dipoles

First we will consider a simple dipole to gain an appreciation for some of the basics of PM-based systems, using a similar approach to [14]. Returning to our derivation for the magnetic fields in a DC electromagnet, we found that the integral of **B** along a closed path, P, that bounds a surface S is given by the current flowing through that surface:

$$\oint_P \frac{\mathbf{B}}{\mu_r \mu_0} \cdot d\mathbf{l} = \oint_P \mathbf{H} \cdot d\mathbf{l} = \int_S \mathbf{J} \cdot d\mathbf{S} = NI. \tag{4.49}$$

For our PM case, there are no external currents and so the integral equals zero:

$$\oint_P \mathbf{H} \cdot d\mathbf{l} = 0. \tag{4.50}$$

If we consider the geometry for the dipole of Fig 4.18, integrating around the path shown, then this becomes

$$\oint_{P1} \mathbf{H} \cdot d\mathbf{l} + \oint_{P2} \mathbf{H} \cdot d\mathbf{l} + \oint_{P3} \mathbf{H} \cdot d\mathbf{l} + \oint_{P4} \mathbf{H} \cdot d\mathbf{l} = 0.$$

As for the electromagnet case, we will assume that the steel has huge relative permeability ($\mathbf{H} \sim 0$) and so we can neglect the integrals along paths 2 and 4;

$$\oint_{P1} \mathbf{H} \cdot d\mathbf{l} = -\oint_{P3} \mathbf{H} \cdot d\mathbf{l},$$

which gives us

$$H_m L_m = -H_g L_g, \tag{4.51}$$

where L_m and L_g are the lengths of the PM block and air gap respectively. Remembering that, in air, we have that $H_g = B_g/\mu_0$, so that

$$B_g = -\mu_0 \frac{H_m L_m}{L_g}. \tag{4.52}$$

We will also assume that any flux leakage from the steel away from the air gap is negligible and so the total flux flowing across the air gap will be the same as that flowing through the PM;

$$B_m A_m = B_g A_g, \tag{4.53}$$

where A_m and A_g are the cross-sectional areas of the PM block and steel pole respectively. Substituting our result for B_g into this gives

$$B_m = -\mu_0 \frac{H_m L_m A_g}{L_g A_m}. \tag{4.54}$$

The relationship between B_m and H_m also, separately, depends upon the material properties (recall Fig 4.17) such that $B_m = \mu_0 H_m + B_r$ for the ideal PM material (a good approximation for the materials discussed earlier) which is perfectly linear, with gradient μ_0, in the second quadrant. This effectively gives us two simultaneous equations which we can use to solve for B_m,

$$B_m = \frac{B_r L_m A_g}{L_g A_m (1 + \frac{L_m A_g}{L_g A_m})}. \tag{4.55}$$

So, in our simplified dipole scenario, the magnetic field within the PM is determined by the physical size and shape of the PM and the air gap, as well as the remanent field of the PM itself. In the simplest case where the size and shape of the PM and air gap are equal ($L_m = L_g$ and $A_m = A_g$), then $B_m = B_g = B_r/2$ and $H_m = -H_g = -B_r/2\mu_0$. To increase the field in the air gap we could double the length of the PM ($L_m = 2L_g$) and then we would get $B_m = B_g = 2B_r/3$. If we want to increase the field further, then we could reduce the steel pole area to concentrate the flux, for example with $L_m = 2L_g$ and $A_m = 2A_g$, then the field in the air gap equals B_r, whilst in the PM it equals $B_r/2$. We should remember that these scenarios are somewhat idealistic, but they do demonstrate how the air gap fields relate to the material properties, as well as the size and shape of the PM block and air gap. By manipulating these parameters, we are changing the working point of the material, this is often shown graphically which can be instructive. The two simultaneous equations which we solved to find B_m are plotted in Fig 4.19. One line represents the intrinsic material properties and the other, called the *load line*, represents the physical layout of the system under consideration. The intersection of the two lines is called the *working point*. The safest region to operate in, in terms of avoiding unwanted demagnetization effects, is when the working point is nearer to the vertical axis (small negative **H**) since at large negative **H** the material is closer to, or possibly in, the non-linear region of the **BH** curve and any changes, perhaps caused by physically moving parts of the system or by temperature changes, can then be irreversible. The most efficient working point to choose is when the **BH** product in the second quadrant is maximised and the maximum magnetic energy is being utilised. For the ideal material this occurs at $B_r/2$.

The easiest PM-based dipole to design and build is one which has a fixed field, with no requirement for any magnetic field adjustability at all, except via manual intervention such as by changing the physical position of the PM or steel pole, perhaps. The actual design of a PM dipole depends, as for other magnet types, on the exact specification and constraints. There is considerable flexibility in the options available. When PM dipoles have been implemented in accelerators, some effort has been put into coping with (the relatively small) temperature effects. The issue being that due to the PM material properties (see Table 4.1), the dipole field will naturally reduce if the temperature increases. A good solution to this issue is to build a *passive* compensation scheme into the magnet itself. This can be achieved by using a second material within the magnet which also has a temperature dependence but in the opposite direction, so the two effects can be made to compensate for each other. A successfully demonstrated solution uses a Ni-Fe alloy which has permeability that varies significantly in the room temperature region [15]. The magnet design includes volumes of this material adjacent to the PM so that some of the magnetic flux is shunted or short-circuited away from the air gap. If the temperature increases, the PM naturally delivers less flux, but simultaneously, the Ni-Fe alloy permeability decreases and so less flux

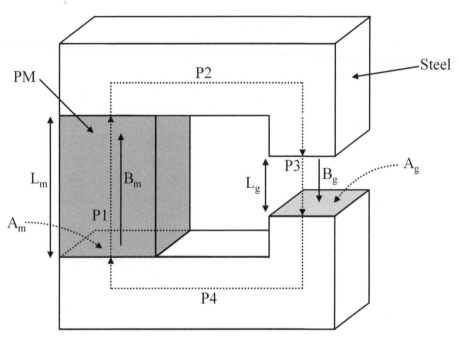

FIGURE 4.18   A simple dipole magnet design driven by a PM block.

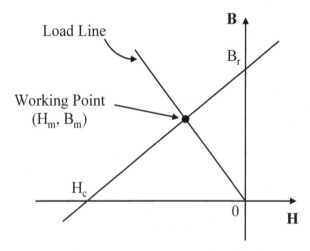

FIGURE 4.19   Graph showing the load line and working point of the PM.

is short-circuited and if the volume of alloy is chosen correctly, then the two effects will compensate each other over the working temperature range of the magnet. Improvements in the dipole field change, with temperature of around two orders of magnitude have been demonstrated for an NdFeB-based dipole, compared to the original level of $-0.11\%/^\circ C$, over a temperature range of more than $10^\circ C$ [16]. It should be noted that inclusion of this passive temperature compensation scheme will lower the maximum magnetic field that is achievable since flux is being short-circuited.

The development and upgrade of storage ring light sources towards diffraction-limited sources, in particular, has recently led to an increased interest in the widespread application of fixed-field PM-based dipoles to take advantage of their compactness compared with electromagnetic equivalents. A second consideration is that dipoles with an optimised longitudinal field variation can offer superior electron or photon beam properties to the facility and that generation of this variation lends itself more naturally to a PM design [17, 18]. One study has shown that a 2 m long PM-based dipole with optimal longitudinal field profile was three times lighter than the equivalent electromagnetic version [19].

If small magnetic field adjustability is required for a particular application (a few percent, say) then it would be reasonable to include electromagnetic coils in the system to provide this range of variation. It should be noted that since the PM material has a relative permeability ~ 1, then it effectively acts like an additional air gap within the system. This means that the coils are less effective than we would normally expect compared to a standard electromagnet which only has an air gap in the region of interest. This can mean that quite powerful coils are required for relatively small adjustments.

If large magnetic field variation is needed from the dipole magnet (more than ten percent, say) then the only really sensible solution is to physically move parts of the system. Solutions exist which move steel components [17] and others that move the PM [20]. These two example designs are presented schematically in Fig 4.20, although many other design solutions are possible. Dipole field variations of a factor ten have been demonstrated although even more variation would be possible for both concepts. The forces that are present in strong magnet systems can be very significant and so the motion system needs to be able to cope with these whilst simultaneously not affecting the precise location of the poles to ensure sufficient field quality is maintained as the field is adjusted.

Assembly of PM-based magnets raises new challenges compared to electromagnets. The most significant extra challenge is, of course, handling of the PM material which cannot be 'turned off'. Great care has to be taken at all stages of assembly to ensure that the attractive forces between the PM and the steel yoke and other magnetic items, such as fasteners and parts of any motion system, are considered. Assembly by hand is generally impossible as the forces are far too high to cope with, and so special fixtures must be designed and built so that the items can be brought together to build the magnet in a safe and controlled manner. Non-magnetic tools, typically made from a Cu-Be alloy, must be used at all times.

PM-Based Quadrupoles

Quadrupoles with fixed gradient strength have been realised in a number of different formats. The most common format used is the Halbach type, which simply consists of a ring of PM blocks [21]. In fact, this format can be used to create any multipole type, simply by adjusting the remanent field direction of the blocks to suit the type desired. The number of segments per ring, which is independent of the multipole order itself, is for the magnet designer to choose, typically being a compromise between complexity and achievable gradient and field quality. There are also hybrid versions which include steel in various configurations [22]. Examples of a PM-only version and a hybrid version are shown in Fig 4.21.

PM quadrupoles with variable gradient have been developed by several groups [22] with

(a)   (b)

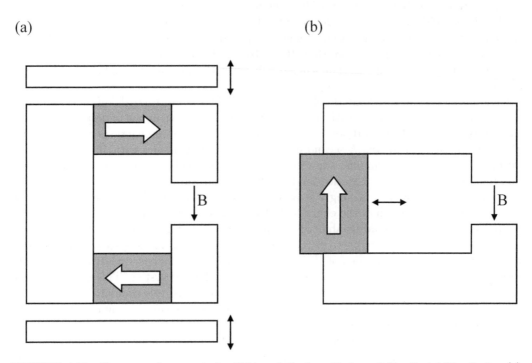

FIGURE 4.20 Two example concepts for PM-based dipoles with large field adjustability. Option (a) has two steel plates which move symmetrically as a pair. As the plates approach the PM, shown in grey, they short circuit the flux and so reduce the field at the beam. Option (b) slides the PM in and out of the steel yoke region to alter the flux path and reluctance through the steel and so the field at the beam.

(a)   (b)

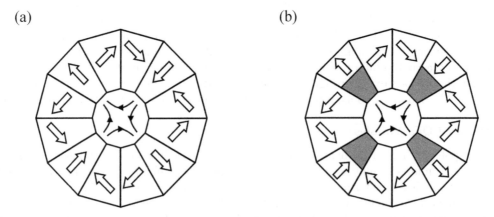

FIGURE 4.21 Two example concepts for fixed gradient PM quadrupoles. Option (a) is a classic Halbach design with 12 PM segments with magnetization direction shown by the arrows. Option (b) is a hybrid variant with steel poles shown in grey.

many different designs being produced. Each design has been optimised to achieve different characteristics and so there is not one design which meets all needs. Some designs can change the gradient by more than an order of magnitude, others are optimised for maximum gradient, others for large good field region, and so on. As for the dipoles, if very modest gradient adjustability is needed then coils can be used. If significant adjustment of the gradient is needed, then some parts of the quadrupole must be physically moved to alter the gradient. Typically, PM blocks are moved linearly or rotated (to alter the magnetization direction) to adjust field levels. For linear motion systems, where the blocks are moved away from the beam axis to lower the gradient, the minimum gradient is determined by how far the blocks are moved and in principle the gradient could be zero if they are moved far enough away. Other examples use an outer steel shell to short circuit the PM blocks to lower the gradient more rapidly for the same physical motion. Three different examples for adjustable quadrupoles are sketched in Fig 4.22.

PM-Based Undulators

The use of PM-based dipoles and quadrupoles is certainly not yet mainstream, although it is becoming more popular as stronger fields from more compact magnets are required, as more complex field shapes are required, and also to reduce the electrical power consumption of accelerator facilities. However, there is one area of accelerator magnets where PMs definitely are the mainstream and that is in the application of undulators for generating light from relativistic electron beams. Storage ring light sources are one of the most common advanced accelerator applications globally, and free-electron lasers are also now developing at a pace. Both of these types of light source rely on undulator magnets to generate the light in these world-leading X-ray sources for researchers. The vast majority of undulators have been, and continue to be, based upon PMs. An undulator is essentially a device which generates a periodic magnetic field, most commonly the field is in the vertical plane and it varies sinusoidally in the longitudinal direction, such that as an electron travels through the magnet it oscillates horizontally from side to side about the beam axis, emitting light in the forward direction. Undulators are built to enhance the light through constructive interference, much like a periodic diffraction grating, and more details on the properties of the light that is generated are given in Chapter 6. The wavelength of light which is observed in the forward direction depends strongly on the electron energy, but also on the period of the magnetic field and peak strength of the field. In essence, high magnetic fields and short magnet periods are optimal as these create the shortest wavelengths possible for a given electron energy. Magnetic fields of the order of 1 T at periods of only 20 to 30 mm are typical. These levels are impossible for electromagnets (unless they are superconducting!) since the physical space available for the coils with sufficient Ampere-turns is just not available. It is at these small dimensions that PMs really excel.

The simplest undulator magnet uses blocks of PMs laid out in two arrays, one above the electron beam, and one below the electron beam (see Fig 4.23 (a)). The magnetization direction of the blocks rotates by 90° each time and the vertical field generated is a very good approximation to a pure sine wave. Another very common design is shown in Fig 4.23 (b) where steel poles are employed in a *hybrid* configuration. For the design which only employs PM blocks (so-called 'pure PM' undulator) it is possible to derive the magnetic field at the electron beam analytically. Assuming that the PM block heights are equal to half the period length, the peak on-axis field, B_{y_0}, is given by

$$B_{y_0} = 1.72 B_r e^{-\pi g/\lambda_u}. \tag{4.56}$$

The inclusion of steel poles, with non-linear permeability behaviour, in the design means that an analytical solution is no longer possible and so magnet design codes are needed to

(a)

(b)

(c)

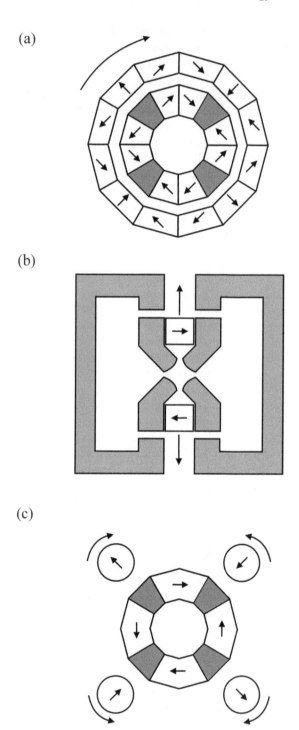

FIGURE 4.22 Three example concepts for variable gradient PM quadrupoles, all shown at their maximum field gradient position. Option (a) is a concentric pair of Halbach rings with the outer ring rotating to adjust the quadrupole gradient [23]. Option (b) has a pair of PM blocks which are driven apart symmetrically about the beam axis and the outer steel shell short-circuits the field to lower the quadrupole gradient more rapidly [24]. Option (c) is a Halbach type with extra PM cylinders which are rotated in unison to alter the quadrupole gradient [25].

(a)

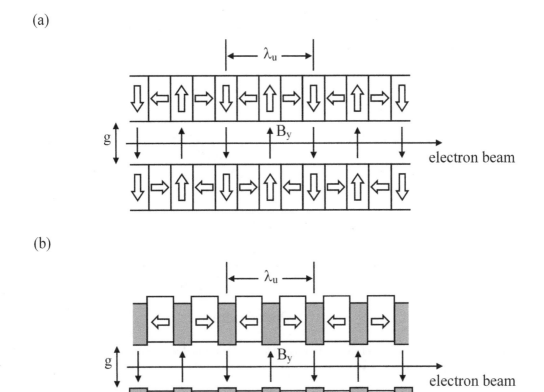

(b)

FIGURE 4.23 Two common magnet designs for undulator magnets which generate sinusoidal vertical field variation in the longitudinal direction. Type (a) uses PM blocks only with magnetization direction varying as shown by the arrows. Type (b) is a hybrid variant with steel poles shown in grey.

model the design to accurately predict the fields that are achieved. An empirical equation, equivalent to the pure PM one, for the peak field in an undulator using PM material (NdFeB) with remanent field of 1.25 T, is given by [26]

$$B_{y_0} = 3.60 \exp\left(-4.45 \frac{g}{\lambda_u} + 0.67 \frac{g^2}{\lambda_u^2}\right). \tag{4.57}$$

This equation is said to be valid over $0.3 < g/\lambda_u < 3.0$. Many similar empirical equations have been generated for different remanent fields, for undulators which generate fields in both transverse planes (elliptical undulators), and also for cryogenic devices. An excellent summary of these various equations is provided in [26]. An example comparison of the peak magnetic fields achievable in a hybrid and pure PM undulator, as a function of period, is shown in Fig 4.24.

4.6 Superconducting Magnets

The application of superconducting (SC) materials to accelerator magnets opens up new possibilities that would otherwise not be available to us. In particular, the generation of multi-Tesla dipole fields is essential for high-energy physics-focused accelerators to reach the

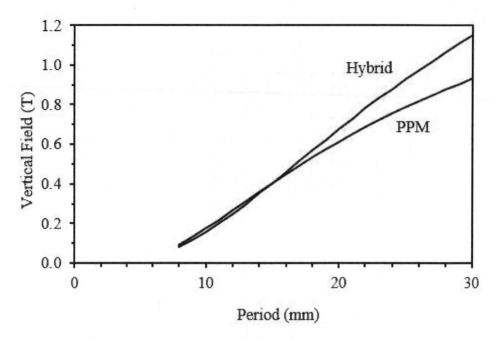

FIGURE 4.24   A comparison of the fields achievable in a hybrid and a pure PM undulator assuming a remanent field of 1.25 T and a magnet gap of 8 mm.

highest possible proton energies in a (relatively!) small circumference facility. SC materials have zero electrical resistance in DC operation and so do not suffer from resistive heating. As a consequence they can carry extremely high current densities that enable dipoles of order 10 T to be fabricated. The main disadvantage of SC magnets is that the materials are only SC at very low temperatures, with accelerator magnets typically operating at 1.9 or 4.2 K. This increases the complexity and the cost significantly and so SC magnets are only used when there is no clear alternative. The understanding, application, and engineering of SC materials and magnets is a specialist topic that many people have spent whole careers on. This section can only be a brief introduction to the topic, highlighting key features and differences to conventional magnets. There are some excellent textbooks on the subject (e.g. [27, 28]) as well as the proceedings of specialist accelerator schools (e.g. [29]) that the reader is invited to study for more details.

4.6.1   Superconducting Materials

The tried and tested SC material of choice is niobium-titanium (NbTi) because it is by far the easiest SC material to work with. It is ductile, easy to insulate, can be readily formed into wires of suitable dimension, is relatively inexpensive, and generally quite forgiving. It can be wound into coils as easily as we wind copper wire. Like all 'Type II' SCs it will remain SC so long as it is operated below its characteristic *critical surface* of temperature, magnetic field at the conductor, and current density. At a fixed operating temperature (e.g. 4.2 K) the surface simply reduces to a line of current density against a magnetic field which defines the SC boundary of the material. At 4.2 K and with a field at the conductor of 6 T, the maximum current density for NbTi is around 2000 A/mm^2 [30]. If, instead, the material is cooled further to 1.9 K, then the same current density can be attained at up to 9 T. Remember that this is the magnetic field that the conductor is experiencing, not the field in the air gap of the magnet. In general, the dipole field at the beam is larger than that at the SC wire itself. The critical surface defines the *limiting boundary* at which the SC becomes normal conducting (resistive) and so it makes good sense for magnet designers to give themselves some safety margin in order to be able to build reliable magnets that will operate at their defined specification for year after year. The more aggressive magnet designs, which try to push the state of the art, might choose to work at around 90% of the limit (the LHC dipoles run at 86% for their nominal field value of 8.3 T [31]) whereas less demanding SC magnets routinely built by industry (e.g. for Magnetic Resonance Imaging systems) are more likely to operate at around 50% of the limit.

If higher-strength magnets are required, that cannot be fabricated by only using NbTi, then another option is to employ Nb_3Sn which, due to a more favourable critical surface, is able to operate at high current densities at much higher magnetic fields. Whereas NbTi can sustain around 2000 A/mm^2 at 6 T at a temperature of 4.2 K, Nb_3Sn can sustain 2700 A/mm^2 at 12 T or 1450 A/mm^2 at 15 T [30], with even more impressive performance possible at 1.9 K. Unfortunately, Nb_3Sn is much more difficult to work with and so is only used when absolutely necessary. Nb_3Sn is a brittle intermetallic compound that is created by raising the constituents to high temperature (typically 650 to 700°C) for many hours and then brought back down to room temperature in a controlled manner (the exact recipe varies grade by grade and is provided by the SC supplier). The primary problem with this particular SC material is that due to its nature, the SC itself is rather brittle and so winding coils with the material causes major degradation to the SC properties. The way around this fragility is to wind the coils before creating the SC material itself. This is called the *wind and react* approach. The unreacted wire is ductile and can be wound in a similar manner to NbTi. However, once the coil is wound, it must then be heat treated for it to become SC and hence useful. The required reaction process adds additional complications to the engineering (coping with the large thermal expansion and subsequently handling of the coil in the brittle state) and precludes the use of common electrical insulating coatings which are unable to withstand the high temperatures. In short, the use of Nb_3Sn adds extra risk to the magnet fabrication process but it does offer a route to significantly higher magnetic fields.

In addition to these two materials there are several high-temperature superconducting (HTS) materials which are being actively applied to accelerators in some niche areas, such as for current leads in the transition range between room temperature and the magnet coils at 4.2 K or below, or being prototyped into coils for evaluation. Examples of these HTS materials are MgB_2, Bi-2212, and REBCO (rare earth barium copper oxide). Further details on the relevant properties of these materials are available in [32].

One point to note is that the current density mentioned earlier is the value within the SC itself. However, in practical situations the SC material is not the only material present, and since the wire is then typically formed into a multi-wire cable, which necessarily includes physical gaps between the wires, and then formed into a coil, with further gaps, the average current density carried by the space occupied by the coil cross section can be much less than the actual peak value within the SC. To account for this *filling factor* in their calculations, magnet designers quote the *engineering current density*, which is simply the average current density flowing through the coil cross section.

A second point to note is that the makeup of an SC wire is actually quite complex. It is formed of a large number of narrow filaments of continuous SC strand held in a copper matrix which supports all the filaments. The number of filaments in a single wire can range from tens to thousands and they can be only a few μm thick in some cases. The copper is very important as it not only supports the SC, it also conducts the current and the heat when the SC is no longer in the SC state. This is a very dangerous and unwanted state because if high currents are flowing the wire will heat up rapidly and melt, so it is essential that there is enough copper within the wire to transport the heat and the current temporarily whilst the high current is removed or diverted as quickly as possible. The wire is typically formed into a multi-wire *Rutherford cable* which has a transposition (an optimised wire 'twist') built in and a good packing factor. There are often twenty to thirty wires that form a cable. The advantage of working with a cable is that large currents can then be carried (thousands of amperes) and the winding becomes easier to handle (far fewer turns per coil).

4.6.2   Coil-Dominated Magnets

We use SC materials to generate dipole fields well in excess of 2 or 3 T. For example, the LHC main dipoles have a nominal operating field of 8.3 T (using NbTi) and the High Luminosity upgrade to the LHC is planning to use 11 T dipoles (using Nb_3Sn). With such high fields, well in excess of the magnetic saturation fields of magnetic steels, it no longer makes sense to use steel poles to shape or enhance the fields. Instead, SC magnets are primarily coil based and rely upon enormous currents flowing to generate the required field levels. This is quite a different approach to that discussed earlier in Section 4.4 and leads to a radically different magnetic design.

Following the approach of [28] it is easy to show that pure multipole fields (i.e. dipole, quadrupole, sextupole, etc.) can be generated in the xy plane by an arrangement of currents flowing parallel to the s (beam) direction. In the idealised case, the current flows in the s direction on the edge of a circle (which is in the xy plane) and so maps out a cylinder. The required current distribution for a pure multipole as a function of the azimuthal angle, θ, is given by

$$I(\theta) = I_0 \cos m\theta. \tag{4.58}$$

A pure dipole is generated inside the cylinder when $m = 1$, a quadrupole when $m = 2$, and so on. This type of magnet is therefore referred to as a *cos-theta* magnet and two example ideal cases are sketched in Fig 4.25. Of course, fabrication of such a design is not really practical and so approximations to the ideal case are made using so-called *sector coils*. In these designs the current density, J, within the wire or cable is constant and the coil geometry is set to approximate the $\cos\theta$ requirement. A simple sector coil for a dipole magnet is shown in Fig 4.26. The inner radius of the coil is r, the coil width is w, and the coil half angle is α. For this geometry the dipole field can be calculated analytically to be [30]

$$B = \frac{2J\mu_0 w}{\pi} \sin\alpha, \tag{4.59}$$

where μ_0 is the permeability of free space. We can see from this equation that the magnetic field at the beam scales with current density and coil width, but does not depend upon the radius. It can be shown [28] that this simple sector coil generates not just a dipole field but also higher-order multipoles (sextupole, decapole, and so on). Furthermore, if the angle, α, is selected to be 60°, then the sextupole term actually cancels to zero. However, the remaining multipole terms are generally considered to not be acceptable (the decapole is still a few percent of the main field, for example) and so this simple sector coil arrangement is not a good enough approximation to the $\cos\theta$ ideal in practice. To overcome this, the solution is to include more degrees of freedom in the design by, for example, adding additional layers or by breaking the coil into more parts, or a combination of the two. This concept is illustrated in Fig 4.27 and has been used successfully by the very-high-field SC magnets employed in accelerators like the LHC. The field quality achieved in practice by this type of magnet is just as high as it is with the iron-dominated, lower-field, magnets discussed earlier.

(a)

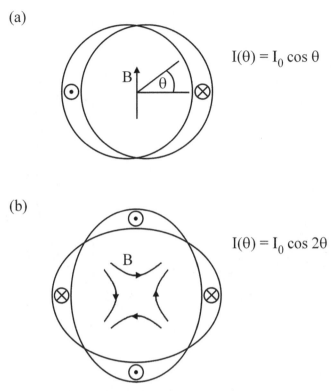

$$I(\theta) = I_0 \cos \theta$$

(b)

$$I(\theta) = I_0 \cos 2\theta$$

FIGURE 4.25 Examples of ideal cos-theta magnets for generating pure multipole fields. The current is flowing into and out of the paper with the distribution as noted in the equation next to each type. (a) is a dipole, which has peak current at the mid-plane and zero current top and bottom, and (b) is a quadrupole, which has peak current in four locations and zero current in four locations.

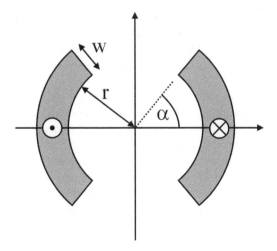

FIGURE 4.26 A simple sector coil approximation to an ideal cos-theta dipole. The current density is uniform within the coils, shown in grey. The current direction is into the page on the right (cross in a circle) and out of the page on the left (dot in a circle).

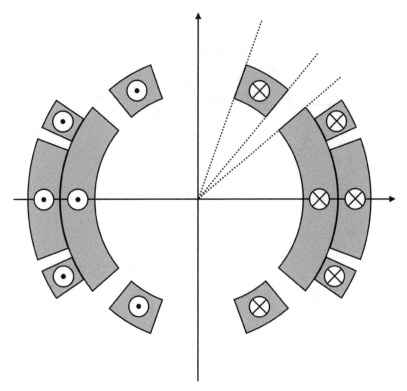

FIGURE 4.27 An illustration of how additional degrees of freedom can be added to the sector coil concept by adding extra layers and splitting layers into parts. The dotted lines highlight just three of the angles which the magnet designer can optimise to ensure the magnet field quality is sufficient.

Inclusion of Steel

Whilst steel is not used to shape the field using pole pieces in these very-high-field magnets, it still has a useful role to play in confining the field to the magnet and so preventing stray fields which can otherwise be quite disruptive to an accelerator facility. To achieve this magnetic shielding, a steel yoke surrounds the coils in the form, at least conceptually, of a thick hollow cylinder. This steel cylinder must be of sufficient thickness that the steel is not saturated. A simple calculation for the 8.3 T LHC dipoles shows that the steel must be at least 100 mm thick [30].

The steel cylinder also has the additional benefit of acting as a virtual coil because of image currents within the yoke. These image currents, which make up the virtual coil, increase the magnetic field within the region of interest. Since the virtual coil has a much larger cross section than the actual coils, the image current density is reduced and the impact on the field is similarly reduced. Nevertheless, the steel yoke surrounding the LHC main dipole coils increases the field by 17%, and this increase seems to be relatively typical for such magnets. The effect of the steel yoke on the field quality should also be taken into account since inner and outer shells will be inverted in the virtual case.

There are some circumstances when it makes sense to also include steel poles to shape the fields in an SC magnet. This type of design is called *superferric*. For the steel to determine the field shape it must not be fully saturated, and so this implies lower magnetic fields than considered above, perhaps in applications such as correction dipoles or higher-order multipoles.

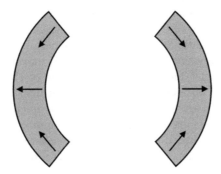

FIGURE 4.28 Sketch showing the direction the forces are acting on a pair of dipole sector coils.

Practical Considerations

We already noted in Section 4.4 that the forces on a coil depend upon the current density and magnetic field at the coil. Since both of these parameters are very large in an SC magnet it is no surprise that the forces acting on the coils can be enormous. If part of a coil moves as an SC magnet is powered then the energy released can be sufficient to cause a *quench* (become resistive) since the heat capacity of materials close to absolute zero is very low and so even very small releases of energy can be enough to raise the temperature locally above the critical surface. Displacement of a coil or part of a coil will also impact the field quality. Typically, the tolerance on the placement of wires in such a magnet is to within 0.1 mm [30] and so even small movements can be quite detrimental.

Handling of the forces exerted on the coils in high-field SC magnets is therefore a major challenge. In some cases the forces are so high that they are beyond the material yield strength and so plastic deformation is a real concern that must be addressed. First, we will consider the direction of the forces and then the countermeasures that are used to make high-field SC magnets viable. If we first apply the right-hand rule to a simple solenoid magnet we will see that the magnetic force is pushing the coil outwards radially away from the axis, creating a *hoop stress* in the coil. Now, if we consider the cos-theta dipole arrangement, we find that there is a radial force in the midplane pushing the coil away from the axis and that the parts of the coil away from the midplane are pushed towards it, as shown schematically in Fig 4.28. In the beam direction, the forces are acting to stretch the coil longitudinally, so overall the forces are trying to expand the coil, much like the solenoid case. Of course, the magnitude and direction of the force within any particular part of the coil depends upon the magnetic field strength and direction and so this rather simple picture presented here is actually much more complex in the real world. A more detailed analysis is presented in [33]. For an example 5 T dipole, the horizontal force is estimated to be 1 MN per longitudinal metre [28].

The solution employed to enable such forces to be handled by the sensitive SC windings is to apply a pre-stress to the coils to counter the magnetic forces. A radial inward compression force is generally applied by mounting the coils inside a pair of stainless steel or aluminium 'collars' which are mechanically pressed around the coils and then secured with dowel rods to maintain this pre-stress on the coils. It should be remembered that this assembly activity takes place at room temperature but that the magnet is operated cold and so the different thermal contractions of the selected materials will also alter the pre-stress levels and must be taken into account.

Another practical consideration is making sure the magnet can cope with the rapid transition between SC and normal conducting that occurs when the magnet quenches. A quench will take place when the SC material passes through the critical surface of field, temperature, and current density, and is thought to generally be caused by a local release of energy due to the friction associated with very small movements of a wire or cable. A quench is a risk to the magnet integrity since a portion of the coil has become resistive and so can heat up rapidly considering the large current densities employed. Local failure of a winding is a very real possibility. For these reasons quench protection is taken very seriously and when a quench is detected, an electronic circuit will turn off the power supply as quickly as possible and the stored energy will be diverted into a secondary circuit which can handle the energy safely. Quench protection systems can be quite complex in detail but are essential for magnet protection since if a coil fails it is effectively scrap.

A further practical consideration is the phenomena known as *training*. This describes the process by which an SC magnet very often improves in performance after successive quench events. It is common for an SC magnet to not reach the design magnetic field when it is first powered, and instead it will reach some intermediate level and then quench. At the second time of powering up, a good magnet will then quench at a higher field and so on. It is as if the magnet is learning (or training) how to cope with higher and higher currents and fields. The explanation for this behaviour is that small motions in the windings are taking place to cause the quench and that the coil is then locally in a more stable position. Good-quality magnets will retain a memory of being trained and so when they are warmed up to room temperature and then cooled back down, they will not need to be trained again. The number of quenches needed to attain design specification is hard to predict but ten or twenty would not be unusual.

4.6.3   SC Undulators

Short period SC undulators can generate higher magnetic fields than the permanent magnet (PM) undulators discussed in Section 4.5, but PM undulators remain the mainstream solution with only a handful of SC examples being used routinely in accelerator-based light sources [34, 35, 36]. The reason that SC undulators are still not the first choice option is in large part due to the extremely successful track record of PM undulators and their ongoing improvement rather than any particular deficiency with SC undulators. Regardless of the progress being made with PM devices, there is still a significant benefit to be gained from using SC materials instead, and it is for this reason that several groups are actively developing short-period, high-field SC undulators [37]. The handful of examples that have been installed into light sources perform extremely well in terms of reliability and stability and there is no reason to doubt that SC undulators will grow in popularity in the future. Indeed, there seems to be a growing view that free-electron laser-based light sources might see the first major installation of these devices in large numbers [38]. As well as increased magnetic field, SC undulators are believed to be several orders of magnitude more resistant to radiation damage than PMs, which is especially important for high bunch repetition rate free-electron lasers.

The magnetic design of SC undulators is very straightforward in concept, with most teams adopting very similar approaches, as illustrated in Fig 4.29. Two physically independent arrays of SC windings are fabricated on steel yokes and arranged in such a way that current flows transversely to the electron beam in an alternating arrangement such as to create the required periodic field. The two arrays are held apart by a non-magnetic fixture and are connected in series. Compared to the SC dipoles and quadrupoles discussed earlier, the forces and quench protection arrangements are much easier to cope with. However, the mechanical tolerances on the wire placement, yoke dimensions, and array separation, of

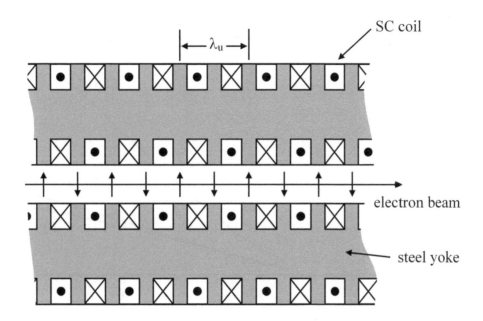

FIGURE 4.29 Sketch showing the side view of a section of a typical SC undulator design for generating vertical magnetic fields.

better than a few tens of μm, are difficult to achieve over the one- or two-metre length of the devices.

As for the hybrid PM undulators which include steel poles, it is not possible to calculate analytically the magnetic field generated by an SC undulator because of the non-linear behaviour of the steel. Instead, scaling laws have been generated which have then been cross-checked against 3D magnetostatic simulations [26, 39, 40] to provide an estimate for the peak field in an NbTi undulator;

$$B_{y0} = (0.3282 + 0.0678\lambda_u - 1.053.10^{-3}\lambda_u^2 + 5.85.10^{-6}\lambda_u^3)e^{-\pi(\frac{g}{\lambda_u}-0.5)}. \qquad (4.60)$$

A comparison of the peak magnetic field achievable in an SC undulator fabricated with NbTi, that is operating at 80% of critical current density, compared against PM-based undulators is given in Fig 4.30. The figure clearly demonstrates the very significant advantage that the SC undulator has over the other options. In this example, the magnet gap is set for all devices at 8 mm. The aperture available for the electron beam is less than the magnet gap in a standard undulator, of either type, due to the need for a beam vacuum chamber within the magnet gap. In the PM case it is now common to put the magnets *inside* the vacuum system to remove the need for this beam vacuum chamber and so increase the magnetic field experienced by the beam as the magnet gap can be reduced for the same beam aperture. It should also be possible to have the SC undulator as part of the beam vacuum system, and so gain a similar benefit in the future, and at least one group is actively pursuing this option [41]. Similarly, the fields could be enhanced in the future by switching to Nb$_3$Sn or one of the HTS materials, this is an active field that is developing quickly.

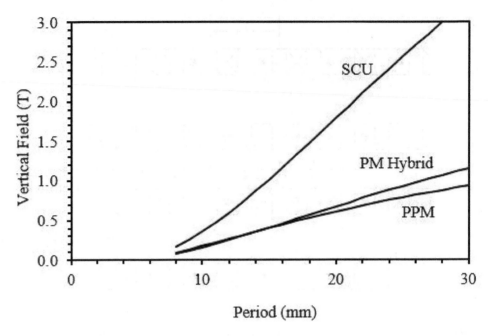

FIGURE 4.30 A comparison of the fields achievable in an SC magnet using NbTi against the PM alternative assuming a remanent field of 1.25 T and a magnet gap of 8 mm.

Exercises

1. What magnetic field is required to bend a beam of protons with a kinetic energy of 800 MeV onto an arc of radius 7 m?

2. If instead of protons we wanted to bend a beam of electrons, of the same kinetic energy on the same arc, what magnetic field would then be required?

3. The LHC has a magnetic dipole field of 8.33 T for protons of kinetic energy of 7 TeV. What is the bend radius of these protons within the dipole magnet?

4. If we instead stored electrons at 7 TeV kinetic energy in the LHC (ignoring any synchrotron radiation effects) what magnetic field would we have to set to ensure that they travel on the same bend radius?

5. We want to design a normal conducting electromagnetic dipole with a magnetic field of 0.8 T and a gap between the poles of 40 mm.

    (a) If we assume that the steel has infinite permeability, how many Ampere-turns are required in each coil of the dipole?

    (b) If we set the number of turns in each coil to be 20, what will be the current flowing through the conductor?

    (c) We choose the conductor cross section to be 10 mm x 10 mm. What will be the current density flowing through the conductor if it is solid, with no integral water cooling channel?

    (d) The current density is sufficiently large that a direct water cooling channel is required. We decide to limit the current density to 10 A/mm². Calculate what cross section is now available for the water cooling channel and, assuming it is a circular channel, what the diameter of this channel will be.

(e) Now, assuming that our dipole has a pole width of 100 mm and is 1 m long, estimate the energy stored in the magnet and then the inductance.

(f) Finally, estimate the magnetic force between the two pole surfaces.

6. We decide to also consider a permanent magnet dipole with the same peak field of 0.8 T and gap between the poles of 40 mm. Again the pole will be 100 mm wide and 1 m long.

(a) Show that when the permanent magnet is used at peak efficiency, and the field within the material is $B_r/2$, that the magnetizing force, H, within the material is $H_c/2$.

(b) Our selected permanent magnet material has $B_r = 1.2$ T. Assuming the material is ideal, with relative permeability of one, what cross section, A_m (refer to Fig 4.18), should the permanent magnet block have for it to operate at maximum efficiency?

(c) If we choose the permanent magnet block to also be 1 m long, like the steel yoke, calculate the required height of the block, L_m.

(d) If you are very keen, repeat this calculation of the permanent magnet block volume for a few alternative magnetic field levels within the block to satisfy yourself that maximum efficiency does correspond with minimum required volume of material.

References

1. Jack Tanabe. *Iron Dominated Electromagnets Design, Fabrication, Assembly and Measurements*. World Scientific, 2005.

2. Andrzej Wolski. Maxwell's equations for magnets. *CERN Accelerator School on Magnets, CERN-2010-004*, 2010.

3. Klaus Halbach. First order perturbation effects in iron-dominated two-dimensional symmetrical multipoles. *Nuclear Instruments and Methods in Physics Research Section A: Accelerators, Spectrometers, Detectors and Associated Equipment*, 74, 1969.

4. L. Pellegrino. Experience with long term operation with demineralized water systems at DAFNE. In *Proceedings of EPAC2004, Lucerne, Switzerland*, 2004.

5. Cherrill M. Spencer. Improving the reliability of particle accelerator magnets: Learning from our failures. *IEEE Transactions on Applied Superconductivity*, 24(3):1–5, June 2014.

6. K. Marinov, J.A. Clarke, N. Marks, and S. Tzenov. Design, tests and commissioning of the EMMA injection septum. *Nuclear Instruments and Methods in Physics Research Section A: Accelerators, Spectrometers, Detectors and Associated Equipment*, 701:164 – 170, 2013.

7. M.J. Barnes, J. Borburgh, B. Goddard, and M. Hourican. Injection and extraction magnets: septa. *CERN Accelerator School on Magnets, CERN-2010-004*, 2010.

8. G.W.Foster, K. Bertsche, J.-F Ostiguy, B. Brown, H. Glass, G. Jackson, M. May, D. Orris, and Dick Gustafson. Permanent magnet design for the Fermilab main injector recycler ring. In *Proceedings of PAC1995, Dallas, USA*, 1995.

9. James T Volk. Experiences with permanent magnets at the Fermilab recycler ring. *Journal of Instrumentation*, 6(08):T08003, aug 2011.

10. C. Benabderrahmane, P. Berteaud, M. Vallau, C. Kitegi, K. Tavakoli, N. Bchu, A. Mary, J.M. Filhol, and M.E. Couprie. $Nd_2Fe_{14}B$ and $Pr_2Fe_{14}B$ magnets characterisation and modelling for cryogenic permanent magnet undulator applications. *Nuclear Instruments and Methods in Physics Research Section A: Accelerators, Spectrometers, Detectors and Associated Equipment*, 669:1–6, 2012.

11. M. Vallau, C. Benabderrahmane, F. Briquez, P. Berteaud, K. Tavakoli, D. Zerbib, L. Chapuis, F. Marteau, O. Marcouill, T. El Ajjouri, J. Vtran, G. Sharma, C. Kitegi, M. Tilmont,

J. Da Silva Castro, M.-H. N'Guyen, N. Bchu, P. Rommelure, M. Louvet, J.-M. Filhol, A. Nadji, C. Herbeaux, J.-L. Marlats, and M.-E. Couprie. Development of cryogenic undulators with PrFeB magnets at SOLEIL. *AIP Conference Proceedings*, 1741(1):020024, 2016.

12. F.-J. Borgermann, C. Brombacher, and K. Ustuner. Properties, Options and Limitations of PrFeB-magnets for Cryogenic Undulators. In *Proc. 5th International Particle Accelerator Conference (IPAC'14), Dresden, Germany, June 15-20, 2014*, pages 1238–1240, July 2014.

13. Ben Shepherd. Radiation damage to permanent magnet materials: A survey of experimental results. *CERN, CLIC-Note-1079*, 2018.

14. Peter Campbell. *Permanent Magnet Materials and Their Application*. Cambridge University Press, 1994.

15. K. Bertsche, J.-F. Ostiguy, and W.B. Foster. Temperature considerations in the design of a permanent magnet storage ring. In *Proceedings of PAC1995, Dallas, USA*, 1995.

16. S. H. Kim and C. Doose. Temperature compensation of NdFeB permanent magnets. In *Proceedings of PAC1997, Vancouver, Canada*, 1997.

17. Takahiro Watanabe, Tsutomu Taniuchi, Shiro Takano, Tsuyoshi Aoki, and Kenji Fukami. Permanent magnet based dipole magnets for next generation light sources. *Phys. Rev. Accel. Beams*, 20:072401, Jul 2017.

18. J. Citadini, L. N. P. Vilela, R. Basilio, and M. Potye. Sirius - Details of the new 3.2 T permanent magnet superbend. *IEEE Transactions on Applied Superconductivity*, 28(3):1–4, April 2018.

19. J. Chavanne and G. Le Bec. Prospects for the use of permanent magnets in future accelerator facilities. In *Proc. 5th International Particle Accelerator Conference (IPAC'14), Dresden, Germany, June 15-20, 2014*, number 5, pages 968–973.

20. A.R. Bainbridge, J.A. Clarke, N.A. Collomb, M. Modena, and B.J.A. Shepherd. The ZEPTO Dipole: Zero Power Tuneable Optics for CLIC. In *Proc. of International Particle Accelerator Conference (IPAC'17), Copenhagen, Denmark, May, 2017*, number 8, pages 4338–4341, May 2017.

21. K. Halbach. Design of permanent multipole magnets with oriented rare earth cobalt material. *Nuclear Instruments and Methods*, 169(1):1–10, 1980.

22. Amin Ghaith, Driss Oumbarek, Charles Kitgi, Mathieu Vallau, Fabrice Marteau, and Marie-Emmanuelle Couprie. Permanent magnet-based quadrupoles for plasma acceleration sources. *Instruments*, 3(2), 2019.

23. T. Mihara, Y. Iwashita, M. Kumada, and C. M. Spencer. Variable permanent magnet quadrupole. *IEEE Transactions on Applied Superconductivity*, 16(2):224–227, June 2006.

24. J. A. Clarke, N. A. Collomb, B. J. A. Shepherd, D. G. Stokes, A. Bartalesi, M. Modena, and M. Struik. Novel tunable permanent magnet quadrupoles for the CLIC drive beam. *IEEE Transactions on Applied Superconductivity*, 24(3):1–5, June 2014.

25. F. Marteau, A. Ghaith, P. N'Gotta, C. Benabderrahmane, M. Vallau, C. Kitegi, A. Loulergue, J. Vtran, M. Sebdaoui, T. Andr, G. Le Bec, J. Chavanne, C. Vallerand, D. Oumbarek, O. Cosson, F. Forest, P. Jivkov, J. L. Lancelot, and M. E. Couprie. Variable high gradient permanent magnet quadrupole (QUAPEVA). *Applied Physics Letters*, 111(25):253503, 2017.

26. F. Nguyen, A. Aksoy, A. Bernhard, M. Calvi, J. A. Clarke, H. M. Castaneda Cortes, A. W. Cross, G. Dattoli, D. Dunning, R. Geometrante, J. Gethmann, S. Hellmann, M. Kokole, J. Marcos, Z. Nergiz, F. Perez, A. Petralia, S. C. Richter, T. Schmidt, D. Schoerling, N. Thompson, K. Zhang, L. Zhang, and D. Zhu. XLS deliverable D5.1 Technologies for the CompactLight undulator. *XLS-Report-2019-004*, 2019.

27. Martin Wilson. *Superconducting Magnets*. Oxford University Press, 1983.

28. K.-H. Mess, P. Schmuser, and S. Wolff. *Superconducting Accelerator Magnets*. World Scientific, 1996.

29. Superconductivity for accelerators. *CERN Accelerator School, CERN-2014-005*, 2014.

30. E. Todesco. Magnetic design of superconducting magnets. *CERN Accelerator School Superconductivity for Accelerators, CERN-2014-005*, 2014.

31. L. Rossi. Manufacturing and testing of accelerator superconducting magnets. *CERN Accelerator School Superconductivity for Accelerators, CERN-2014-005*, 2014.

32. R. Flukiger. Superconductivity for magnets. *CERN Accelerator School Superconductivity for Accelerators, CERN-2014-005*, 2014.

33. F. Toral. Mechanical design of superconducting accelerator magnets. *CERN Accelerator School Superconductivity for Accelerators, CERN-2014-005*, 2014.

34. S. Casalbuoni, A. Cecilia, S. Gerstl, N. Glamann, A. Grau, T. Holubek, C. Meuter, D. Saez de Jauregui, R. Voutta, C. Boffo, Th. Gerhard, M. Turenne, and W. Walter. Recent developments on superconducting undulators at ANKA. *Proceedings of IPAC2015, Richmond, USA*, page 2485, 2015.

35. Y. Ivanyushenkov, C. Doose, J. Fuerst, K. Harkay, Q. Hasse, M. Kasa, D. Skiadopoulos, E. Trakhtenberg, Y. Shiroyanagi, and E. Gluskin. Development and performance of 1.1-m long superconducting undulator at the Advanced Photon Source. *Proceedings of IPAC2015, Richmond, USA*, page 1794, 2015.

36. J Bahrdt and E Gluskin. Cryogenic permanent magnet and superconducting undulators. *NIM A*, 907:149–168, 2018.

37. J.A. Clarke and T.W. Bradshaw. Superconducting undulator workshop report. *ICFA Beam Dynamics Newsletter*, 65:148, 2014.

38. P. Emma, N. Holtkamp, H.-D. Nuhn, D. Arbelaez, J. Corlett, S. Myers, S. Prestemon, R. Schlueter, C. Doose, J. Fuerst, Q. Hasse, Y. Ivanyushenkov, M. Kasa, G. Pile, E. Trakhtenberg, and E. Gluskin. A plan for the development of superconducting undulator prototypes for LCLS-II and future FELs. *Proceedings of FEL2014, Basel, Switzerland*, page 649, 2014.

39. S.H. Kim. A scaling law for the magnetic fields of superconducting undulators. *Nuclear Instruments and Methods in Physics Research Section A: Accelerators, Spectrometers, Detectors and Associated Equipment*, 546(3):604 – 619, 2005.

40. E.R. Moog, R.J. Dejus, and S. Sasaki. Comparison of achievable magnetic fields with superconducting and cryogenic permanent magnet undulators – A comprehensive study of computed and measured values. *Advanced Photon Source, Argonne National Laboratory ANL/APS/LS-348*, 2017.

41. J.A. Clarke, K. Marinov, B. J. A. Shepherd, N.R. Thompson, V. Bayliss, J. Boehm, T. Bradshaw, A. Brummitt, S. Canfer, M. Courthold, B. Green, T. Hayler, P. Jeffery, C. Lockett, D. Wilsher, S. Milward, and E. C. M. Rial. Optimization of superconducting undulators for low repetition rate FELs. In *Proc. of International Free Electron Laser Conference (FEL'17), Santa Fe, NM, USA, August, 2017*, number 38, pages 411–414, 2017.

5
Single Particle Motion

Now that we understand how to accelerate particles in an accelerator and how to produce the magnetic fields that steer, focus and manipulate the bunches, we can turn our attention to the dynamics of the transverse motion. We shall learn what these cavities and magnets do to a charged particle beam and how we can design magnet layouts to achieve the goals of our machine. We'll look at the basic equations governing the motion of charged particles in an EM field and the consequences for accelerator builders. In this chapter we shall focus on single particle motion, so the interaction between particles (collective effects) is considered in Chapter 7.

We shall start by some general consideration of the motion, observing that the rate of oscillations in the transverse plane is larger compared to the longitudinal plane in a strong focusing accelerator. * After a discussion of various ways of 'doing' dynamics we shall consider Hill's equation and explore the consequences, leading to linear single particle dynamics and the Courant-Snyder formalism. Following this we turn our attention to the real-life situation, when things are not quite ideal. This starts with magnetic imperfections, focusing on the linear case, and moves to a beam of charged particles with a spread in momentum. This gives us dispersion, chromaticity and momentum compaction. Finally we consider the motion of beams of non-interacting particles and an introduction to non-linear dynamics.

---

*This is not necessarily true in weaker focusing or lower-energy accelerators.

5.1   Preliminary Considerations

Let's start with a thought experiment! If you have ever visited a particle accelerator, cast your mind back to that visit. Or close your eyes and imagine you are flying towards, and then inside a particle accelerator. Ideally this would be a particle accelerator ring. Perhaps you have a favourite one! Once you are there, take a good look. What observations do you make and what would impress you? After you have thought about this, we will tell you our observations.

So, what did you come up with? We thought the following features were worthy of note:

- The particles spend a long time in the ring. In many rings this can be many hours. The fact that the particles do not move to very large amplitude and touch the machine aperture means their motion is stable. In the LHC the protons travel around the ring over 11,000 times per second and stay in there for many hours, whilst in Diamond the revolution frequency is 533.8 kHz.

- The machine repeats itself, i.e. it is periodic. We can see this from the layout of the magnets.

- The particle motion does not have the same periodicity as the machine on a particle-by-particle basis but the envelope of the motion follows the machine periodicity.

You may have gotten the first one, though the second two are less obvious. These are observations we will study and explain using beam dynamics.

There is a huge range of particle accelerators, from the very small to the very large Large Hadron Collider. These can often be classified by a small number of high-level parameters, and doing so is useful to compare machines and get a feel for scale and purpose. The first way to classify machines is in terms of the type of particle accelerated and its geometry. For example, the CLARA accelerator at Daresbury is a linear (single pass) electron accelerator and the Large Hadron Collider is a circular (many pass) proton accelerator. Following this, the beam energy, in terms of MeV, GeV or TeV for example, gives the energy scale of the accelerator and the current (or bunch charge) gives the scale of the number of particles accelerated or stored. The design beam energy at the LHC is 7 TeV and each as-designed beam stores 0.5 amperes of proton current. At Diamond the electrons go around the 562 m ring 534,000 times a second. The maximum beam current is 300 mA. Following this there are a myriad of accelerator parameters used to discuss, compare and classify the accelerators. For example, the colliding beam luminosity in the case of a collider and the beam lifetime in the case of a storage ring. The calculation and evolution of these parameters is something we can compute using single and multi-particle dynamics. We perform beam dynamics calculations and modelling to understand the motion of particles in linear and circular accelerators, to understand the fundamentals of existing machines, optimise and commission accelerators, design new machines, e.g. a new collider, and design novel machines, e.g. a non-scaling FFA. So the science of beam dynamics is central to making and operating particle accelerators. How do we do this? The fundamental tool of a person engaged in beam dynamics is knowing how to calculate the motion of a charged particle in a real electromagnetic field, which includes motion in magneto-static configurations, what happens in a time-dependent field, computing charged particle optics, understanding the approximations used, how the particles interact with the surroundings, and whether the particles in a bunch interact with themselves. For now, we shall concern ourselves with single, non-interacting particles, starting initially with static magnetic fields and bringing in time-dependent electric fields later in the chapter. We shall then worry about what it means to have many interacting particles in Chapter 7.

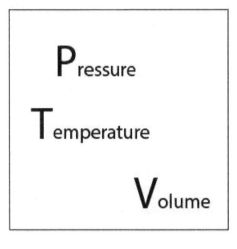
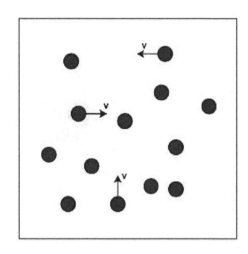

FIGURE 5.1 The global view of a gas in terms of macroscopic variables such as pressure (P) and temperature (T) (left) and the local view of a collection of gas in terms of gas molecules (right).

The most basic question we can ask is: how do I represent the beam passing through the accelerator in my beam dynamics language? This leads us to a hierarchy of beam descriptions and in the course of our analysis of beam dynamics we will use different, but related, descriptions of the beam. A useful way to think of this is in terms of a microscopic or a macroscopic description, with the latter only using a few, global parameters. A useful analogy is a box of gas of some substance, as shown in Figure 5.1. We can think of this system as being described by several numbers: the pressure (P), the temperature (T), the volume (V), the number of moles (n), etc. An equation of state relates these quantities together, and, in the case of an ideal gas, we have the ideal gas law

$$PV = nRT, \tag{5.1}$$

which relates the state variables to each other, R being the ideal gas constant, and tells us how they change. This gives a description of the gas in terms of a few variables.

Our gas is also made up of a collection of gas molecules, each with a position and a momentum in every degree of freedom of the system. Each molecule has a speed v and a kinetic energy (translational energy). This is a microscopic view of our gas in a box, and an equally valid way of thinking about the box of gas. The two pictures are related in a fundamental way

$$U = \frac{3}{2}kT, \tag{5.2}$$

with U (the average kinetic energy) directly proportional to the macroscopic temperature of the gas T, k is the Boltzmann constant. Hence the microscopic (particle) view and the macroscopic view are related, as they should be, as we're talking about the same box of gas. Both views can be useful to understand the system. It is common in physical systems to have several different, but equivalent, views of the same situation, e.g. physics of an ideal gas, quantum mechanics, with wave and matrix formulations. We have this situation in beam dynamics.

The first view is the global view where we assume a ring or beam line exists as an object and study the global properties of the system. For example, the stability of the beam or the number of oscillations per turn (tune, which we shall discuss later in this chapter).

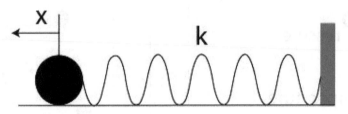

FIGURE 5.2   The standard problem of a harmonic oscillator in one dimension, where the restoring force is proportional to the distance from the equilibrium position.

Then we have a local view, where we worry about the details of the machine and think about individual particles. We need to think about which frame of reference is best and what the fields look like in this frame. We then can ask how a single particle moves in this system. As an aside, in this book we shall use the words machine, ring and lattice. By machine we mean any complete accelerator system, for example the LHC or Diamond. By ring we mean a closed arrangement of dipoles forming a repeating path for the particle beam, and by lattice we mean any general arrangement of magnets appearing inside an accelerator.

So we have different ways of looking at a beam. There are also several ways of doing particle dynamics. These ways are equivalent to each other and all can be used to solve dynamical problems. The three formulations of dynamics are 1) Newtonian dynamics 2) Lagrangian dynamics, and 3) Hamiltonian dynamics. The one you should choose depends on the kind of problem you are solving. In accelerator physics we tend to use Newtonian and Hamiltonian dynamics, and each one has its own merits.

5.2   The Dynamics of a Simple Harmonic Oscillator

In this section we shall see that the transverse motion of a particle in an accelerator is very similar to the motion of a simple harmonic oscillator and we can learn a lot from thinking about this similarity.

Let's work in one dimension, denoting the position from some equilibrium point by $x(t)$ and the velocity by $\dot{x}(t)$, as shown in Figure 5.2. If the restoring force is given by $F(t) = -kx(t)$, called Hooke's law, where k is the spring constant, then the coordinate $x(t)$ obeys

$$\ddot{x}(t) + \omega_0^2 x(t) = 0, \tag{5.3}$$

where we write $\omega_0^2 = k/m$, with m denoting the mass of the oscillating particle. This is just using Newton's second law with Hooke's law. Note that in reality Hooke's law is only approximately true for a real spring and experimental measurements show that most springs have higher-order non-linear terms in the restoring force. Our intuition, and experience of masses on a spring, tell us the solution should be oscillating (or diverging), which we can see by substituting a trial solution $x(t) = \exp(\lambda t)$ where λ is some constant, into the equation of motion. This gives $\lambda_{1,2} = \pm i\omega_0$ and a general solution of

$$\begin{aligned} x(t) &= A\cos(\omega_0 t) + B\sin(\omega_0 t) \\ &= C\sin(\omega_0 t + \phi). \end{aligned} \tag{5.4}$$

The amplitude C and phase ϕ depend on the initial conditions, unlike the natural angular frequency of oscillation ω_0, which depends on the spring and the mass that is oscillating. Note if we replace the sign of the force in Hooke's law we replace the sine and cosine in the solution by sinh and cosh, and obtain hyperbolic diverging solutions (if you are not familiar with sinh and cosh, spend some time reading about them, as we shall use these functions again). These ideas will come back for our description of beam dynamics, and the observation that we can obtain both oscillating solutions or diverging solutions by a restoring force proportional to distance from an equilibrium point and by flipping the sign in the force. This makes physical sense as the force now pushes the mass to large values of x for $x > 0$ and the force pushes the mass to smaller values of x for $x < 0$.

It is a standard plan of attack when solving differential equations to rewrite a second-order differential equation as two first-order differential equations and, if we did so, we could write the solution in terms of two constants, or invariants, of the system. The first one is linked to the total energy, thus determining the size of the motion (and can be linked to amplitude), and the second one appears as a phase in the harmonic function describing the motion. These constants are both linked to initial conditions, and where we release the mass (or for an accelerator later in this chapter, our particle).

We shall see the appearance of such invariants in our study of transverse motion, and they shall prove to be very important in our study of beam dynamics. The language of invariants is particularly powerful in physics and engineering and gives a useful framework to understand and predict motion. The invariant of the motion using Newton's formulation of dynamics emerged after a bit of work. This structure is very clear if we tackle dynamics problems using an alternate formulation first proposed by Hamilton, which we discussed when we thought about formulations of dynamics in the previous sections. As we saw, we shall take the approach of Newton and use forces in this book to make the physical results clear, but let's take a short diversion and consider our harmonic oscillator from the point of view of Hamilton. In this framework the central object is the Hamiltonian, which contains the physics of the system and is formulated in terms of coordinates and their corresponding canonical momenta. Together these form something called a conjugate pair, and we have one pair for each dimension of the system. For our one-dimensional oscillator we have the position x and the canonical transverse momenta p_x. Once we have the Hamiltonian (which we will do in a moment) we use Hamilton's equations to figure out the motion,

$$
\begin{aligned}
\frac{\mathrm{d}x}{\mathrm{d}t} &= \frac{\partial H}{\partial p_x}, \\
\frac{\mathrm{d}p_x}{\mathrm{d}t} &= -\frac{\partial H}{\partial x}.
\end{aligned}
\tag{5.5}
$$

Notice that we have two first-order differential equations to solve, instead of a single second-order differential equation in Newton's approach.

The Hamiltonian for the oscillator is given by

$$
H = \frac{p_x^2}{2m} + \frac{1}{2}m\omega_0^2 x^2,
\tag{5.6}
$$

where m is still the mass of the particle. This is really just the sum of the kinetic and potential energies and, in the presence of only forces that are constant in time we can prove this sum of terms is conserved. This is easily done by taking the total derivative of H and using Hamilton's equations, and we shall leave this as an exercise for the keen reader. It means that

$$
\frac{\mathrm{d}H}{\mathrm{d}t} = 0,
\tag{5.7}
$$

and H is an invariant. Now, directly applying Hamilton's equations gives

$$
\begin{aligned}
\frac{\mathrm{d}x}{\mathrm{d}t} &= \frac{p_x}{m}, \\
\frac{\mathrm{d}p_x}{\mathrm{d}t} &= -m\omega_0^2 x,
\end{aligned}
\tag{5.8}
$$

which is what we obtained using Newton's approach with forces. We will not use Hamiltonians for the study of transverse motion but there are many good references [1] and the approach is very useful for studying non-linear motion.

5.3   Hill's Equation

We have just seen that we can obtain either converging (oscillating) or diverging solutions by using Hooke's law of a restoring force for an oscillator, which states that the restoring force is proportional to the distance from the equilibrium position. This is exactly the behaviour we shall see in our focusing quadrupole elements and we will show this can be used to obtain stable behaviour in both transverse planes. Quadrupoles have four magnetic poles and are the building blocks of our focusing lattices. The linearly rising magnetic field will give rise to the focusing of our beam.

This feature will emerge from our fundamental equation of transverse motion, called Hill's equation. The equation for the horizontal motion, with coordinate x, is

$$
x''(s) + \left(k_x(s) + \frac{1}{\rho(s)^2} \right) = 0,
\tag{5.9}
$$

and the equation of the vertical motion, with coordinate y, is

$$
y''(s) + k_y(s) = 0.
\tag{5.10}
$$

In these equations s is the longitudinal distance along our accelerator beamline, $k_{x,y}(s)$ denotes the momentum-normalised focusing strength and ρ is the bending radius, as defined in the figure. We shall define these quantities more carefully later. Note we've written out the explicit dependence of the functions on s, which can be dropped for brevity once we know what is going on.

It's worth spending some time to understand the features of these equations, as they will govern our study of transverse dynamics, at least linearly. What does this last statement mean? Before we answer that, let's think about coordinate systems. We shall not present a complete derivation of Hill's equations for accelerator physics – this can be found in all the standard textbooks and we learn nothing significantly useful if we present it in this book. For very good treatments see [2, 3]. In this chapter we work with transverse coordinates, so in the horizontal plane we use x, which is the horizontal position and we use y for the vertical position. The question then is, what are x and y relative to? To understand this we need to look at the coordinate system.

The coordinate system we use is shown in Figure 5.3 and forms the basis for the analysis in this chapter. We are developing the equations of motion in a linear or circular machine, and our equations work in both situations. For the circular case, the curvature is provided by a set of dipoles, which define a curved trajectory through the tunnel. The local curvature is denoted by ρ and the distance along this curve in the laboratory frame is denoted by s. Our coordinate system is often called a co-moving system and will move along the reference trajectory defined by the dipoles at the same speed as some reference particle. We then define all quantities, for example transverse positions x and y, with respect to this reference

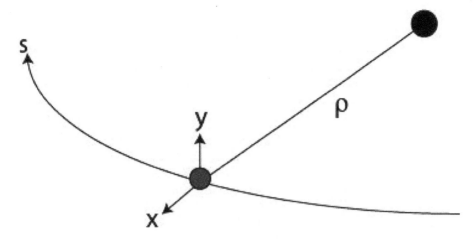

FIGURE 5.3 Our co-moving coordinate system for describing the location of our particle in the accelerator. The bending arises from the dipoles forming the geometry of the machine.

particle. The curved reference trajectory is normally called the orbit, and the coordinate system moves with a reference particle around the design orbit defined by the dipoles. The co-moving system has the consequence that we won't see the curve in the dipoles explicitly, only the focusing around the reference particle due to the bending. For the linear machine case, there is no focusing terms from the bending.

So our coordinates represent deviations with respect to the design (ideal) orbit, and we assume these deviations will be small (x is normally around millimetres). The assumption that quantities like x are small will make our equations linear; we will discuss lifting this constraint later. For coordinates relative to this design orbit we use the position and slope $dx/ds = x'$, and δ will denote deviations from the reference particle momenta. A transverse position vector in this frame then is

$$\mathbf{R} = x\mathbf{x} + y\mathbf{y}, \tag{5.11}$$

where $\mathbf{x}$ and $\mathbf{y}$ are unit vectors in the co-moving frame and

$$r = \rho + x. \tag{5.12}$$

Let's sketch out the derivation. Once we've defined our co-moving coordinate system and understand what it means to differentiate position vectors, we are able to write down the left-hand side of Newton's second law. We've one side of Newton's second law and we need the other side. What is the force? This is the right-hand side, which we shall equate to the left-hand side when we figure it out.

In the presence of electric and magnetic fields we use the Lorentz equation (or Lorentz force law), already seen in Chapter 2,

$$\mathbf{F} = q\left(\mathbf{E} + \mathbf{v} \times \mathbf{B}\right), \tag{5.13}$$

where $\mathbf{v}$ is the velocity of the charged particle with charge q. This physical law tells us, in vector notation, the force on a charged particle moving with velocity $\mathbf{v}$ from an electric field $\mathbf{E}$ and magnetic field $\mathbf{B}$. For our purposes we shall assume the velocity in the longitudinal direction is far bigger than the transverse velocity, so the transverse velocities are small, as are quantities like x'. Equating this Lorentz force to $m \cdot \ddot{x}$ gives a set of equations of the horizontal and vertical motion. There is also an equation for the relative longitudinal motion but we'll disregard this and approach motion in this plane in another way. It turns out this longitudinal motion is far slower than the transverse motion, which we shall see in Section 5.9 and means we treat these two kinds of motion differently. We can also make another assumption, that is, we are interested for now in dipole and quadrupole fields, and so the field which appears in the Lorentz force law can be written as

$$\mathbf{B} = B_{0y}\mathbf{y} + g\left(x\mathbf{y} + y\mathbf{x}\right). \tag{5.14}$$

In this equation, B_{0y} is the dipole field which generates the curved coordinate system, and the second term is the quadrupole field we impose on the beam. The curl-free nature of the free-space Maxwell equations means $g = \partial B_x/\partial y = \partial B_y/\partial x$.

We can now linearise these equations, meaning dropping any terms in any of the variables of order two and greater in any of the variables, i.e. we cross out any term looking like x^2, xx', x'^2, x^3 and so on. This is an approximation and means our equations describe linear motion and are valid for small values of the variables (so that $x^2 \ll x$ etc). We also expand the momentum deviation δ, which appeared in the denominator due to Newton's second law, and we write

$$\frac{1}{1+\delta} \sim 1 - \delta + \mathcal{O}(\delta^2). \tag{5.15}$$

We drop terms of δ^2 and higher, and so assume the quantity δ is small. For now we shall also drop terms containing δ, and restore them when we talk about dispersion. Later, when we think about something called chromaticity, we shall restore terms like $x \cdot \delta$. But for now our Hill's equation to describe linear, on-momentum motion is

$$x''(s) + \left(k_x(s) + \frac{1}{\rho(s)^2}\right) = 0, \tag{5.16}$$

in the horizontal plane and

$$y''(s) + k_y(s) = 0 \tag{5.17}$$

in the vertical plane. We have defined some notation, and defined

$$k_x = \frac{g}{B\rho} + \frac{1}{\rho^2}, \tag{5.18}$$

and the equivalent for the vertical plane,

$$k_y = -\frac{g}{B\rho}. \tag{5.19}$$

Note the vertical plane only has a contribution from the quadrupoles through g, and there is a minus sign difference between the planes for the g terms.

So we have our equations of motion. Let's think about their features. To do this we write both equations very compactly as one equation by defining some notation. Let $u = x, y$ for the variables, and wrap up the second term in each equation into a single function. Hence Hill's equations become

$$u'' + Ku = 0, \tag{5.20}$$

where $K = g/(B\rho) + 1/\rho^2$ in the horizontal plane and $K = g/(B\rho)$ in the vertical plane. We can see now these equations look exactly the same as our harmonically oscillating mass on a spring. Imagine for a moment there was no $1/\rho^2$ term in the equation for K in the horizontal plane. This would mean the quadrupole gradient sign sets the value of the restoring force, so that a positive g would focus in the horizontal plane and defocus in the negative plane. Therefore negative g would do just the opposite. We also see the implication

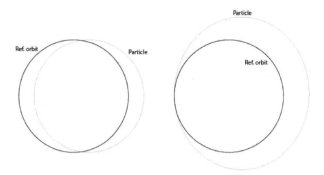

FIGURE 5.4 The natural focusing of particles in a constant magnetic field for clockwise moving particle. The left particle has a small transverse offset compared to the reference particle. The right particle has a small momentum offset compared to the reference particle.

of Maxwell's equations – a quadrupole designed to focus in one plane must defocus, or create a diverging trajectory, in the other plane. We can control the spring constant but with some constraints! We shall explore the role of focusing in the coming pages, but perhaps you can start to imagine what a series of magnets would look like if we need to somehow have confined motion in both planes.

Let's think about that $1/\rho^2$ term. It's acting like a focusing magnet but only in the plane of the bending, in this case the horizontal plane. This is called natural, or body, focusing and arises from the effect of the dipole magnets defining the reference trajectory. The natural focusing arising through bending can be visualised using the analysis of Figure 5.4. This figure shows the reference orbit as the darker line, and a particle moving with respect to this orbit as the lighter line. The left-hand figure shows what happens when the general particle has a small transverse position offset. If you follow it around the ring it oscillates around the reference orbit, with one complete oscillation per turn. The right-hand figure shows the general particle having a small momentum offset, and showing stable behaviour. Note there is no natural focusing in the vertical plane as we don't bend in this plane. For a straight beamline this term is absent. As an aside, we should also mention that the length through which a magnet acts on the beam is often longer than its physical length of material due to field lines curving at each magnet end. This is called the effective length of a magnet.

So Hill's equations describe how the transverse coordinates x and y evolve as a function of distance through the magnetic lattice, and look like linear harmonic oscillator equations. We can use this fact to solve them easily, which means writing down explicit functions $x(s)$ and $y(s)$. Of course we can solve Hill's equations in many ways, including numerically, but let's pursue the approach most commonly taken in the literature. Let's consider the horizontal motion and take the case $K > 0$. We know the solution is built of harmonic functions and contains two unknown constants, so let's guess at

$$x(s) = c_1 \cos(\sqrt{K}s) + c_2 \sin(\sqrt{K}s), \tag{5.21}$$

where c_1 and c_2 are the constants fixed from the initial conditions. We can take the derivative

$$x'(s) = -c_1 \sqrt{K} \sin(\sqrt{K}s) + c_2 \sqrt{K} \cos(\sqrt{K}s), \tag{5.22}$$

and substitute into Hill's equation to quickly check this is indeed a solution. To find the constants we note that $x(0) = x_0$ and $x'(0) = x_0'$, giving

$$\begin{aligned} c_1 &= x_0, \\ c_2 &= \frac{x_0'}{\sqrt{K}}, \end{aligned} \tag{5.23}$$

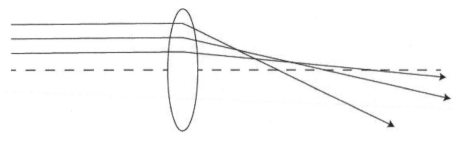

FIGURE 5.5   The differing quadrupole kicks obtained for different particle transverse offsets.

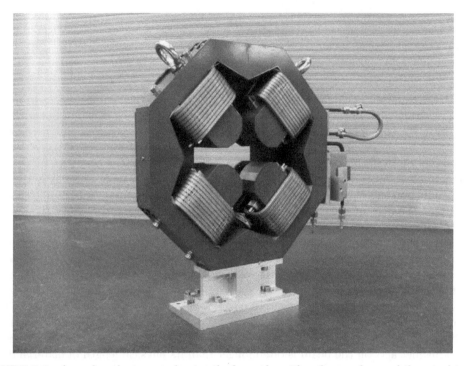

FIGURE 5.6   A quadrupole magnet, showing the four poles with coils wound around them to drive the magnetic field. © STFC

and so

$$x(s) = x_0 \cos(\sqrt{K}s) + \frac{x_0'}{\sqrt{K}} \sin(\sqrt{K}s). \qquad (5.24)$$

This has a first derivative of

$$x'(s) = -x_0 \sqrt{K} \sin(\sqrt{K}s) + x_0' \cos(\sqrt{K}s), \qquad (5.25)$$

so the kick given by the quadrupole magnet points back towards the origin (focusing) and gets bigger the further the particle is away from the origin. This focusing effect for off-axis particles is shown in Figure 5.5 and a real quadrupole can be seen in Figure 5.6.

The equations evolving x and x' can be written as a matrix equation, wrapping two equations into one and using linear algebra to express the linear nature of our system. Hence we can write

$$\begin{pmatrix} x \\ x' \end{pmatrix}_s = M_{\text{quad}} \cdot \begin{pmatrix} x \\ x' \end{pmatrix}_0 \qquad (5.26)$$

for the evolution of the vector formed by x and x' from position $s = 0$ to position $s = s$ in a nicely compact way. This is just another way to write our two separate equations evolving x and x' that we just discussed. The matrix M_{quad} is given, for a focusing quadrupole, by

$$M_{\text{F}} = \begin{pmatrix} \cos(\sqrt{K}s) & \frac{1}{\sqrt{K}} \sin(\sqrt{K}s) \\ -\sqrt{K} \sin(\sqrt{K}s) & \cos(\sqrt{K}s) \end{pmatrix}. \tag{5.27}$$

In this linear formalism, the particle is represented by a point in (x, x') space, known as trace-space. The matrix M_{F} acts on these trace-space vectors to evolve the particle in s. What happens if $K < 0$? Now we have the equation of motion

$$x'' - |K|u = 0, \tag{5.28}$$

which has a diverging solution which can be written in terms of sinh and cosh functions. So we write

$$x(s) = c_1 \cosh(\sqrt{K}s) + c_2 \sinh(\sqrt{K}s), \tag{5.29}$$

where, as before, c_1 and c_2 are the constants fixed from the initial conditions. We can solve for the constants and write as a matrix equation as we did for the focusing case, giving

$$M_{\text{D}} = \begin{pmatrix} \cosh(\sqrt{|K|}s) & \frac{1}{\sqrt{|K|}} \sinh(\sqrt{|K|}s) \\ +\sqrt{|K|} \sinh(\sqrt{|K|}s) & \cosh(\sqrt{|K|}s) \end{pmatrix} \tag{5.30}$$

for the defocusing quadrupole.

So a given quadrupole magnet focuses on one plane and defocuses in the other plane, by virtue of Maxwell's equations. By convention we say a horizontally-focusing quadrupole is a 'focusing' quadrupole, conventionally known as an 'F-quadrupole'. The focusing strength is related to the gradient of the magnetic flux density B by

$$k = \frac{g}{B\rho} = \frac{q}{p} \frac{dB_y}{dx}. \tag{5.31}$$

A quadrupole with a positive sign for dB_y/dx is horizontally-focusing, whereas a negative dB_y/dx is horizontally-defocusing; the latter is conventionally known as a 'D-quadrupole'.

A drift space is a region of the beam line with no electromagnetic fields. We can figure out the evolution equations for x and x' by either simple geometry or taking a limit of the quadrupole matrices for $K \to 0$. Either way we find the variables change as

$$\begin{aligned} x(L) &= x_0 + x_0' \cdot L, \\ x'(L) &= x_0', \end{aligned} \tag{5.32}$$

where the drift space has length L and (x_0, x_0') are the particle coordinates on entry to the drift space. This can be written as a matrix

$$M_{\text{drift}} = \begin{pmatrix} 1 & L \\ 0 & 1 \end{pmatrix}, \tag{5.33}$$

telling us how a particle evolves in a drift, with a very clear geometrical interpretation. A useful approximation for the quadrupole matrix we already know (e.g. M_{F}) with finite length and called the thick lens matrices, is the limit when the focal length, f, of the quadrupole lens is long compared to its length, l. Hence we consider

$$f = \frac{1}{Kl} \gg l, \tag{5.34}$$

which we find by letting $l \to 0$ whilst keeping the product Kl constant; the product Kl is known as the integrated strength. This gives the quadrupole focusing matrix in this limit as

$$M_{\text{thin}} = \begin{pmatrix} 1 & 0 \\ -\frac{1}{f} & 1 \end{pmatrix}. \tag{5.35}$$

This is the matrix for a horizontally-focusing quadrupole in the thin-lens approximation with a focal length of f. The kick towards the axis for a particle with a non-zero position with respect to the axis is clear. The matrix for the defocusing case is obtained by the transformation $f \to -f$. The thin-lens approximation is useful for quick calculations of beamlines.

We now have a linear matrix formalism for the evolution of a coordinate vector (x, x'), and know the matrix for a focusing quadrupole, M_{quad} is given by

$$M_{\text{F}} = \begin{pmatrix} \cos(\sqrt{K}s) & \frac{1}{\sqrt{K}}\sin(\sqrt{K}s) \\ -\sqrt{K}\sin(\sqrt{K}s) & \cos(\sqrt{K}s), \end{pmatrix}, \tag{5.36}$$

a defocusing quadrupole,

$$M_{\text{D}} = \begin{pmatrix} \cosh(\sqrt{|K|}s) & \frac{1}{\sqrt{|K|}}\sinh(\sqrt{|K|}s) \\ +\sqrt{|K|}\sinh(\sqrt{|K|}s) & \cosh(\sqrt{|K|}s), \end{pmatrix} \tag{5.37}$$

a drift,

$$M_{\text{drift}} = \begin{pmatrix} 1 & L \\ 0 & 1 \end{pmatrix}, \tag{5.38}$$

and a thin lens quadrupole

$$M_{\text{thin}} = \begin{pmatrix} 1 & 0 \\ -\frac{1}{f} & 1 \end{pmatrix}. \tag{5.39}$$

In a real accelerator we have lots of these elements arranged one after the other, as shown in Figures 5.7 and 5.8, with focusing quadrupoles, defocusing quadrupoles and intervening drift spaces. Take a close look at both these figures and try to spot the beamline elements we are discussing in this chapter. We shall look at lattices more closely later in this book.

Each element is represented by a matrix, at least in our linear approximation. The question arises: how do we transform through these sequences of elements? The answer is intuitive and we shall not prove it. We multiply the matrices of each element to give an overall transfer matrix through the system, beginning with the start of the beamline on the right and pre-multiplying by the next element seen by the beam. So imagine we have a beamline consisting of a focusing quadrupole, followed by the drift space, followed by a defocusing quadrupole and followed by a drift space. This is the order of elements seen by the beam. The overall matrix for the transformation of the particle by the system is given by

$$M_{\text{cell}} = M_{\text{drift}} \cdot M_{\text{D}} \cdot M_{\text{drift}} \cdot M_{\text{F}}. \tag{5.40}$$

Note the focusing quadrupole matrix sits on the right of the series of matrices in this expression; the overall transfer matrix is obtained by multiplying the individual transfer matrices in reverse order. We have also called the composite system a cell, for reasons which will become clear. The overall motion of a particle at the start of this system (defined as $s = 0$) to the end (defined as $s = 1$) is

$$\begin{pmatrix} x \\ x' \end{pmatrix}_1 = M_{\text{cell}} \cdot \begin{pmatrix} x \\ x' \end{pmatrix}_0. \tag{5.41}$$

FIGURE 5.7 A beamline section for a so-called transfer line between two accelerators, showing an arrangement of focusing and defocusing quadrupoles (with four poles), dipoles (with two poles), and intervening drift spaces, mounted on a common supporting girder. © STFC

FIGURE 5.8 A beamline, showing an arrangement of focusing quadrupoles, defocusing quadrupoles and intervening drift spaces. © STFC

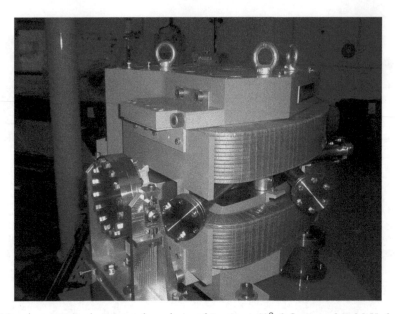

FIGURE 5.9 A sector dipole magnet, here designed to give a $60°$ deflection of 35 MeV electrons. Note the curve of the coils and poles, so that the electrons enter and exit perpendicular to the magnet end faces; this gives no edge-focusing effect. Rectangular magnets are also used, which do give edge focusing. © STFC

We would encourage the reader to think now how they would turn this formalism into a simple particle evolution code (called a tracking code in the field). Under what circumstances would your code give valid results? When you get some time, use your favourite programming language to write this code.

The beamline pictured in Figure 5.7 also contains elements which bend the beam around the trajectory of the machine, and are the dipoles we used to define the curved reference trajectory. Do these have a matrix? The bending effect to define the reference trajectory is already included in our equation in the co-moving coordinate system but, as we discussed previously, this bending introduces some natural transverse focusing. This can be described by a matrix. To obtain it, we start from the matrix for a focusing quadrupole, Equation 5.27, in terms of K and note for a pure bending element

$$K = \frac{1}{\rho^2}, \tag{5.42}$$

where again ρ is the bending radius. Hence we obtain the following matrix for a dipole of length l

$$M_{\text{dipole}} = \begin{pmatrix} \cos\theta & \rho\sin\theta \\ -\frac{1}{\rho}\sin\theta & \cos\theta \end{pmatrix}, \tag{5.43}$$

where $\theta = l/\rho$ is the bend angle of the dipole. The geometric (natural) focusing is now clear. Note the matrix in the non-bending plane is a drift. An example of an accelerator dipole is shown in Fig 5.9.

The consequence of Maxwell's equations with no sources (specifically curl $\mathbf{B} = 0$), discussed in Chapter 2, means a horizontally focusing quadrupole is defocusing in the vertical plane, and vice versa. This follows from applying this Maxwell equation to the field of the quadrupole in free space and is due to the electromagnetic nature of the devices. However, all is not lost, and we can build systems of quadrupoles which overall focus in both planes by alternating polarity of quadrupoles. This alternating gradient, or strong focusing, principle

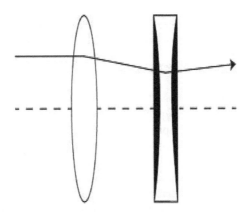

FIGURE 5.10   The path of a particle through a system of alternating quadrupole magnets. The first lens is focusing in this plane, and the second lens is defocusing (denoted by the shaded concave parts of the lens). The particle receives net focusing in both planes.

was first proposed by Nicholas Christofilos in 1949, who patented rather than published the result. A group at Brookhaven National Laboratory – Ernest Courant, M. Stanley Livingston and Hartland Snyder – independently discovered the same principle three years later when trying to solve an operational problem on the Cosmotron accelerator [4]. Today the strong focusing principle is central to the design of many particle accelerators.

Consider the system of a thin-lens focusing quadrupole (focal length f_1) separated by a drift (length d) from a defocusing quadrupole (focal length f_2). If we compute the overall matrix of this system and look at the (2,1) element, this will give us the reciprocal of the system's overall focal length, by comparison with the thin lens matrix for a quadrupole. If we do this we obtain

$$\frac{1}{f} = \frac{1}{f_1} + \frac{1}{f_2} - \frac{d}{f_1 f_2}. \tag{5.44}$$

If we choose $f_1 = -f_2 = f_x$ then the leading terms cancel and we obtain overall focusing in both planes at the same time, with focal length $f = f_x^2/d$. This is a very pleasing and surprising feature, and is the bedrock of many modern accelerators. We can understand this result by thinking of ray tracing and reference to the rays in Figure 5.10. Think it through yourself by visualising test rays at various transverse offsets.

We have brought in the idea of a map in the form of a matrix, and this idea needs a bit of explanation. The matrix M is that map that brings an initial state vector $X(s_0)$ to a final state vector $X(s_1)$, so that

$$X(s_1) = \mathrm{M} \cdot X(s_0). \tag{5.45}$$

For the linear case, the map can be represented as a matrix and the matrix representation is equivalent to the linear system. For non-linear systems, matrices do not work anymore and we need to find new representations of the maps, for example Taylor maps or Lie maps. For further discussion we refer the reader to Wolski's textbook [1].

We know how to combine matrices (linear maps) with the rule

$$\mathrm{M}(s_2|s_0) = \mathrm{M}(s_2|s_1) \cdot \mathrm{M}(s_1|s_0), \tag{5.46}$$

again noting the order. Matrix algebra is not commutative, so we cannot switch the position of matrices in our expressions, or equivalently the order of beamline elements matters. But it is associative and we can form matrix sub-groups (provided we maintain the order of the matrices!). One particularly useful map is the one-turn map. If we start at location s in a ring of circumference C, then the one-turn map is defined as one turn around the ring

$$M(s + C|C). \tag{5.47}$$

This means the map for N revolutions of the ring is found from N applications to a given particle state vector of the one-turn map. We'll come back to this idea when we discuss beam stability. The one-turn map in fact applies to any system, including a linear beamline, with periodicity. In this case it is called the one-period map and gives the transformation through one period.

We've basically re-written our equations in terms of matrices. Is this useful? Yes, as it means we can use all the formal machinery of linear algebra, e.g. matrix multiplication, eigenvalues, eigenvectors, traces, similarity transforms, plus quickly see the impact on the beam of a series of elements. All this is very powerful and useful! Alas a lot of these ideas are beyond this introductory textbook.

The computer realisation of these ideas gives birth to a simple beam-tracking code. These accelerator codes simply assume a piecewise-continuous representation of the accelerator structure, with the order of elements the same as the real beam line being modelled. However, the number of matrices is not the same as the number of elements. This is because of edge focusing, which gives a kick to the beam at the entrance and exit faces of a rectangular dipole. We can write this kick as a matrix, and we'll cover it later in the chapter. But be aware the numbers of matrices in a computer code is always more than the number of elements!

Now we understand what an element does to our particle, we can track single particles through a composite system and, assuming the particles do not interact, many particles.

5.4   The Courant-Snyder Formalism

We've written down Hill's equation for linear beam motion, determined that we can solve it, and written the solution using matrices and linear algebra. Hill's equation is a second-order differential equation for a system with periodic focusing properties and we saw it is a little like an oscillating mass on a spring with a spring constant that changes with time. In fact, the variable spring constant $k(s)$ for our accelerator in the quadrupole gradient and depends on the magnetic properties of the ring. If this ring has periodicity L, then so does the function $k(s)$,

$$k(s + L) = k(s). \tag{5.48}$$

Hence we can expect a kind of quasi-harmonic oscillation, where the frequency and amplitude depend on the location in the ring and show periodicity similar to that of the function $k(s)$. All this means is that we can now follow the motion of particles through our beamline in terms of the transverse coordinates as a function of distance through the machine.

The Courant-Snyder formalism, the subject of this section, solves Hill's equations with an ansatz based on this intuition and parameterises the beam motion into a neat formalism. It also leads to a macroscopic description of the beam and the famous β-function of accelerator lattice design. We assume a solution of Hill's equation inspired by our intuition about the position-dependent amplitude and phase, namely

$$x(s) = \sqrt{2A\beta(s)} \cos(\psi(s) + \psi_0). \tag{5.49}$$

This initially looks strange, so let's pick it apart. $\beta(s)$ has the physical meaning of an amplitude of the motion, which depends on the position s around the accelerator. $\psi(s)$ is a position-dependent phase appearing inside our oscillating function and A is an overall constant. Because Hill's equation is linear, the constant does not appear in it. We'll see later that A is special and is called the single-particle emittance. We note that emittance is generally used as a quantity describing entire beams and we, at the moment, are concerned with single particles. The factor of 2 is for later convenience in the definition of the emittance. We choose to use 'single-particle emittance' as opposed to 'action', to avoid confusion with more formal treatments involving Hamilton's equations. Our treatment is not rigorous enough to use the word 'action' and we hope readers with knowledge of the action will forgive us; Andy Wolski's textbook gives more details [1]. Our ansatz is, in essence, a parameterisation of the anharmonic motion of our particle, with a maximum amplitude that changes with s through the machine. We use β here to mean the β-function, and whilst the same symbol is used for the relative velocity $\beta = v/c$, it is usually clear from the context which quantity is being referred to.

The variable $\beta(s)$ is the key quantity in the Courant-Snyder formalism and has many names: the 'beta function', the beam envelope function, the Courant-Snyder β-function, the amplitude function and so on. It is always chosen to be positive. We'll see that it represents the focusing properties of a lattice, and a small β-function means a tightly-focused lattice. The periodicity of the magnetic system is very important, and this will mean

$$\beta(s + L) = \beta(s), \tag{5.50}$$

for some periodicity L. So the β-function follows the repeating structure of the beamline focusing elements.

If we take the derivatives of the Courant-Snyder ansatz and substitute into the equation of motion, we find we get two terms: one proportional to cosine and one proportional to sine. This is a good exercise to do and we shall leave this to the reader. The coefficients of these terms must vanish separately, and we eventually obtain two differential equations

$$\frac{1}{2}(\beta\beta'' - \frac{1}{2}\beta'^2) - \beta^2\psi'^2 + \beta^2 k = 0 \tag{5.51}$$

and

$$\beta'\psi' + \beta\psi'' = (\beta\psi')' = 0. \tag{5.52}$$

The second equation can be integrated immediately and, choosing a constant of integration, gives

$$\beta\psi' = 1. \tag{5.53}$$

Now we have an equation for the phase function

$$\psi(s) = \int_0^s \frac{\mathrm{d}s}{\beta(s)}. \tag{5.54}$$

This position-dependent phase (known as the phase advance) is related to an integration of the β-function along the beam line, and knowing the β-function means we can compute the phase function. We can now eliminate the phase function from the first of the differential equations to get a differential equation for the β-function

$$\frac{1}{2}\beta\beta'' - \frac{1}{4}\beta'^2 + \beta^2 k = 0. \tag{5.55}$$

So knowing the distribution of focusing strengths along a beam line determines $\beta(s)$, although we rarely (i.e. never under normal circumstances) solve this equation in practice. Finally, we define the two functions (with $\beta(s)$, called the lattice functions),

$$\alpha(s) = -\frac{1}{2}\frac{\mathrm{d}\beta(s)}{\mathrm{d}s} \tag{5.56}$$

and

$$\gamma(s) = \frac{1 + \alpha(s)^2}{\beta(s)}. \tag{5.57}$$

Once the β-function is known, and hence α and γ, the motion of a single particle is completely specified by specifying the single-particle emittance and the initial phase factor of the particle. So we have

$$\begin{aligned}
x(s) &= \sqrt{2A\beta(s)} \cos(\psi(s) + \psi_0) \\
x'(s) &= -\frac{\sqrt{2A}}{\sqrt{\beta(s)}} \left[\alpha(s) \cos(\psi(s) + \psi_0) + \sin(\psi(s) + \psi_0)\right],
\end{aligned} \tag{5.58}$$

where the second equation is the derivative of the first. We can combine these two equations to give the quantity

$$\beta x' + \alpha x = -\sqrt{2A\beta(s)} \sin(\psi(s) + \psi_0), \tag{5.59}$$

which means we can write an expression which is invariant for a particle

$$x^2 + (\beta x' + \alpha x)^2 = 2A\beta. \tag{5.60}$$

By expanding the bracket and using the definitions of α and β we obtain

$$\gamma x^2 + 2\alpha x x' + \beta x'^2 = 2A. \tag{5.61}$$

This is a very important equation, so let's look at it carefully. For every point in the accelerator we have a value of the functions $\alpha(s)$, $\beta(s)$ and $\gamma(s)$. They depend on the lattice (through the focusing function $k(s)$) and are different for every point. At a particular point, if we combine the particle position and angle with these lattice functions we get an invariant, which was the single-particle emittance A we first saw in the solution to Hill's equations in the Courant-Snyder formalism. As the particle moves to the next location in the accelerator, where we have different lattice functions, the particle has a different position and angle. However if we form this combination of quantities again, Equation 5.61, at the new location we get the same value as before. In other words, the single-particle emittance is a constant of the motion and always has the same value at every point.

You may have seen this equation before in geometry. If not, imagine there was no xx' term. What would it look like in the (x, x') plane? It would be a circle, with equation

$$\gamma x^2 + \beta x'^2 = 2A. \tag{5.62}$$

The conserved quantity

$$\gamma x^2 + 2\alpha x x' + \beta x'^2 = 2A \tag{5.63}$$

actually describes an ellipse in the (x, x') plane, with ellipse parameters described by the values of α, β and γ. β controls the extent along the x-axis, γ controls the extent along the x' axis and α tells you how upright the ellipse is. The area of the ellipse is given by

$$\text{area} = \pi 2A, \tag{5.64}$$

so the area transcribed by the particle as it moves in (x, x') space is constant, since A is a constant. In general an ellipse may be described in the (x, y) plane as

$$c_1 x^2 + 2c_2 xy + c_3 y^2 = c_4, \tag{5.65}$$

with area $\pi c_4 / \sqrt{c_1 c_3 - c_2^2}$. For our ellipse in the (x, x') plane, we can find the points of intersection by setting $x = 0$ or $x' = 0$ and obtain

$$x = \sqrt{\frac{A}{\gamma}}, \quad x' = \sqrt{\frac{A}{\beta}}. \tag{5.66}$$

The maximum values of x and x' as the particle moves around the ellipse can be found by rearranging and differentiating, to obtain

$$x_{\text{extreme}} = \pm\sqrt{A\beta}$$
$$x'_{\text{extreme}} = \pm\sqrt{A\gamma}. \tag{5.67}$$

Therefore, for a fixed A, the parameter $\beta(s)$ controls the size of the particle's excursions in position space, and the parameter $\gamma(s)$ controls the size of the particle's excursions in angular space. When we come to talk about beams, which are collections of particles, we'll see they are measures of the beam's spatial size and angular divergence. When one is small, the other must be big and vice versa, since they are intrinsically linked.

If you recall, the lattice parameters are functions of the focusing of the lattice so every point in the lattice has a value of the lattice functions. Hence every point in the lattice has its own orientation of the ellipse. A given particle has its own value of the single-particle emittance, thus setting the area of the ellipse it moves around. To see what is going on clearly, let's play a mind game: we sit at one location in the ring and watch a single particle, turn after turn after turn. So every time the particle comes past us we write down its position and angle. This can be done with a simple computer code, and we generate the coordinates of the particle turn after turn,

$$(x_1, x'_1), \quad (x_2, x'_2), \quad (x_3, x'_3), \quad (x_4, x'_4), \cdots \tag{5.68}$$

where (x_i, x'_i) are the particle coordinates on the ith turn. All of these points lie on the perimeter of the ellipse, as they must since our fixed point has fixed values of α, β and γ, and A is invariant. Note the particle jumps around the ellipse and does not move around it continuously. If you wrote the tracking code in the previous section you could try this exercise for some stable lattice.

The β-function is a key quantity in the Courant-Snyder formalism. By definition we take β to be a position function of position s in the machine, and it carries the same periodicity that the lattice itself carries. It is determined by the focusing properties of the lattice, and is a function which is routinely computed in the design and operation of particle accelerators. It is maximum in a focusing quadrupole and minimum in a defocusing quadrupole. Let us now look at some examples.

The β-functions in each plane of the long straight section of the Large Hadron Collider are shown in Figure 5.11. We can see the periodic solution in the arc, and the small β-functions in the middle of the plot (usually denoted β^* in colliders), which correspond to the interaction point where collisions take place. We shall discuss the mini-beta principle soon. Note the large β-function spikes, which correspond to large particle excursion. The section which smoothly joins the arc β-function to the minimum is called the matching section. And we can measure the β-function too. Generally in science we can measure quantities if we change something they depend on in a systematic way. Hence careful changes of quadrupole currents allow β-functions to be reconstructed. We shall discuss this more later. So we have a formalism for linear beam motion in terms of the Courant-Snyder parameters. These quantities are central to linear beam dynamics and are used to design accelerators all around the world. Let's study them some more. It turns out that it is possible to write the transfer matrix between two points in a lattice in terms of the Courant-Snyder parameters at each of the two points and the phase advance between the points.

Let us now write this general transfer matrix. To begin with, we return to the Courant-Snyder form of the solution to Hill's equation; note that it depends on two constants and write this ansatz in a slightly different form,

$$x(s) = c_1 \sqrt{\beta(s)} \cos \psi(s) + c_2 \sqrt{\beta(s)} \sin \psi(s), \tag{5.69}$$

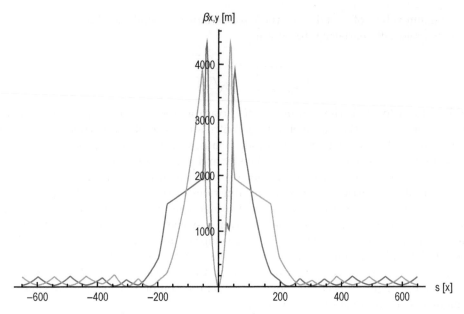

FIGURE 5.11 The LHC β-functions in the long straight section, 600 m either side of the interaction point. At $s = 0$ we have the tight focusing of the interaction point, and the large β-functions in the quadrupoles around this point arise from strong focusing. The periodic β-functions in the periodic arc can be seen at large values of $\pm s$.

where c_1 and c_2 are constants yet to be determined. If we define the initial conditions at the point '0' to be $\beta(0) = \beta_0$, $\alpha(0) = \alpha_0$ and $\psi(0) = \psi_0$ and write the initial particle coordinates to be x_0 and x_0' then we can fix the unknown constants to be

$$
\begin{aligned}
c_1 &= \frac{x_0}{\sqrt{\beta_0}}, \\
c_2 &= \sqrt{\beta_0}\,x_0' + \frac{\alpha_0}{\sqrt{\beta_0}}x_0.
\end{aligned}
\tag{5.70}
$$

We see the expression for $x(s)$ is linear in x_0 and x_0',

$$
x(s) = \sqrt{\frac{\beta(s)}{\beta_0}}\left[\cos\psi(s) + \alpha_0\sin\psi(s)\right]x_0 + \sqrt{\beta_0\beta(s)}\,x_0'\sin\psi(s).
\tag{5.71}
$$

Taking the derivative of this expression, we can cast this equation into a convenient matrix form as it's linear, to get

$$
\begin{pmatrix} x \\ x' \end{pmatrix}_{s_1} = M(s_1|s_0)\cdot\begin{pmatrix} x \\ x' \end{pmatrix}_{s_0}
\tag{5.72}
$$

where we have

$$
M(s_1|s_0) = \begin{pmatrix} \sqrt{\frac{\beta_1}{\beta_0}}(\cos\psi + \alpha_0\sin\psi) & \sqrt{\beta_1\beta_0}\sin\psi \\ \frac{\alpha_0-\alpha_1}{\sqrt{\beta_1\beta_0}}\cos\psi - \frac{1+\alpha_1\alpha_0}{\sqrt{\beta_1\beta_0}}\sin\psi & \sqrt{\frac{\beta_0}{\beta_1}}(\cos\psi - \alpha_1\sin\psi) \end{pmatrix}.
\tag{5.73}
$$

The subscripts 0 and 1 refer to the beginning and end of the transfer map and ψ is $\psi(s_1) - \psi(s_0)$. This means the transfer matrix between two points is purely determined by the lattice functions at each point and the phase advance between the points. This is a remarkable and very useful result.

The one-turn (strictly one-period) map is a very important quantity. Starting with our expression for the transfer matrix between two points, $M(s_1|s_0)$, we observe that the map for one turn of the ring means we come back to the same position. Hence we get

$$\begin{aligned}
\beta_1 = \beta_0 &= \beta \\
\alpha_1 = \alpha_1 &= \alpha \\
\gamma_1 = \gamma_0 &= \gamma
\end{aligned} \tag{5.74}$$

and we have

$$M(s+L|s) = \begin{pmatrix} \cos\Psi + \alpha\sin\Psi & \beta\sin\Psi \\ -\gamma\sin\Psi & \cos\Psi - \alpha\sin\Psi \end{pmatrix}, \tag{5.75}$$

where the phase advance over the period is $\Psi = \psi_1 - \psi_0$. This describes the transformation over one period of the accelerator lattice, and is called the one-turn, or one-period map. It is very important and tells us lots about the beam motion. We shall use it very shortly to examine beam motion stability.

Before we turn our attention to the information contained in the one-turn map, let's figure out how to calculate the lattice functions from it. If we multiply all the matrices for all the elements in the ring together, we obtain the total matrix for one turn of the machine (again, strictly, one period), which we write as

$$M = \begin{pmatrix} m_{11} & m_{12} \\ m_{21} & m_{22} \end{pmatrix}. \tag{5.76}$$

We can get the one-turn phase from the trace of this matrix by comparing it to the form we have for the one-turn map in terms of lattice functions, obtaining

$$\Psi = \arccos\left(\frac{m_{11} + m_{22}}{2}\right). \tag{5.77}$$

Note that we get only the ψ part of $2\pi n + \psi$. We can get the lattice functions from the other matrix elements as

$$\begin{aligned}
\beta &= \frac{m_{12}}{\sin\Psi}, \\
\alpha &= \frac{m_{11} - m_{22}}{2\sin\Psi}, \\
\gamma &= -\frac{m_{21}}{\sin\Psi}.
\end{aligned} \tag{5.78}$$

So we have a route to the lattice functions through the one-turn map. We compute this object and this gives the lattice functions at that point. This is how codes such as MADX [5] work.

Note for the phase over the period or turn to be real-valued, and using our expression for the phase above, we see that the absolute value of the trace $(m_{11} + m_{22})$ of M must be equal to or less than 2, i.e. $|Tr(M)| \leq 2$.

Imagine we know the one-turn map at one location, say s. Is there a way to figure it out at another location, say s', provided we know the transfer matrix M for s to s'? The answer is yes. They are related to each other by a similarity transform, and so

$$M(s'+C|s) = M(s'|s) \cdot M(s+C|s) \cdot M^{-1}(s'|s). \tag{5.79}$$

We shall state this without proof. Similarity transforms come from matrix theory and lead to all manner of nice properties such as identical eigenvalues and traces before and after the transformation. Let's be concrete and denote the matrix M (from s to s') by

$$M = \begin{pmatrix} m_{11} & m_{12} \\ m_{21} & m_{22} \end{pmatrix}. \tag{5.80}$$

Let's use this to figure out how the lattice functions transform from place to place if we know the transfer matrix. Starting with the similarity transform, we can express the one-turn maps in terms of the lattice functions at the locations s and s' as our standard expression,

$$M(s'|s) = \begin{pmatrix} \cos\Psi + \alpha\sin\Psi & \beta\sin\Psi \\ -\gamma\sin\Psi & \cos\Psi - \alpha\sin\Psi \end{pmatrix}. \tag{5.81}$$

Performing the similarity transformation we can obtain expressions for the lattice functions at position $s = 1$, given those at $s = 0$, as

$$\begin{pmatrix} \alpha_1 \\ \beta_1 \\ \gamma_1 \end{pmatrix} = \begin{pmatrix} m_{11}m_{22} + m_{12}m_{21} & -m_{11}m_{21} & -m_{12}m_{22} \\ -2m_{11}m_{12} & m_{11}^2 & m_{12}^2 \\ -2m_{21}m_{22} & m_{21}^2 & m_{22}^2 \end{pmatrix} \cdot \begin{pmatrix} \alpha_0 \\ \beta_0 \\ \gamma_0 \end{pmatrix}. \tag{5.82}$$

So knowing M, we can transform the lattice functions to any point in the beam line. Needless to say, this expression is important and very useful.

So we know how particles evolve (transform) in accelerator elements. How do the lattice parameters transform? Let's look at a drift space of length L, with an incoming particle described by x_0 and x_0', the incoming lattice parameters are β_0, α_0 and γ_0, and all quantities evolving to position 1. The transfer matrix is

$$M(1|0) = \begin{pmatrix} 1 & L \\ 0 & 1 \end{pmatrix}, \tag{5.83}$$

and the particle evolves as $x_1 = x_0'L + x_0$ and $x_1' = x_1$. Recall we are evolving from $s = 0$ m to $S = 1$ m. What about the lattice parameters? They evolve as

$$\begin{aligned} \beta_1 &= \beta_0 - 2\alpha_0 L + \gamma_0 L^2 \\ \alpha_1 &= \alpha_0 - \gamma_0 L \\ \gamma_1 &= \gamma_0. \end{aligned} \tag{5.84}$$

Note the quadratic term in the evolution of β, which we shall return to soon.

Several times we have used the phase advance for one turn of a ring (period),

$$\Psi = \int_s^{s+L} \frac{ds}{\beta(s)}, \tag{5.85}$$

i.e. given by an integral over the β-function. We often call the phase advance for one turn of a ring the tune (denoted ν or Q), and express it in units of $2 \times \pi$.

$$\nu = \frac{\Psi}{2\pi} = \frac{1}{2\pi} \int_s^{s+L} \frac{ds}{\beta(s)}. \tag{5.86}$$

There is one tune for each plane, including the longitudinal plane, and it's a very important function for beam dynamics. Note we can evaluate the tune at any point in the ring and always get the same answer (a property not shared by α, β and γ). This is because the trace is invariant under similarity transformations.

Note that all our equations in this section assume the beam motion is linear in the transverse coordinates. With the caveat in mind, we can start to build complicated arrangements of magnets. This art is called lattice design.

5.5 Lattice Design

In this section we shall construct our lattices from some basic building blocks, made of the magnets we have met so far. The art of lattice design for machines around the world works in this way, and we shall see how small chunks of magnet layout are created and combined to give larger structures. While what we do applies equally to circular and linear machines, we often use circular machines as the example, and will point out along the way how linear machines are designed.

Let's start with the basic bending we need, which means dipoles. Synchrotrons, and sometimes high-energy particle colliders, are circular machines, so we need plenty of dipoles in the lattice to bend the particles around the tunnel or ring. This creates the design orbit of the machine, and the middle of our moving coordinate system. Then, once the design orbit is sorted out, we need to design the magnetic lattice, and position the quadrupoles and higher-order magnets. This process is called lattice design. Fig 5.7 shows a section of an accelerator lattice, showing the arrangement of dipoles and quadrupoles in the lattice – lattice design is deciding the placement and strength of these elements.

Our first task to figure out the geometry of the ring, and define the curved reference orbit using a layout of dipole magnets. This forms the fundamental footprint of the machine and defines our coordinate system for future analysis. The use of dipoles to form the bending of an accelerator is a very important application. In circular machines, the dipoles are needed to form the ring geometry and must add to a total bend angle of 2π.

Consider a particle bending through angle $d\theta$, with arc length ds and bend radius ρ, such that $\theta = ds/\rho$. For a weak bending magnet we can approximate ds as dl, where dl is an element of the length of the magnet. The integral over the magnet length gives the total bend angle,

$$\theta = \frac{\int B dl}{B\rho} \tag{5.87}$$

which we need to be 2π for a circular tunnel to get all of the way round the circumference of the machine's footprint.

For an example, consider the LHC. This is a two-beam circular proton synchrotron at CERN, Geneva. Here we have 1232 dipoles, each of 14.3 m length, and each beam has a design momentum of 7 TeV/c. The rigidity of this beam was discussed in Chapter 2, where we defined the beam rigidity as

$$(B\rho) = p/q. \tag{5.88}$$

For the LHC we need $NlB = 2\pi \cdot p/q$ to complete the ring so the required field is 8.3 T, which is of course the design strength of the LHC dipoles.

We should say that dipoles are also very important in linear beam lines, to give the right angle of beam delivery. They are also used in a transfer line to form dog-legs and chicanes designed to manipulate beams, especially for bunch compression. For example, a dog-leg in an electron machine to perform bunch compression. Other uses include removal of background particles in a linear collider and separation of beams in the LHC.

The dipoles are now defined and the basic machine geometry fixed. Now we need to concern ourselves with linear beam focusing and the quadrupoles. For this we need the principle of alternating gradient. Recall that two quadrupoles of opposite polarity could provide focusing in both planes at the same time. This fantastic result is one of our fundamental building blocks – the FODO cell – and allows us to construct periodic, stable structures. The FODO cell consists of a horizontally focusing quadrupole (F), a space (O), a defocusing quadrupole (D) and a space (O), giving an alternating gradient layout. We can repeat the FODO cell to make a FODO channel of arbitrary length. Note the drift space (O) can contain nothing, a bend, some diagnostics, an RF cavity or even an entire experiment in some cases.

To understand the beam dynamics in a FODO cell we need to compute the one period map, giving the linear motion over one FODO cell. To do this we simply multiply the matrices of the components of the cell together, conventionally starting in the middle of one of the quadrupoles, which means we start and end with a quadrupole matrix of half strength. This is not so strange and ultimately means we find the maximum and minimum β-function points, which occur at symmetry points of the cell in the middle of each quadrupole.

Recall our matrices describing the action of the linear accelerator elements on the beam. For the focusing quadrupole we had

$$M_{\mathrm{F}} = \begin{pmatrix} \cos(\sqrt{K}s) & \frac{1}{\sqrt{K}} \sin(\sqrt{K}s) \\ -\sqrt{K} \sin(\sqrt{K}s) & \cos(\sqrt{K}s) \end{pmatrix}, \tag{5.89}$$

and for the defocusing quadrupole we had

$$M_{\mathrm{D}} = \begin{pmatrix} \cosh(\sqrt{|K|}s) & \frac{1}{\sqrt{|K|}} \sinh(\sqrt{|K|}s) \\ +\sqrt{|K|} \sinh(\sqrt{|K|}s) & \cosh(\sqrt{|K|}s) \end{pmatrix}. \tag{5.90}$$

We also need the matrix for a drift,

$$M_{\mathrm{drift}} = \begin{pmatrix} 1 & L \\ 0 & 1 \end{pmatrix}, \tag{5.91}$$

and the matrix for a thin-lens quadrupole,

$$M_{\mathrm{thin}} = \begin{pmatrix} 1 & 0 \\ -\frac{1}{f} & 1 \end{pmatrix}. \tag{5.92}$$

Using the thick lens elements, we can multiply these matrices in sequence, in 'FODO' order,

$$M_{\mathrm{FODO}} = M_{\mathrm{F}/2} \cdot M_{\mathrm{drift}} \cdot M_{\mathrm{D}} \cdot M_{\mathrm{drift}} \cdot M_{\mathrm{F}/2}. \tag{5.93}$$

Let's take some real numbers for a real machine to give a feeling for quantities. Let us take our quadrupole strengths to be $K = \pm 0.54102$ m^{-2}, the quadrupole lengths to be $l_q = 0.5$ m and the separation distance to be $L = 2.5$ m. This gives, if we do the maths

$$M_{\mathrm{FODO}} = \begin{pmatrix} 0.707 & 8.206 \\ -0.061 & 0.707 \end{pmatrix}. \tag{5.94}$$

This is the one-period map of the FODO cell, and contains a lot of information on the stability and the focusing properties of our lattice. First of all, we can ask if the FODO cell stable? For this we need the (absolute) trace of the one-turn map to be less than or equal to 2. Here it is 1.415. So this FODO cell will give stable dynamics in this plane and the particle motion is bounded. Make sure you can repeat these calculations.

What is the betatron phase advance per cell? Recall that

$$\Psi = \arccos\left(\frac{m_{11} + m_{22}}{2}\right), \tag{5.95}$$

and so the phase advance per cell is 45°. This is a 45° cell.

What are the lattice functions at the point of the one-turn map? For us, this is in the middle of the focusing quadrupole. Well, we use

$$\beta = \frac{m_{12}}{\sin \Psi},$$
$$\alpha = \frac{m_{11} - m_{22}}{2 \sin \Psi},$$
$$\gamma = -\frac{m_{21}}{\sin \Psi}, \qquad (5.96)$$

and find that $\beta = 11.611$ m and $\alpha = 0$. (For this case, what does the ellipse look like?) What does MAD compute? Try it yourself (www.cern.ch/mad)! Or construct your own code.

We can also make our life easier and compute the matrix for our FODO cell using the thin-lens matrices. Again, starting from the middle of Q_F we have

$$M_{\text{FODO}} = \begin{pmatrix} 1 & 0 \\ -\frac{1}{2f} & 1 \end{pmatrix} \begin{pmatrix} 1 & L \\ 0 & 1 \end{pmatrix} \begin{pmatrix} 1 & 0 \\ \frac{1}{f} & 1 \end{pmatrix} \begin{pmatrix} 1 & L \\ 0 & 1 \end{pmatrix} \begin{pmatrix} 1 & 0 \\ -\frac{1}{2f} & 1 \end{pmatrix}. \qquad (5.97)$$

Doing the mathematics, we end up with the matrix in terms of L and f

$$M_{\text{FODO}} = \begin{pmatrix} 1 - \frac{L^2}{2f^2} & 2L(1 + \frac{L}{2f}) \\ -\frac{L}{2f^2}(1 - \frac{L}{2f}) & 1 - \frac{L^2}{2f^2} \end{pmatrix}, \qquad (5.98)$$

which contains lots of information. Straight away we can ask, for what parameters is the FODO cell going to give stable motion? This means

$$\mid \text{Tr}(M) \mid \leq M \to |f| \geq \frac{L}{2}. \qquad (5.99)$$

We can also write the cell phase advance in terms of the parameters

$$\cos \Psi = \frac{1}{2} \text{Tr}(M) = 1 - \frac{L^2}{2f^2}. \qquad (5.100)$$

Our stability equation seems to say motion is stable when focusing is weak (long focal lengths)! Strong quadrupoles aren't always the way to go to get stable motion and controlled β-functions.

Now we can compute the lattice functions in the cell. Note that β in the focusing and defocusing quadrupoles are maximised there, and this maximum depends solely on the cell length and phase advance. Using

$$\beta = \frac{m_{12}}{\sin \Psi},$$
$$\alpha = \frac{m_{11} - m_{22}}{2 \sin \Psi}, \qquad (5.101)$$

we obtain in the focusing quadrupole

$$\beta_F = \frac{2L(1 + L/2f)}{\sin \Psi}, \quad \alpha_F = 0. \qquad (5.102)$$

The expression for β_D can be obtained in a similar way.

So we build our ring out of dipoles and FODO cells. What about an experiment or a region free of magnets for diagnostics? We need to stop focusing for a while, so we should ask what will happen? Remember we derived the expression for the evolution of the β-function in a drift as

$$\beta_1 = \beta_0 - 2\alpha_0 L + \gamma_0 L^2, \qquad (5.103)$$

showing what happens to our beta function in a drift. At a symmetry point $\alpha = 0$ and $\gamma = 1/\beta$ giving the increase of the β-function after the symmetry point as

$$\beta(s) = \beta^* + \frac{s^2}{\beta^*}, \tag{5.104}$$

where we denote the β-function at the symmetry point as β^*. This is very bad for accelerator designers! What happens can be understood in terms of the ellipse. The area of the ellipse is constant, so squeezing β means we increase γ, so the beam rapidly diverges after it leaves the symmetry point. This is an example of Liouville's theorem, which states that the the area occupied by a beam in phase space is constant as it moves through the accelerator. We saw our ellipse area was constant, which is a consequence of Liouville's idea.

Fig 5.11 is the region around the ATLAS experiment in the LHC. Here we have a waist (a minimum in the β-function) at the ATLAS interaction point at $s = 0$ m, and sitting at $s = \pm 22$ m are strong quadrupoles (in fact a triplet of quadrupoles) to make the beam waist. Around these we have matching quadrupoles to match the β-function back into the periodic solution in the LHC arc FODO cells. A problem with large β-functions in the triplet quadrupole is that this implies a large aperture requirement due to large beam sizes, as well as other problems we shall see later.

Accelerator design starts with defining the geometry of the machine, attending to the dipoles and then using a linear approximation to analyse the linear dynamics. For a modern accelerator the machine parameters are defined, with many constraints such as cost, where tunnels can go, what the accelerator is specified to achieve and so on. The gross parameters such as energy, luminosity, radiation output, etc. are defined. This is the first step and defines the global properties of the accelerator and its broad aims. At this stage the user community should be involved to ensure the machine will meet the user need.

Next, we need to consider magnetic technology to define the maximum dipole and quadrupole strengths. This defines the geometry of the machine. Then the linear lattice is then constructed based on the fundamental building blocks. The linear lattice should fulfill the accelerator physics criteria and provide global quantities such as circumference, emittance, betatron tunes, magnet strengths, and some other machine parameters. Design codes such as MADX [5], ASTRA [6] and GPT [7, 8] (the list is nearly endless!) are used for the determination or matching of lattice functions and parameter calculations. Periodic cells are needed in a circular machine. The cell can be the kind we have looked at, namely FODO, or many others we can come to later after we have discussed dispersion. Next combined-function or separated-function magnets are selected and matching or insertion sections are introduced to get the desired machine functions in an experimental region or an undulator, for example. There is more to do, but we shall come back to this recipe for accelerator design once we've learned some new concepts.

As an aside, let's consider an open-ended design problem. Design a lattice with four identical FODO cells, with each cell containing a thin lens, a dipole, another thin lens and another dipole. The machine should store protons with a total energy of 3 GeV per proton, with a bend radius of around 80 m. Choose suitable values for the quadrupole strengths, drift lengths and bending radii so that the motion is stable. Plot the beta functions and dispersion in both planes and calculate the ring tunes. What is the momentum compaction factor? You could use MAD, or any other suitable code or programming language.

We shall finish this section with one more matrix we need to know. So far we've defined our linear matrix formalism and figured out matrices for drift spaces, quadrupoles and dipoles. The latter matrix is a purely focusing effect in the plane of the bending, with the bending effect of the dipole contained in the co-moving coordinate system. This means the beam changes in angle when it passes through the dipole. When we come to build this magnet, we have a choice to make. The first choice is called a sector dipole, where the beam is perpendicular to the entrance and exit faces and the magnet follows the curved trajectory of the reference orbit. A second choice, which is easier to fabricate, is a rectangular dipole where the entrance and exit faces are parallel to each other and so the curved trajectory of the beam makes an angle with the entrance and exit faces, normally written as ψ in the literature. The impact of this entrance and exit angle is that the beam receives a small focusing kick, essentially due to the larger or lesser amount of magnetic field seen by the beam. This is called edge focusing. The particle is bent through an angle of

$$\Delta\theta = \frac{x_0 \tan \psi}{R}, \tag{5.105}$$

where R is the bend radius of the dipole and x_0 is the transverse position of the particle under consideration. Hence we can write the coordinate transformation as $x = x_0$ and

$$x' = x_0' + x_0 \frac{\tan \psi}{R}, \tag{5.106}$$

or as a matrix as

$$M_{\text{edge}} = \begin{pmatrix} 1 & 0 \\ \frac{\tan \psi}{R} & 1 \end{pmatrix}. \tag{5.107}$$

In the vertical plane there is also a focusing effect, which is given by

$$M_{\text{edge}} = \begin{pmatrix} 1 & 0 \\ -\frac{\tan \psi}{R} & 1 \end{pmatrix}. \tag{5.108}$$

So we see a positive ψ causes horizontal defocusing and vertical focusing.

We should mention at this time that magnetic fields extend beyond the physical extent of the magnet with a non-linear character that is not fully included in the effective length. These fringe fields can disturb the beam in strong magnets and can be very important to the dynamics. For full details see [9, 10, 11].

5.6 Errors and Misalignments

In our analysis we started with an arbitrary field and made an expansion, which we called the multipole expansion. When we used the expansion in our discussion of Hill's equation, we only kept the first two terms, equivalent to the constant and linear terms in a Taylor series. These correspond to dipole and quadrupole fields. The complete multipole expansion for the transverse fields looks like

$$\begin{aligned} B_y + iB_x &= \sum_{n=1}^{\infty} C_n z^{n-1} \\ &= \sum_{n=1}^{\infty} C_n (x + iy)^{n-1}, \end{aligned}$$

as we saw in Chapter 4. Here, C_n are the multipole coefficients. In this expression we have our dipole and quadrupole fields, plus higher-order terms like the sextupole, octupole and so on. Linear beam dynamics is the study of the dipole and quadrupole terms, and non-linear dynamics is the domain of the higher-order terms.

To realise these fields in a real accelerator we build the magnets which present the required multipoles to the beam, and to do this we need to specify some field quality. These magnets will never be perfect and the design and construction of them will lead to multipole coefficients different from the design values and the addition of extra multipoles within the

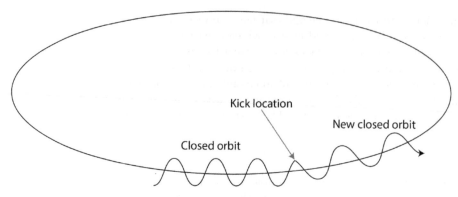

FIGURE 5.12 Closed-orbit distortion from a single dipole kick, showing the change in the motion of the particle after the kick to a new closed orbit about the ring.

constraints from symmetry and fabrication tolerances. Therefore these magnets will have mostly the field component we want, but will have small contributions of higher-order field components. This is what we saw in the chapter on magnets. What do these unwanted multipole terms do to the beam? In the design process, most synchrotrons specify a field quality of one to ten parts in 10,000, which may not seem much but can cause many problems for our beam. We need to be able to calculate the resulting motion. We also need to align magnets correctly, otherwise this will lead to further unwanted effects on the beam. For example, a quadrupole can be misaligned either horizontally or vertically, so generating an additional dipole field in the beam. A further origin for unwanted fields on the beam is from a power supply to a dipole or a quadrupole that may vary over time, thus producing a field which is not precisely what is required. The bottom line is our lattice is never as we designed it and we need to deal with field errors for every machine we attempt to build and operate.

5.6.1   Closed-Orbit Distortion

The design orbit defined by all of the dipoles in the ring is also known variously as the reference orbit or reference trajectory. For an ideal machine this is the trajectory that goes through the middle of each magnet, closes upon itself in a circular machine, and is the reference orbit to which we define the particle coordinates. This is the desired situation but is not achieved in real rings. If there is a small dipole kick at some location, arising from any of the reasons we have just discussed like a quadrupole misalignment or a power supply error, the beam will feel an extra kick and this orbit will distort. This distortion will run around the entire ring or along the entire beamline. This is shown in Figure 5.12, where we see the orbit change resulting from a kick at a fixed location, denoted "kick location". An important consequence of this is that a small kick at some location will be seen everywhere in the beamline or ring!

This closed-orbit distortion defines a position-dependent orbit offset around the ring, which can be seen in the figure. In effect, the particles no longer oscillate around the design orbit but around a new closed orbit, meaning the particles oscillate not about the middle of every magnet but some other orbit $x(s)$, where

$$x(s) = x_\beta(s) + x_{\text{CO}}(s). \tag{5.109}$$

Here $x_\beta(s)$ denotes our betatronic oscillations around the ideal orbit and $x_{\text{CO}}(s)$ denotes the position-dependent shift of this reference trajectory. This new orbit $x_{\text{CO}}(s)$ must obey the periodicity of the ring. How do we find it? Well, a short but not terribly enlightening calculation involving inclusion of a kick in our formalism gives

$$x_{CO}(s) = \theta \frac{\sqrt{\beta(s)\beta(s_0)}}{2\sin\pi\nu}\cos(\pi\nu - |\psi(s) - \psi(s_0)|), \qquad (5.110)$$

where a dipole kick of angle θ is located at location s_0. The betatronic phase at s is denoted $\psi(s)$ and the tune of the machine is denoted by ν.

Note that our expression for the closed-orbit distortion has a denominator containing the sine of the tune multiplied by π. This means that if the tune is an integer, the argument of the sine becomes a multiple of π, and so this factor diverges and gets very big. This means the closed-orbit distortion, proportional to this quantity, gets very large. This is an example of resonance, where the machine tune is such that harmful beam behaviour occurs.

Let's think physically what that means. Imagine the tune was set to 2 in a machine, meaning the beam made one complete betatron oscillation every turn of the machine. Then the particle would encounter a particular dipole error at one point in the machine every turn, and at the same point in its betatron oscillation. This means the effect of the dipole error adds up turn after turn after turn, pushing particles to very large excursions transversely. This is clearly bad. We avoid this by minimising magnetic dipole errors and staying away from dangerous values of the tune. Here, because of our dipole error, we should avoid integer tune values.

We'll soon see there are many other resonances which occur at other tune values. We're about to analyse the next kind of magnetic error – quadrupole errors – which will mean we need to stay away from half-integer tune values to avoid harmful behaviour.

It will turn out that generally resonances occur when the tunes of the machine in both planes satisfy the condition

$$m\nu_x + n\nu_y = p, \qquad (5.111)$$

where m, n, p are integers. This contains the condition for our dangerous integer and half integer tune values, and much more besides. The order of the resonance is given by $m + n$. Note that this condition not only includes constraints on either the horizontal or vertical tune to avoid resonance, but also resonance conditions that mix the horizontal and vertical tune. These are called coupling resonances, with the lowest-order coupling resonance having the condition

$$\nu_x \pm \nu_y = p, \qquad (5.112)$$

and this is known as the linear coupling resonance. This resonance, driven by non-linear elements, couples both transverse planes together and leads to the exchange of motion and beam emittance between the planes. Further terminology is a sum resonance, which is a positive sign between ν_x and ν_y in Equation 5.111, and a difference resonance, which is a negative sign between ν_x and ν_y in Equation 5.111. A structural resonance is the case of the integer p corresponding to the superperiodicity of the machine, as these resonances are especially strongly driven and hence dangerous.

A very common plot is the resonance condition plot, where we plot all the resonance conditions on a plot of (ν_x, ν_y). This plot is shown in Figure 5.13. Each condition corresponds to a line on the plot, at some order. The tunes of the machine in each plane are chosen to avoid these resonance lines, and this tune point is known as the working point of the machine.

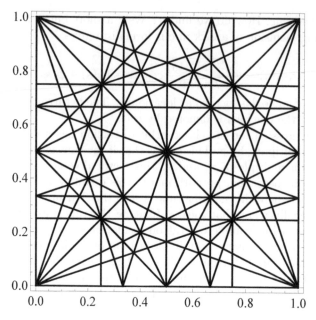

FIGURE 5.13 Diagram of the resonance conditions of a circular machine. The lines correspond to the solutions of the resonance condition described in the text and represent harmful conditions for the beam. The lines are plotted in the space of (ν_x, ν_y).

5.6.2 A Quadrupole Error

Now imagine we had an extra quadrupole in our ring, or a quadrupole field error. This would change the focusing, so every quantity associated with focusing will change. This will perturb the beam away from the design, and cause:

1. change in the tune of the machine, and

2. change in the β-function of the machine all round the ring, known as β-beat.

Let's calculate it. Imagine our quadrupole error had integrated strength $KL = +q$. This means it has a matrix which kicks the x' of the particle and looks like

$$\begin{pmatrix} 1 & 0 \\ -q & 1 \end{pmatrix}. \tag{5.113}$$

If we represent the rest of the machine by the one-turn map,

$$M(s_0 + L|s_0) = \begin{pmatrix} \cos\Psi + \alpha\sin\Psi & \beta\sin\Psi \\ -\gamma\sin\Psi & \cos\Psi - \alpha\sin\Psi \end{pmatrix}, \tag{5.114}$$

then the effect on the global dynamics of the machine can be calculated from the matrix product to give a new one turn map

$$M(s_0 + L|s_0) = \begin{pmatrix} \cos\Psi + \alpha\sin\Psi & \beta\sin\Psi \\ -\gamma\sin\Psi & \cos\Psi - \alpha\sin\Psi \end{pmatrix} \begin{pmatrix} 1 & 0 \\ -q & 1 \end{pmatrix}. \tag{5.115}$$

This gives, if we spend five minutes doing the matrix multiplication,

$$M(s_0 + L|s_0) = \begin{pmatrix} \cos 2\pi\nu + \alpha_0 \sin 2\pi\nu - q\beta_0 \sin 2\pi\nu & \beta_0 \sin 2\pi\nu \\ -\gamma_0 \sin 2\pi\nu - q(\cos 2\pi\nu - \alpha_0 \sin 2\pi\nu) & \cos 2\pi\nu - \alpha_0 \sin 2\pi\nu \end{pmatrix}. \tag{5.116}$$

This is the perturbed one-turn map, and all symbols with a 0 subscript represent the values of the unperturbed machine. If we denote the tune and lattice functions of the perturbed machine by a subscript p, then the one-turn map looks like

$$M(s_0 + L|s_0) = \begin{pmatrix} \cos 2\pi\nu_p + \alpha \sin 2\pi\nu_p & \beta \sin 2\pi\nu_p \\ -\gamma \sin 2\pi\nu_p & \cos 2\pi\nu_p - \alpha \sin 2\pi\nu_p \end{pmatrix}. \tag{5.117}$$

Equating the traces of these two matrices gives

$$2\cos 2\pi\nu_p = 2\cos 2\pi\nu - q\beta_0 \sin 2\pi\nu \tag{5.118}$$

which relates the unperturbed and perturbed tune. We note if q is small, then the perturbed tune is close to the unperturbed tune, so $\nu \simeq \nu_p$. Let's assume the tune shift is small, and write $\nu_p = \nu + d\nu$. If we then expand the cosine function using a standard identity

$$2\cos 2\pi(\nu + d\nu) = 2\cos 2\pi\nu \cdot \cos 2\pi d\nu - 2\sin 2\pi\nu \cdot \sin 2\pi d\nu. \tag{5.119}$$

To simplify this expression we recall the quadrupole error in our lattice is small and so the tune shift $d\nu$ is small. Hence we can assume that $\cos 2\pi d\nu \simeq 1$ and $\sin 2\pi d\nu \simeq 2\pi d\nu$, giving the important result

$$q\beta_0 = 4\pi d\nu, \tag{5.120}$$

and so the tune shift from a small quadrupole of strength q is

$$\Delta\nu = \nu_p - \nu = \frac{q\beta_0}{4\pi}. \tag{5.121}$$

This is a very important result. Note the following important features:

1. The perturbed tune increases if $q > 0$, which corresponds to a focusing quadrupole i.e. more focusing means more oscillations. So we get a positive tune shift for increased particle focusing.

2. This means a pure quadrupole field error would shift the tune one way in one plane and the other way in the other plane. However, note that we can also get tune shifts from space charge, beam-beam effects and electron clouds, which can cause same-sign tune shift in both planes.

3. The effect of the quadrupole error is proportional to the local β-function. This is a common feature that the β-function magnifies local field errors.

If we have a distribution of quadrupole errors around the ring, $k(s)$, the approximate tune shift can be calculated from

$$\Delta\nu = \frac{1}{4\pi} \oint ds\beta(s)k(s). \tag{5.122}$$

We note this can also be used to measure the β-functions. To do this, we vary a single quadrupole in the ring, and measure the tune, as the response of the beam is proportional to the β-function. In general, the β-function tells you how sensitive the beam is to perturbations. For example, for LHC luminosity upgrades, we may have to live with very large β-functions in the arcs of the LHC. This means the proton beams will be more sensitive to field errors.

What about the change in beta function due to our quadrupole error q at s_0? Skipping the derivation (which is short and not particularly enlightening), we obtain

$$\frac{\Delta\beta}{\beta} = -\frac{q\beta_0}{2\sin 2\pi\nu} \cos(2\pi\nu + 2|\psi(s) - \psi(s_0)|). \tag{5.123}$$

Note the β-perturbation is a function of s, so is a 'beta wave' around the ring The distortion oscillates at twice the betatron frequency, which is why it's called a β-beat. Note also the strength of the distortion is proportional to the quadrupole error and also the beta function at the position of the quadrupole error. The β-beat measured in the LHC is shown in Figure 5.14.

Finally, we have a $\sin 2\pi\nu$ term in the denominator. This means the expression will get very large whenever the tune approaches a half-integer. This is resonance again, and means large particle amplitudes are driven for half-integer machine tunes.

5.7   Off-Momentum Particles

5.7.1   General Considerations

So far we have considered beam motion when the particles have the design momentum p. We refer to these particles as on-momentum particles, and this defines the ideal, on-momentum motion. However, in general, a particle's momentum will be $p \pm$ (something small), where the beam consists of a group of particles with some distribution of momenta. In fact when we come to talk about longitudinal dynamics, it's a necessary consequence of longitudinal stability that we have a range of momenta in the beam. And so we need to think about what happens when we have particles which are not at the momentum for which we designed our machine.

So let's write, for our momentum,

$$p + \Delta p = p(1 + \delta), \tag{5.124}$$

where $\delta = \Delta p/p$ and parameterises the deviation of a given particle's momentum away from the design momentum. Now we can explore the consequence of non-zero δ. So how does this change our picture?

Let's think about dipoles and bending first. Imagine we design our accelerator and figure out we need a certain dipole field strength to bend a particle of a certain momentum around a bend in the tunnel. This momentum is the one we design our accelerator for and so is called the design momentum. We send our design particle with the design momentum into this dipole and it bends through the right angle. This defines the machine geometry. Now imagine we send through a particle with slightly less momentum. What will happen? Well, the dipole field strength of the magnet is fixed so the particle will get a change in its angle which is greater than the design particle. Hence the trajectory, or orbit, of this off-momentum particle will be different. Imagine we send through a particle with slightly more momentum. Now the particle will get a change in its angle which is less than the design particle due to the fixed field. Hence the trajectory, or orbit, of this off-momentum particle will also be different. This change in orbit for particles with differing momenta is called dispersion because particle beams with a spread of momentum get dispersed by a dipole.

Let's now consider our particle with some momentum less than the design momentum passing through a quadrupole. What happens now? Well, the quadruple is designed to focus particles with the design momenta to a single point, and so our particle will see too much field, be over-focused and so not be focused to the correct point. Similarly, particles with too much momentum will not be focused enough. The quadrupoles, through their k distribution, fix the focusing of the lattice and so the β-function of the lattice and the tune will change. The change in these quantities is said to arise from chromaticity, or momentum-dependent focusing.

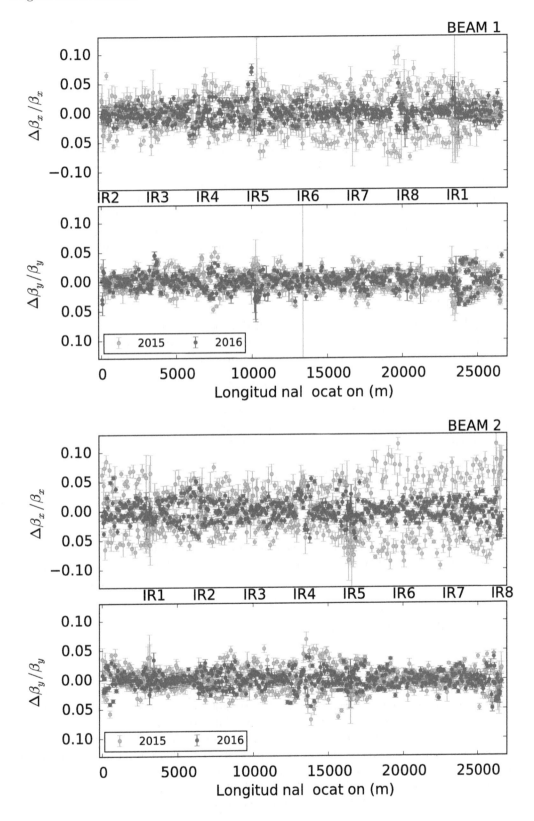

FIGURE 5.14 The β-beat in the LHC as measured in 2015 and 2016. In a perfect machine $\Delta\beta/\beta = 0$ at all locations. Used with kind permission from [12].

Let's be more quantitative. Recall our derivation of Hill's equation gave us

$$x''(s) + \left(k_x(s) + \frac{1}{\rho(s)^2}\right) = 0, \tag{5.125}$$

in the horizontal plane and

$$y''(s) + k_y(s) = 0 \tag{5.126}$$

in the vertical plane. In these equations, earlier in this chapter, we dropped δ where it appeared. We need to figure out how to modify these equations to retain the effects of a particle being off-momentum. We do this by retaining terms of the coordinate δ we didn't previously retain. Let's explore them one by one. We know that we build accelerator beamlines by using magnets, with dipoles and quadrupoles being our basic building blocks, and we shall find the retained terms affect all of our elements. We shall also find that our particle path length will change for off-momentum particles, an effect known as momentum compaction and crucial for the longitudinal motion in our accelerators. We shall consider each case one by one.

5.7.2   Dispersion

First of all, let's consider trajectory change, called dispersion. In this case, the derivation of Hill's equations needs to be modified to retain terms in the expansion that are linear in δ, which arise from the expansion of the momentum in terms of the momentum deviation δ,

$$m\gamma v_s = p(1 + \delta). \tag{5.127}$$

In this equation, m is the particle mass, γ is the relativistic gamma function and v_s is the particle speed along the reference trajectory. The complete derivation is left to the reader [3], but is straightforward, and if we do this we obtain the revised Hill's equations

$$x''(s) + \left(k_x(s) + \frac{1}{\rho(s)^2}\right) = \frac{\delta}{\rho}, \tag{5.128}$$

in the horizontal plane and

$$y''(s) + k_y(s) = 0 \tag{5.129}$$

in the vertical plane. Note the vertical plane equation is unmodified as the bending, in our analysis, is purely in the horizontal plane. We still have the definitions

$$k_x = \frac{g}{B\rho} + \frac{1}{\rho^2}, \tag{5.130}$$

and the equivalent for the vertical plane,

$$k_y = -\frac{g}{B\rho}. \tag{5.131}$$

The new Hill's equation in the horizontal plane is the inhomogeneous equation of motion, like before but with a non-zero right-hand side term not containing x or its derivative. This is the inhomogeneous term and leads to dispersion. The extra term on the right-hand side, proportional to δ, will drive the horizontal motion of an off-momentum particle, which we shall call horizontal dispersion, or simply dispersion. Note there is no dispersion-driving term in the vertical plane as there is no bending for our derivation.

The general solution for the horizontal motion of a particle is given by the sum of two terms: the betatron motion term $x_\beta(s)$ and an off-momentum dispersion term

$$x(s) = x_\beta(s) + x_h(s). \tag{5.132}$$

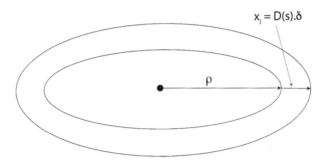

$$x_i = D(s).\delta$$

$$\rho$$

FIGURE 5.15 The orbit offset of dispersion, showing the shift from the on-momentum orbit by $D(s).\delta$.

We can think of $x_\beta(s)$ as a closed-orbit term, around which $x_h(s)$ oscillates. This follows from the theory of differential equations, where we add the special solution obtained from the driving term to the homogeneous version of the differential equation. Essentially, dispersion is a shift of the closed orbit around which the betatron oscillations occur, as shown in Figure 5.15.

To analyse and understand our dispersive motion it is convenient to define a special trajectory, $D(s)$, which is that trajectory followed by a particle that has $\delta = 1$. This trajectory, while physical, has no particles following it as they would be lost due to the large transverse deviation, but is a tool to parameterise the motion. So let's consider our newly defined dispersion function $D(s)$. This is actually a physically allowed orbit, and the one a particle with $\delta = 1$ has should this particle exist. As this $D(s)$ is a physical orbit it is focused by the lattice, meaning both dispersion and dispersive motion is focused by the lattice. The motion of the particle is the sum of our old $x(s)$ and the dispersion, so that

$$x(s) = x_\beta(s) + \delta \cdot D(s). \tag{5.133}$$

One way of viewing this equation is thinking of the dispersive term as a closed orbit around the accelerator, and a particle oscillates around this dispersive orbit through the usual betatron oscillations. This is like a dipole error closed-orbit distortion. What are typical values? Well, x_β is typically a few mm, values of $D(s)$ might be < 1 metre, and δ might typically be 0.001.

So how do we calculate $D(s)$? We need to find a solution to the inhomogeneous Hill's equation and add it to the general solution of the homogeneous equation, so we need to solve

$$x''(s) + \left(k_x(s) + \frac{1}{\rho(s)^2}\right) = \frac{\delta}{\rho}. \tag{5.134}$$

To calculate $D(s)$, consider motion in a dipole (so no gradients) and we have $\delta = 1$ for the trajectory corresponding to $D(s)$. Therefore $D(s)$ is a solution of the resulting inhomogeneous equation

$$D''(s) + \frac{1}{\rho(s)^2}D(s) = \frac{1}{\rho}. \tag{5.135}$$

We have already solved the homogeneous equation (with the right-hand side equal to 0) as this is the matrix we already found for a dipole. Now we need to find a particular solution of the inhomogeneous equation and add this solution (D_I) to the solution of the homogeneous equation. Since the right-hand side is a constant, then a valid choice of a particular solution is a constant: we can try a constant as a solution

$$D_I = C \tag{5.136}$$

and we can readily find that $D = \rho$ by inserting this into the equation of motion. This means our general solution is

$$D(s) = A\cos(s/\rho) + B\sin(s/\rho) + \rho, \tag{5.137}$$

and its derivative is

$$D'(s) = -\frac{A}{\rho}\sin(s/\rho) + \frac{B}{\rho}\cos(s/\rho). \tag{5.138}$$

We can find A and B from the initial conditions by noting that $D(s = 0) = D_0$ and $D'(s = 0) = D_0'$ and so we have equations to evolve the dispersion function D through the dipole, which are

$$
\begin{aligned}
D(s) &= D(0)\cos(s/\rho) + D'(0)\rho\sin(s/\rho) + \rho(1 - \cos(s/\rho)), \\
D'(s) &= -\frac{D(0)}{\rho}\sin(s/\rho) + D'(0)\cos(s/\rho) + \sin(s/\rho).
\end{aligned}
\tag{5.139}
$$

These equations are linear and readily written as a matrix equation,

$$
\begin{pmatrix} D(s) \\ D'(s) \\ 1 \end{pmatrix} =
\begin{pmatrix}
\cos(s/\rho) & \rho\sin(s/\rho) & \rho(1 - \cos(s/\rho)) \\
-\frac{1}{\rho}\sin(s/\rho) & \cos(s/\rho) & \sin(s/\rho) \\
0 & 0 & 1
\end{pmatrix}
\begin{pmatrix} D(0) \\ D'(0) \\ 1 \end{pmatrix}.
\tag{5.140}
$$

Note the upper-left 2×2 matrix is just the transfer matrix for a dipole we have already derived. This means the dispersion function obeys the matrix equations we know already; in a dipole, dispersion is also produced (or driven). The dispersion function in a quadrupole obeys the quadrupole transfer matrix, and so the dispersion function is focused in a quadrupole in the normal way. However, there is no extra dispersion driven in a quadrupole, and so M_{13} and M_{23} are zero in the matrix.

Finally, as the motion is given as the sum of the betatron motion and the dispersion

$$x(s) = x_\beta(s) + D(s)\delta. \tag{5.141}$$

The general motion of a particle can be written as a 3×3 matrix equation,

$$
\begin{pmatrix} x(s) \\ x'(s) \\ \delta \end{pmatrix} =
\begin{pmatrix}
M_{11} & M_{12} & D \\
M_{21} & M_{22} & D' \\
0 & 0 & 1
\end{pmatrix}
\begin{pmatrix} x(0) \\ x'(0) \\ \delta \end{pmatrix}.
\tag{5.142}
$$

For a short sector dipole with bending angle θ small compared to 1,

$$\theta = \frac{l}{\rho} \ll 1, \tag{5.143}$$

we can write this matrix in the simpler form

$$
\begin{pmatrix}
1 & l & l\theta/2 \\
0 & l & \theta \\
0 & 0 & 1
\end{pmatrix}.
\tag{5.144}
$$

This is useful for quick calculations and corresponds to having a thin-lens kick for an off-momentum particle. A quadrupole has no driving term for the dispersion and the 3×3 map is given by

$$
\begin{pmatrix}
M_{11} & M_{12} & 0 \\
M_{21} & M_{22} & 0 \\
0 & 0 & 1
\end{pmatrix}.
\tag{5.145}
$$

In this expression, the unity factor in the (3,3) element simply expresses the invariance of δ (the momentum is unchanged). When we considered β-functions, we looked at the form they took in our basic lattice building block – the FODO cell. What happens to the dispersion in a FODO cell? Consider a FODO cell with thin-lens quadrupoles. Now that we know dispersion is driven by dipoles, we can calculate the dispersion function in the same way we computed the β-function. Let's find the dispersion at the middle of the F-quadrupole, so we have a magnetic arrangement (with B denoting a dipole)

$$\frac{\text{QF}}{2} \cdot \text{B} \cdot \text{QD} \cdot \text{B} \cdot \frac{\text{QF}}{2}. \tag{5.146}$$

Looking at only the horizontal motion we find the one-cell map can be constructed from half-quadrupole maps, full quadrupole maps, and the thin-lens dipole map. Multiplying these five matrices together,

$$M = \begin{pmatrix} 1 & 0 & 0 \\ -\frac{1}{2f} & 1 & 0 \\ 0 & 0 & 1 \end{pmatrix} \begin{pmatrix} 1 & L & L\theta/2 \\ 0 & 1 & \theta \\ 0 & 0 & 1 \end{pmatrix} \begin{pmatrix} 1 & 0 & 0 \\ \frac{1}{f} & 1 & 0 \\ 0 & 0 & 1 \end{pmatrix} \times$$
$$\begin{pmatrix} 1 & L & L\theta/2 \\ 0 & 1 & \theta \\ 0 & 0 & 1 \end{pmatrix} \begin{pmatrix} 1 & 0 & 0 \\ -\frac{1}{2f} & 1 & 0 \\ 0 & 0 & 1 \end{pmatrix}, \tag{5.147}$$

we arrive at

$$M = \begin{pmatrix} 1 - \frac{L^2}{2f^2} & 2L(1+\frac{L}{2f}) & 2L\theta(1+\frac{L}{4f}) \\ -\frac{1}{2f}(1-\frac{L^2}{2f^2}) & 1 - \frac{L^2}{2f^2} & 2\theta(1-\frac{L}{4f}-\frac{L^2}{8f^2}) \\ 0 & 0 & 1 \end{pmatrix}. \tag{5.148}$$

We have left the matrix multiplication to the dear reader. Here L is the length of each dipole, θ is the bend angle and f is the quadrupole focal length. The upper 2×2 was obtained before, and now we have information on the dispersion.

The dispersion in the middle of the focusing quadrupole D_F and its gradient D'_F must satisfy the closed-orbit condition,

$$\begin{pmatrix} D_F \\ D'_F \\ 1 \end{pmatrix} = M \cdot \begin{pmatrix} D_F \\ D'_F \\ 1 \end{pmatrix} \tag{5.149}$$

which leads us to

$$D_F = \frac{L\theta(1 + \frac{1}{2}\sin\frac{\phi}{2})}{\sin^2\phi/2}, \tag{5.150}$$

and $D'_F = 0$ at the symmetry point in the middle of the quadrupole. The dispersion in the middle of the defocusing quadrupole can be found by transforming the dispersion to the middle of this quadrupole.

We've seen how to combine alternating gradient quadrupoles to make a focusing structure in both planes. This is called the FODO cell and is an example of a basic optical building block we use to construct lattices. There are many possible configurations of dipoles and quadrupoles that can give stable motion. We can talk about dispersion-free lattices, which are important in many applications. These allow bending of the beam without generating additional dispersion (known as an achromat). Examples are the Chasman-Green structure, triple-bend achromat. We also can build dispersion suppressors, which match the periodic dispersion in the arc (perhaps made of FODO cells) into a dispersion-free straight. We can also displace the beam transversely without generating dispersion using a sequence of only bends. Sometimes called a geometrical achromat.

Let's look at achromats in more detail. Consider a simple double-bend achromat (DBA) cell with a single quadrupole in the middle of two dipoles that bend in the same direction. The role of the quadrupole is to focus the dispersion halfway through the structure and allow it to be closed (i.e. set to zero at the dipole exit) by the second dipole. We use the thin lens approximation and write down the dispersion matching condition, i.e. we expect some dispersion D_c in the middle of the quadrupole and feed into the system zero dispersion

$$\begin{pmatrix} D_c \\ 0 \\ 1 \end{pmatrix} = \begin{pmatrix} 1 & 0 & 0 \\ -\frac{1}{2f} & 1 & 0 \\ 0 & 0 & 1 \end{pmatrix} \begin{pmatrix} 1 & L_1 & 0 \\ 0 & 1 & 0 \\ 0 & 0 & 1 \end{pmatrix} \begin{pmatrix} 1 & L & L\theta/2 \\ 0 & 1 & \theta \\ 0 & 0 & 1 \end{pmatrix} \begin{pmatrix} 0 \\ 0 \\ 1 \end{pmatrix}. \tag{5.151}$$

Here f is quadrupole focal length, θ and L are the bend parameters and L_1 is the distance between the quadrupole and bend centres. In essence we match to the $D_c' = 0$ condition at the middle of the quadrupole, i.e. the quadrupole turns over the sign of the dispersion generated by the bend and the dispersion is a maximum in the quadrupole centre. The required focal length is

$$f = \frac{1}{2}\left(L_1 + \frac{1}{2}L\right) \tag{5.152}$$

and resulting D_c is hence

$$D_c = \left(L_1 + \frac{1}{2}L\right)\theta. \tag{5.153}$$

Note the dispersion at the quadrupole becomes higher for longer distances and bigger bend angles. This analysis shows what is possible, but in practice we need extra quads for matching and maybe a reduction of the required quad strength by splitting the central quad.

The optical functions (β, α, γ) for a vertical double-bend achromat (DBA) with a quadrupole triplet between them are shown in Fig 5.16. Note the bending is done in the vertical plane and the structure is achometic in this plane. The horizontal dispersion seen in this figure is pre-existing to the achomatic structure. This figure is taken from the lattice of the LHeC collider [13, 14] and shows the action of the triplet to focus the vertical dispersion from the same-sign dipoles.

5.7.3 Momentum Compaction

We have seen that a momentum offset changes the horizontal orbit of a particle through dispersion if we have horizontal bending. Ideally, a machine with only horizontal bends does not generate any vertical dispersion. However, dispersion does generate a longitudinal effect, as the total circumference of an off-momentum particle's trip around the machine will be different to the reference particle. This matters for synchronisation and for longitudinal dynamics. What is this circumference, or path length, error? Consider the situation in Figure 5.17. The path length in this dipole for the ideal particle is given by $\rho\theta$, and the path length for a particle at radius $\rho + x$, where x can come from any source, is $(\rho + x)\theta$. Hence the path length change due to the particle not being on the design orbit is

$$\Delta C = (\rho + x)\theta - \rho\theta = x\theta. \tag{5.154}$$

The change in circumference of the machine, made up of lots of dipoles, is given by an integral over the whole ring

$$\Delta C = \oint \frac{x_{CO}(s)}{\rho(s)} ds, \tag{5.155}$$

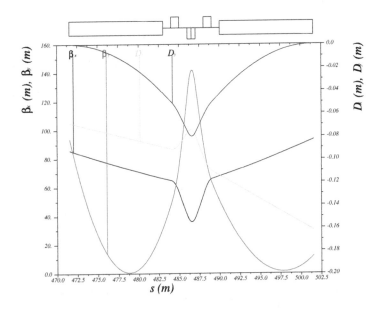

FIGURE 5.16 A double-bend vertical achromat (DBA) structure from a real lattice, showing the optical functions through the structure. The diagram at the top shows the lattice, with blocks for the two vertical dipoles and squares above and below the axis for quadrupoles. Note the central defocusing quadrupole (square below the axis) turns over the labelled vertical dispersion D_y. Also note the action of the quadrupoles on the β-functions.

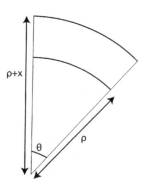

FIGURE 5.17 The origin of momentum compaction, showing the longer orbit travelled at the large radius.

where we know the closed-orbit distortion around the ring $(x_{CO}(s))$. For the case where the closed-orbit distortion is given by a momentum error, we can say

$$\Delta C = \delta \oint \frac{D(s)}{\rho(s)} ds, \tag{5.156}$$

and so the difference in circumference is proportional to the momentum deviation. Note this is because we work with the linear dispersion and in reality the closed-orbit distortion will also depend on higher powers of δ. So we define the *linear* momentum compaction factor

$$\alpha_c = \frac{1}{\delta} \frac{\Delta C}{C} \tag{5.157}$$

or

$$\frac{\Delta C}{C} = \alpha_c \delta. \tag{5.158}$$

In general, we then have an integral around the ring to compute the momentum compaction factor,

$$\alpha_c = \frac{1}{C} \oint \frac{D(s)}{\rho(s)} ds, \tag{5.159}$$

because a ring has many sources of path length deviation. The momentum compaction factor is an important lattice design parameter. A large value means the path length varies a lot for off-momentum particles. This means the particles tend to spread out and the bunch length becomes long. Similarly, a small value means a shorter bunch length. Typically $\langle D \rangle > 0$, so the particles tend to orbit on the outer side of the ring.

In this section we have looked at trajectory changes that depend linearly on the momentum deviation, so that $x = D(s)\delta$. In general we can have an arbitrary dependence of the transverse position on the momentum deviation, and write

$$x = D_1 \delta + D_2 \delta^2 + ... \tag{5.160}$$

where D_1 is the linear dispersion (the kind we have discussed in this chapter so far) and D_2 is called non-linear dispersion, or second-order dispersion. We shall discuss these kinds of ideas more in the section on non-linear dynamics shortly.

5.7.4   Chromaticity

We have seen that dipoles cause orbit changes to particles due to their spread of momentum. This is dispersion. Now let's think about focusing errors due to these off-momentum particles in quadrupoles. Consider some particles of slightly different energy passing through a quadrupole, as shown in Figure 5.18.

Higher-momentum particles have a greater beam rigidity than the reference particle, and so are deflected less when passing through a fixed magnetic field. This means focusing is momentum-dependent and the particle's focusing will change with momentum. Similarly, a lower-momentum particle will be overfocused by the quadrupole field. This means the machine's β-function and tune will depend on momentum deviation. This effect is referred to as chromaticity. If the machine tunes depend on the momentum deviation, we can write linearly in δ

$$\nu_{x,y} = \nu_{x,y}(0) + \xi_{x,y}\delta \tag{5.161}$$

where we've defined the linear chromaticity for each plane $\xi_{x,y}$. Non-linear chromaticity is an obvious extension, giving shifts to the machine tune dependent on δ to higher powers. This is a topic for a more advanced treatment, but for now it's good to know of its existence.

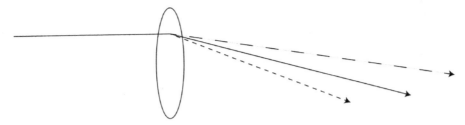

FIGURE 5.18 Different focusing of a quadrupole lens. The solid ray is the nominal ray, for which the quadrupole field is designed. The short-dashed ray is overfocused and the long-dashed ray is underfocused, corresponding to too little and too much momentum respectively.

To analyse linear chromaticity we return to the equations of motion, but this time keeping all terms containing x and δ. We proceed in the same way as we've done before, but when we expand the various terms, we keep the term $x.\delta$ we previously dropped. This generates a chromatic term in our equations of motion

$$x''(s) + \left(k_x(s) + \frac{1}{\rho(s)^2}\right) = \frac{\delta}{\rho} + \left(\frac{2}{\rho^2} + \frac{g}{B\rho}\right)x\delta \qquad (5.162)$$

where we defined as usual

$$k_x = \frac{g}{B\rho} + \frac{1}{\rho^2}. \qquad (5.163)$$

We can think of these chromatic terms as a quadrupole field error of strength

$$\Delta K_x = -\left(\frac{2}{\rho^2} + \frac{g}{B\rho}\right)\delta. \qquad (5.164)$$

A similar analysis in the vertical plane would have found a chromatic perturbation of

$$\Delta K_y = \frac{g}{B\rho}\delta. \qquad (5.165)$$

We already know how to compute the effect of a quadrupole field error. Recall the tune shift from a quadrupole error $k(s)$ in our lattice

$$\Delta\nu = \frac{1}{4\pi} \oint ds\beta(s)k(s), \qquad (5.166)$$

which means we can write down the tune-shift arising from the chromatic perturbation term,

$$\Delta\nu = \frac{1}{4\pi} \oint ds\beta(s)(-1)\left(\frac{2}{\rho^2} + \frac{g}{B\rho}\right)\delta. \qquad (5.167)$$

This expression is linear in the momentum deviation and gives us the tune shift for this focusing error. It is conventional to define the horizontal tune change per unit δ as the horizontal chromaticity

$$\xi_x = -\frac{1}{4\pi} \oint ds\beta(s)\left(\frac{2}{\rho^2} + \frac{g}{B\rho}\right). \qquad (5.168)$$

We call the chromaticity 'natural' as it arises from the quadrupoles which make up the lattice. Any lattice with quadrupoles naturally generates this chromaticity. Similarly, in the vertical plane,

$$\xi_y = -\frac{1}{4\pi} \oint ds\beta(s)\frac{g(s)}{B\rho}. \qquad (5.169)$$

The horizontal β-function is biggest in horizontally focusing quadrupoles (and vice versa), so the natural chromaticity is normally negative in both planes. The linear chromaticity Q' is sometimes written as the linear change in the tune

$$\Delta Q = Q'\delta. \tag{5.170}$$

For a FODO cell we can show that

$$\xi_x = -\frac{\beta_F - \beta_D}{4\pi f}, \tag{5.171}$$

which is a very useful expression and is proportional to the difference in β-functions at the F and D quadrupoles. Chromaticity is naturally generated by any focusing lattice, so when we have non-zero k we have chromaticity and it tends to be negative in both planes x and y.

The optics of the LHC long straight section were shown in Figure 5.11. The chromaticity generated in the strong quadrupoles increases with the β-function, and so large chromaticity is generated in the quadrupoles around the LHC's interaction point. This is an unavoidable consequence of the mini-beta layout.

The chromaticity number tells us how much the tune shifts for a unit shift in the momentum deviation ($\Delta p/p = 1$). So given the beam has an energy spread, it tells us the spread of the tune of the beam. Tune is a finite region in tune space. If we measure the beam's frequency spectrum by a pick-up device and perform a Fourier analysis, we'll see spikes at the fractional part of the tune, and the width of the spike will give an estimate of the chromaticity.

How do we correct chromaticity? Well, it basically comes about when a particle which is slightly off-momentum sees a different quadrupole field than it should and this particle is focused differently from the others. So in essence we need a correcting device which has some kind of transverse position-dependent focusing. A sextupole! A sextupole field has field components given by

$$B_x = Sxy \;,\, B_y = \frac{S}{2}(x^2 - y^2), \tag{5.172}$$

where S defines the sextupole strength, $\mathrm{d}^2 B_y/\mathrm{d}x^2$. Note the field is quadratic in x and y, and also (for the first time) we see products of x and y in our equations, known as coupling. A sextupole couples the beam planes. An off-momentum particle passing through the sextupole has displacement

$$x = x_\beta + D\delta, \tag{5.173}$$

with $y = y_\beta$ in the vertical plane. And so the fields seen by the particle are found by substitution

$$\begin{aligned} B_x &= S(x_\beta + D\delta)y_\beta \\ &= Sx_\beta y_\beta + SD\delta y_\beta \end{aligned} \tag{5.174}$$

and

$$B_y = \frac{S}{2}(x_\beta^2 - y_\beta^2) + SD\delta x_\beta + \frac{S}{2}D^2\delta^2. \tag{5.175}$$

There are many terms here, some helpful and some harmful. The helpful ones for us are

$$\begin{aligned} B_x &= SD\delta y_\beta \\ B_y &= SD\delta x_\beta, \end{aligned} \tag{5.176}$$

where the horizontal dispersion function has made each sextupole into a quadrupole with an effective gradient $S.D.\delta$. We can use these to cancel the natural chromaticity in the lattice and cancel the chromatic tune shift. But it's not all perfect. Remember we ignored plenty of terms in the fields of the sextupoles; some of the terms are good and fix our chromaticity, but some are bad and introduce non-linearities and coupling into our accelerator ring. These terms can harm the beam. It is not possible to represent sextupoles in our linear formalism, and often the best way to understand the impact of sextupole fields is to track particles with matrices, and stopping to be more careful every time a sextupole is encountered. This leads to the study of a machine's dynamic aperture, or what amplitude of particle can survive for many turns. To get stable solutions for the off-momentum particle, we need to put sextupole magnets and RF cavities in the lattice beam line. Such nonlinear elements induce nonlinear beam dynamics and the dynamic acceptances in the transverse and longitudinal planes need to be carefully studied in order to get sufficient tolerance or acceptance (for long beam current lifetime and high injection efficiency). For the modern high-performance machines, strong sextupole fields to correct high chromaticity have large impact on the nonlinear beam dynamics and this is one of the most challenging lattice design issues to deal with. In the real machine, there are always imperfections in the accelerator elements. So, one also needs to consider engineering and alignment limitations or errors, component vibrations, and so on. Correction schemes such as orbit correction and coupling correction need to be developed, involving elements such as dipole correctors, skew quadrupoles and beam position monitors.

So, to close, how can we measure the chromaticity? Generally in science we change something to measure it and so we change the beam momentum and make a linear fit of the tune. For more details see [3].

5.8 Beams of Many Particles, and Emittance

So far we've defined the single-particle emittance A (or action) of a particle,

$$x(s) = \sqrt{2A\beta(s)}\cos(\psi(s) + \psi_0), \tag{5.177}$$

which was the constant in our Courant-Snyder analysis and defines the amplitude of the motion. The motion of an individual particle is then completely specified by its single-particle emittance A and by its initial phase ψ_0. Different particles will have different single-particle emittances and initial phases but they all have the same Courant-Snyder functions, at least for beams with no momentum spread. Therefore, each particle has its own invariant ellipse, with areas fixed by its value of A, around which the particle slides as it moves through the lattice. This is what we saw earlier in this chapter. The particle with $x = x' = 0$ has zero emittance and always stay at $x = x' = 0$. This is called the 'ideal particle', and does not exist in practice. Before we worry about definitions of whole beam emittance (or simply called emittance), let's think more about single particle motion. Dropping the initial phase we get

$$x(s) = \sqrt{2A\beta(s)}\cos(\psi(s)). \tag{5.178}$$

Now recall we wrote the invariant equation as

$$x^2 + (\beta x' + \alpha x)^2 = 2A\beta, \tag{5.179}$$

which can be interpreted in the $(x, \beta x' + \alpha x)$ plane as a circle of radius $\sqrt{2A\beta}$. So if we use these coordinates, particles move on circles in this plane. This can be a very useful concept. If we have the challenge of representing a beam of particles, containing many different values of A, we can now see a possible definition of the overall beam emittance –

we choose one circle, corresponding to one particle, to represent the beam, which includes a certain fraction of all the particles in the beam. If we transform back to an ellipse, this representative ellipse plays the same role. This assumes some distribution of particles in phase space, which we shall come back to and the precise analytical distribution depends on how the beam is prepared or stored and can often be taken to be a simple analytic form.

So, in other words, we always have more than one particle in our beam and so, generally, need to understand how to characterise a beam of particles, each with their own value of A. We can choose one of the particle's emittances to represent the emittance of the entire beam or choose some other number to represent the beam as a whole. For example, 68% of all particles, or 95%, or some definition based on the typical value. This one number is called beam emittance, or emittance.

Very often an expression for the emittance based purely on knowledge of the particle distribution is useful. As an example, when we make simulations we have access to all the positions and angles in the beam and so we can define the RMS emittance as

$$\epsilon_{\text{rms}} = \sqrt{\langle x^2 \rangle \langle x'^2 \rangle - \langle xx' \rangle^2}. \tag{5.180}$$

In this expression we have defined the beam distribution moments as integrals over the particle density $\rho(x, x')$ as

$$
\begin{aligned}
\langle x \rangle &= \int x \cdot \rho(x, x') \mathrm{d}x \mathrm{d}x', \\
\langle x' \rangle &= \int x' \cdot \rho(x, x') \mathrm{d}x \mathrm{d}x', \\
\langle x^2 \rangle &= \int (x - \langle x \rangle)^2 \cdot \rho(x, x') \mathrm{d}x \mathrm{d}x', \\
\langle x'^2 \rangle &= \int (x' - \langle x' \rangle)^2 \cdot \rho(x, x') \mathrm{d}x \mathrm{d}x', \\
\langle xx' \rangle &= \int (x - \langle x \rangle)(x' - \langle x' \rangle) \cdot \rho(x, x') \mathrm{d}x \mathrm{d}x'. \tag{5.181}
\end{aligned}
$$

Note that we can write these expressions as sums over a finite number of particles N in a form such as

$$\langle x^2 \rangle = \frac{1}{N} \sum_{i=1}^{N} x_i^2, \tag{5.182}$$

for when we deal with numerical representations of particle beams (for example, in a beam simulation).

How does this relate to our definition of the single-particle emittance A? We have defined $x(s)$ for a single particle, and so its derivative is (where α is the Courant-Snyder parameter)

$$x'(s) = -\sqrt{\frac{2A}{\beta(s)}} \left(\cos(\psi(s)) + \alpha \sin(\psi(s)) \right). \tag{5.183}$$

We have already observed that this corresponds to the particle moving around an ellipse as the coordinates (x, x') evolve, with the area of the ellipse being specified by A. This motion can also be understood in terms of two alternative variables to (x, x'), namely the size of the ellipse the particle moves around, A, and the angle around the ellipse, ψ. This is entirely equivalent to x and x'. If we use (A, ψ) to describe the particle, then the transformation linking the two descriptions is

$$x = \sqrt{2A\beta}\cos(\psi)$$

$$x' = -\sqrt{\frac{2A}{\beta}}\left(\cos(\psi) + \alpha\sin(\psi)\right). \tag{5.184}$$

The variables A and ψ are known as action-angle variables, as we know very well now, is a conserved quantity for a given particle. We can expect our beam to have a uniform range of values of ψ so the average value of a collection of particles, each with their own value of A would be

$$\langle x^2 \rangle = 2\beta(s)\langle A\cos(\psi)\rangle. \tag{5.185}$$

Here the angular brackets mean we average over all particles in the bunch. If all the particle angles are randomly distributed and uncorrelated with A, then we can write

$$\langle x^2 \rangle = \beta(s)\langle A \rangle \tag{5.186}$$

or, defining $\langle A \rangle = \epsilon$,

$$\langle x^2 \rangle = \beta(s)\epsilon. \tag{5.187}$$

Similarly, we can use the derivative of our expression for x to obtain $\langle xx' \rangle$ and $\langle x'^2 \rangle$, obtaining

$$\langle xx' \rangle = -\alpha(s)\epsilon \tag{5.188}$$

and

$$\langle x'^2 \rangle = \gamma(s)\epsilon. \tag{5.189}$$

Combining these we obtain our expression for the beam emittance ϵ in terms of our beam moments,

$$\epsilon_{\text{rms}} = \sqrt{\langle x^2 \rangle\langle x'^2 \rangle - \langle xx' \rangle^2}. \tag{5.190}$$

The RMS emittance of a beam is useful because we can simply sum over the coordinates of known particles and it coincides with the single-particle emittance of a beam in a circular machine (with all particles sitting on an ellipse with the same single-particle emittance).

Now imagine we had a beam distribution in A and ψ_0, which is just a collection of particles in our machine. For example, imagine we had a bunch of particles uniformly distributed in ψ_0 and Gaussian distributed in (x, x'). The link between A and (x, x') is given by

$$\gamma x^2 + 2\alpha xx' + \beta x'^2 = 2A, \tag{5.191}$$

and so we can write for the particle density

$$\begin{aligned}
\Psi(x, x') &= \frac{1}{N}\exp(-\frac{\epsilon}{\epsilon_{\text{rms}}}) \\
&= \frac{1}{N}\exp\left(-\frac{\gamma x^2 + 2\alpha xx' + \beta x'^2}{2\epsilon_{\text{rms}}}\right) \\
&= \frac{1}{N}\exp\left(-\frac{x^2 + (\alpha x + \beta x')^2}{2\beta\epsilon_{\text{rms}}}\right).
\end{aligned} \tag{5.192}$$

We can fix the normalisation by requiring

$$\int_{-\infty}^{\infty} dx \int_{-\infty}^{\infty} dx' \, \Psi(x, x') = 1 \tag{5.193}$$

to obtain

$$\Psi(x, x') = \frac{1}{2\pi\epsilon_{\text{rms}}} \exp\left(-\frac{x^2 + (\alpha x + \beta x')^2}{2\beta\epsilon_{\text{rms}}}\right). \tag{5.194}$$

If we perform the integration for the second moments of the beam distribution, defined as above, we now obtain

$$\begin{aligned}
\langle x^2 \rangle &= \beta\epsilon_{\text{rms}}, \\
\langle xx' \rangle &= -\alpha\epsilon_{\text{rms}}, \\
\langle x'^2 \rangle &= \gamma\epsilon_{\text{rms}}.
\end{aligned} \tag{5.195}$$

To obtain an expression for the average beam emittance in terms of these quantities, we use

$$\gamma x^2 + 2\alpha xx' + \beta x'^2 = 2A, \tag{5.196}$$

and, taking averages, we obtain

$$\begin{aligned}
2\langle\epsilon\rangle &= \gamma\langle x^2\rangle + 2\alpha\langle xx'\rangle + \beta\langle x'^2\rangle \\
&= \epsilon_{\text{rms}}(2\gamma\beta - 2\alpha^2) \\
&= 2\epsilon_{\text{rms}}.
\end{aligned} \tag{5.197}$$

And so we find our RMS definition of the emittance to be the same as the average value of our single-particle emittances,

$$\langle\epsilon\rangle = \epsilon_{\text{rms}}. \tag{5.198}$$

We can show that the ellipse with a single-particle emittance of ϵ_{rms} corresponds to 68% of the particles in the beam. This is left as a very instructive exercise for the reader.

Finally we close this section with the concept of normalised emittance. The beam emittance we have discussed in this section is also known as the geometric emittance ϵ. If we increase the momentum of the beam (i.e. via an acceleration process) then the transverse velocities remain constant whilst the longitudinal velocities increase; the emittance of the beam reduces $\propto 1/\beta\gamma$ where β and γ are the usual relativistic parameters. This process is known as adiabatic damping [3]. It is useful to introduce the normalised emittance ϵ_N, defined as

$$\epsilon_N = \beta\gamma\epsilon. \tag{5.199}$$

In the absence of other processes, ϵ_N remains constant under acceleration and does not depend on the momentum of the beam; in the case of high-energy electrons where $\beta \simeq 1$ to a good approximation, we have $\epsilon_N = \gamma\epsilon$.

5.9   Longitudinal Dynamics

In this section we shall explore some dynamics of the longitudinal plane, focusing on longitidinal stability from a beam dynamics perspective. The detailed discussion of RF cavities and their fields can be found in Chapter 3.

So far we have studied transverse motion using (x, x') and (y, y'), so motion has been 4D and purely in the transverse planes. Now we need to study the remaining direction, involving the coordinates in the longitudinal direction. This is called synchrotron motion, and we need to worry about energy gain, longitudinal stability and how we focus in accelerating structures. In analogy with our study of transverse motion, we could expect to use s and s' as the longitudinal coordinates, and proceed in much the same way. In fact, instead of s', we use the momentum deviation δ or the energy deviation. But it makes no fundamental difference.

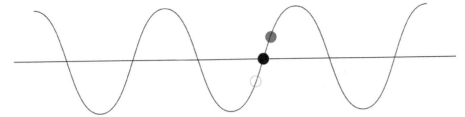

FIGURE 5.19 The RF waveform and particles arriving either at the design time in the waveform, early, or late. The system is designed so that the design particle sees for zero field and an increase in particle energy decreases the arrival time at the next cavity. The filled particle is the synchronous particle, the grey particle arrives too late and the unfilled particle arrives too early.

All accelerators, or at least what are now known as conventional accelerators, use radio-frequency cavities to accelerate. These cavities also provide longitudinal stability, as we shall see, and this is very important in accelerator design and operation. The RF field varies sinusoidally in time, hence only particles arriving at the correct time will get the design acceleration. A real bunch of charged particles has a finite bunch width and hence some of the particles will arrive too early or too late (possibly due to having too much or too little momenta), and hence will experience a different accelerating voltage than the centre of the bunch, dependent on the arrival phase of the bunch. In designing a machine we will choose the phase at which the particles will arrive at the cavity, known as the synchronous phase, ϕ_s.

In the linac, the phase is defined relative to the maximum of a harmonic voltage, and so $\phi_s = 0$ corresponds to maximum acceleration (known as being on-crest). This is because linacs are generally operated close to the maximum in voltage. A different definition is generally used in a circular machine, with $\phi_s = 0$ corresponding to a minimum in the harmonic voltage. Hence $\phi_s=0$ provides zero acceleration.

Taking the linac definition, if the synchronous phase is between 0 and $\pi/2$, then particles arriving late will get more acceleration, and late particles will get more acceleration. The opposite is true if the synchronous phase is between $-\pi/2$ and 0, where early particles will get more acceleration. This is shown in Figure 5.19, where the circles represent particles arriving at different times or phases, for a synchronous phase of $\pi/2$ in the linac definition.

Now we come to a very important principle, and one which makes accelerators work – the principle of phase stability. In order to achieve stable acceleration we would want the time it takes for particles to reach the next RF cavity (or return to the same RF cavity in a synchrotron) to be slightly longer for early particles or shorter for late particles. This will provide a restoring force to particles towards the bunch centre and ensure that particles do not slip in phase so much that they are no longer synchronous with the RF. In the next two sections we shall see how we can achieve this for circular machines and then in linear machines.

5.9.1 Longitudinal Dynamics in Circular Machines

One way of proceeding would be to define longitudinal lattice functions, in perfect analogy to our studies of transverse beam dynamics, and this can be done. However, synchrotron motion is very slow compared to transverse motion and this approach is not the most natural way to do things. However, we can define a synchrotron tune, which turns out to be much less than the transverse tunes and we can write

$$\nu_{x,y} \gg \nu_s, \tag{5.200}$$

where ν_s is our longitudinal, or synchrotron, tune. Because the motion is so slow, we can ignore the s-dependent effects around the ring, and avoid a longitudinal Courant-Snyder formalism.

In circular machines, to get the same magnitude and phase of accelerating field every time a particle travels around the ring and returns to the cavity, the RF frequency ω has to be an integer multiple h of the revolution frequency ω_0

$$\omega_{\mathrm{RF}} = h\omega_0, \tag{5.201}$$

so the beam always sees the correct accelerating field and gains the correct amount of energy. In these equations h is known as the harmonic number. But what if h is slightly wrong, i.e. h=110.0000000001 instead of 110 exactly? Then the next turn the field seen by the particle will be slightly different than what is needed. Then, after many turns, the beam will be increasingly out of phase with the RF system and will no longer be accelerated. We surely need to be tolerant to very small errors in the frequencies as a beam is made up of particles with a spread of phases. This was resolved by the very important principle of phase stability, discovered independently by Edwin McMillan and Vladimir Veksler in 1945 [15, 16]. For stable motion we choose our RF frequency, which fixes the synchronous particle. Now, particles with slight deviation in longitudinal coordinates will oscillate (albeit slowly) around this synchronous particle.

Our cavity is designed to generate a time-dependent longitudinal electric field to transfer energy from this field to the particle, as we discussed in Chapter 3. The RF voltage applied to the particle is sinusoidal in time,

$$V(t) = V_0 \sin \omega_{\mathrm{RF}} t, \tag{5.202}$$

and if we pick the RF frequency to be an integer multiple of the revolution frequency the beam sees the same voltage every time it crosses the cavity. This is called synchronism, and can be written as $\omega_{\mathrm{RF}} = h\omega_0$. So now the cavity is set up so that the particle at the longitudinal centre of the bunch, called the synchronous particle, acquires just the right amount of energy and it sees the same voltage each turn.

$$V(t) = V_0 \sin(\omega_{\mathrm{RF}} t + \phi_0) = V_0 \sin(\phi_s t). \tag{5.203}$$

In the case of no acceleration, the synchronous particle has $\phi_s = 0$, and so it sees a zero of the harmonically varying voltage. Referring to Figure 5.19, consider now another particle arriving at some other phase ϕ. If a particles arrives early, it sees too little voltage, so that $\phi < \phi_s$, and if a particles arrives late it sees too much voltage, so that $\phi > \phi_s$. If we want to accelerate, we choose $0 < \phi_s < \pi$ so that a synchronous particle gains energy on each turn of the machine.

Let's consider our ring, for which the synchronism condition is fulfilled for a phase ϕ_s. This could be accelerating or not, as it doesn't matter here. Consider the sinusoidal RF waveform in Figure 5.20.

The effect of the slope in the voltage function depends on the particles energy. There are two effects to consider, the particles velocity and the path the particle takes around the ring in the accelerator's dipole and quadrupole fields. At low particle energies (compared to the rest energy) the dominant effect is the change in the particle's velocity, and hence the revolution time, while at high energies where the particle is travelling close to the speed of light the change in momentum causes the particle to have a larger or smaller bending radius and so the particle with higher energy will take a longer oscillating path around the ring.

First, let's take the low-energy case. Let eV_s be the energy gain in one cavity for the particle to reach the next cavity with the same RF phase. The points in (energy, phase) space

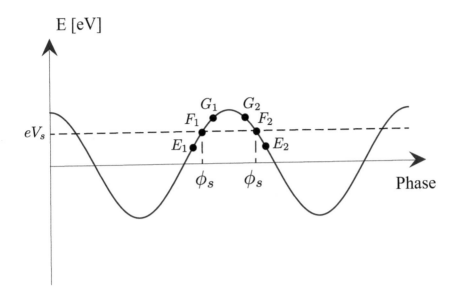

FIGURE 5.20   The RF waveform, showing stable and unstable fixed points.

where this happens are called fixed points, here F_1 and F_2. This is shown in Figure 5.20. Imagine a particle arrives a little later than the synchronous particle. So it sees a slightly later phase of the RF waveform. This is the point G_1. This means it gets a larger energy kick, so has a higher velocity and gets around the ring faster. This means it arrives slightly earlier than it did, and hence moves towards the fixed point F_1.

Similarly, an early particle will see E_1, get a smaller kick and move towards F_1. Hence F_1 is a stable fixed point. Therefore an increase in energy is transferred into an increase in speed, hence a quicker time to the next cavity. E_1 and G_1 will move towards F_1. This is the principle of phase stability. This means the particles oscillate around the synchronous phase, and have a natural spread of momenta.

Let's play the same game for the other fixed point F_2. Here, if we follow the same logic we find the points E_2 and G_2 move away from F_2. Hence we call F_2 an unstable fixed point. So we can classify fixed points as either stable or unstable.

So this works if an increase in energy becomes a decrease in time to the next cavity. What happens if the particle is moving at the speed of light? This means that gaining energy does not increase its speed. For this case, higher energy translates into a longer revolution time, which means F_2 becomes a stable point and F_1 becomes an unstable fixed point. Now E_1 and G_1 will move away from F_1 (which is now unstable), while E_2 and G_2 will go towards F_2 (which is now stable). This arises because particles with lower energy move on an inner dispersive orbit, with a lower revolution time.

So, the stability behaviour changes as the particles accelerate and become relativistic and this change of behaviour – when F_1 and F_2 swap between stable and unstable points is called transition.

Let's look at this more carefully now. Particles with different momenta travel on different paths and we know the revolution time T depends on the circumference, C, taken by a particle and its speed, v,

$$T = \frac{C}{v}. \qquad (5.204)$$

The fractional revolution frequency for a slightly different circumference and speed is therefore given by

$$\frac{\Delta f}{f} = -\frac{\Delta T}{T} = -\frac{\Delta C}{C} + \frac{\Delta v}{v}. \tag{5.205}$$

What this means is the particle arrival time is affected both by a longer path around the machine and also by the particle moving faster. We can relate both these contributions to the fractional momentum deviation,

$$\frac{\Delta f}{f} = -\left(\alpha_c - \frac{1}{\gamma^2}\right)\delta = -\eta\delta. \tag{5.206}$$

Here we have defined the phase slippage factor

$$\eta = \alpha_c - \frac{1}{\gamma^2} = \frac{1}{\gamma_T^2} - \frac{1}{\gamma^2}. \tag{5.207}$$

In this discussion we have used the useful equations

$$\frac{\Delta v}{v} = -\frac{1}{\gamma^2}\frac{\Delta p}{p} \tag{5.208}$$

and

$$\frac{\Delta C}{C} = \alpha_c \frac{\Delta p}{p}. \tag{5.209}$$

The quantity γ_T is called the transition gamma and is related to the momentum compaction factor of the lattice through

$$\gamma_T = \frac{1}{\sqrt{\alpha_c}}. \tag{5.210}$$

Below the transition energy we have $\gamma < \gamma_T$ and so $\eta < 0$. So a higher-momentum particle has a revolution time shorter than that of the synchronous particle and so makes a single turn back to the cavity in a shorter time. This means our fixed point F_1 is stable and F_2 is unstable.

Above the transition energy we have $\gamma > \gamma_T$ and so $\eta > 0$. Now the opposite is true. Higher-momentum particles have a revolution time greater than that of the synchronous particle. This means our fixed point F_1 is unstable and F_2 is stable. At the transition energy the machine is isochronous (same revolution time) for all momenta and all particles circulate with the same period. This is $\eta = 0$. The point of transitioning from below transition to above transition is a dangerous time for the machine, as longitudinal confinement is briefly lost and the RF phase suddenly has to jump from one stable region to another.

Now that we have stable regions in longitudinal phase space, we can start to study the longitudinal dynamics. The derivation of the longitudinal oscillations in a circular machine is beyond the scope of this introductory book, but let's sketch some important ideas. If we give the parameters of the cavity we can compute the motion in the longitudinal phase space. This is called a phase space portrait. In the transverse plane we used the variables x and x', which made sense for this motion, but in the longitudinal place it's more common to use the particle phase difference from the synchronous phase and the relative energy of the particle. We see regular motion around the stable fixed points, and unstable motion elsewhere. The dividing line between stable and unstable motion in this plot is known as the separatrix, shown in Figure 5.21. The separatrix starts from a point very close to (but not exactly at) the unstable fixed point, moves away and forms an 'alpha' or fish shape around the stable fixed point. The area of stable motion enclosed is called the bucket and there is one bucket per RF period. In the LHC the RF system oscillates at 400 MHz, the stable regions (buckets) are separated by 2.5 ns and we fill every 10th bucket with protons.

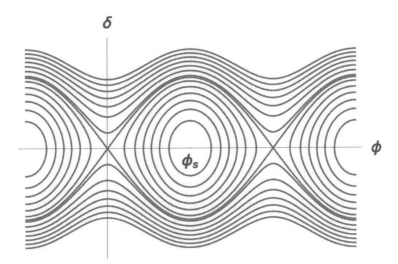

FIGURE 5.21 Longitudinal stable and unstable motion, written in terms of the variables (ϕ, δ). The synchronous phase is ϕ_s. The boundary between stable and unstable motion is called the separatrix.

A complete and quantitative discussion of this topic can be found in [3]. However we can learn something by sketching out the analysis. We proceed by looking at energy balance for one complete turn of the machine, balancing the energy gained by a particle when arriving at the cavity at some varying phase to the energy lost per turn by the particle as it moves around the ring. The change in the particle energy is the difference between these two quantities and this leads to a first-order differential equation for the rate of change of the particle's energy. We can also obtain a differential equation for the rate of change of the particle's arrival phase turn by turn, obtaining

$$\Delta\dot\phi = \frac{2\pi q}{\beta^2 T_0}\left(\alpha_c - \frac{1}{\gamma^2}\right)\cdot\frac{\Delta E}{E}. \tag{5.211}$$

In this equation q is the particle charge, T_0 the revolution time and β is the particles relative velocity. It is an equation for the rate of change of the arrival phase ψ in terms of the energy deviation $\Delta E/E$. What this says is that there are two ways for a particle to pick up a phase difference with respect to the synchronous particle, and both are related to the energy error with respect to this synchronous particle. The first term arises as the circumference of the machine for an off-energy particle is different than the design circumference. We learned all about this when we looked at the momentum compaction. The second term comes from the fact that an off-energy particle has a different speed than the design speed. Both terms are related to the energy deviation and have opposite sign in most cases. The relative size of the two terms determines if a machine is below or above transition. If we combine our equation for the rate of change of arrival phase with the equation for the change of energy we can obtain a second-order differential equation for the energy, and an expression for the synchronous frequency. For full details see [3].

Longitudinal Dynamics in Linacs

In linacs each bunch only passes through each cavity once and the bunches are almost always being accelerated with lots of RF cavities closely spaced together. Turning our attention to a linac, only the particle speed changes matter for longitudinal stability as there are no dipoles and hence no momentum compaction. This means that the subject of longitudinal dynamics is mostly concerned with proton and ion beams up to a few GeV, and the very start of electron linacs up to a few MeV (in the case of electrons they will become relativistic in the first few cells of the first RF cavity they see). The change in particle energy from cell-to-cell means that the relativistic β of the particle beam changes along the length of the linac. Therefore the length of the accelerating cells, numbered 1 to n, increases along the length of the linac. The approximate length of cell $n+1$, to have the particles enter this cell at the same phase as the previous cell, is given by

$$L_{n+1} = \frac{v_n \phi_a}{2\pi f} + \frac{dv_n}{dz}\frac{L_n \pi_a}{4\pi f}, \tag{5.212}$$

where v_n is the velocity of the particle entering cell $n+1$, and hence leaving cell n, with phase advance, ϕ_a, and f is the frequency. Replacing v_n with β_n, which is the ratio of the particle velocity to the speed of light of the particle entering cell $n+1$, we can find [17]

$$L_{n+1} = \frac{\beta_n}{\frac{2f}{c} - \frac{1}{2}\frac{d\beta}{dz}}. \tag{5.213}$$

We may choose to use a longer or shorter cell length to have the synchronous phase vary from cell to cell.

Many linacs that require short relativistic electron bunches will include magnetic chicanes, using four dipole magnets, in order to create a difference in path for high- and low-energy particles, such that a beam with a variation in energy along its length will experience momentum compaction but only inside the chicane. The velocity gradient, $d\beta/dz$, is related to the accelerating gradient and chosen as a compromise between peak electric fields and the length of the structure to reach 1 MeV. It should be noted that due to relativity, $d\beta/dz$ is not constant for a constant accelerating gradient and decreases to zero as the particles gain energy and the particle velocity tends to the speed of light.

Similarly to circular machines below γ_T, low-energy electron linacs and low to intermediate energy proton and ion linacs experience longitudinal bunching. In the case of linacs, the synchronous phase is chosen in the design of the linac, with the RF cell length chosen to vary along the linac with increasing particle velocity matched to the acceleration of a chosen synchronous particle at the desired design gradient. As mentioned previously a particle which arrives early will experience less acceleration and will fall behind, a particle arriving late will get more acceleration and catch up. It hence makes sense to again analyse longitudinal dynamics in terms of the longitudinal phase space discussed previously in this chapter, which is a consideration of particle energy versus particle time or phase with respect to the synchronous particle's energy and time/phase for motion in the linac. As in circular machines, if we take a particle with a displacement in phase or energy from the synchronous particle it will follow a path in longitudinal phase space, with stable particles following closed loops and unstable particles following continuous paths that slip from one RF cycle to the next. This interface in phase space is again known as the separatrix, shown in Figure 5.21. Not every RF bucket will necessarily be filled with particles in a linac as there may be other reasons to want to space bunches out in time as we will see in Chapter 7.

By analysing the Hamiltonian of the acceleration in a linac (assuming smooth continuous acceleration), we can define the maximum deviation from the synchronous particle at the synchronous phase and the maximum phase deviation at the synchronous energy while still providing stable acceleration [18]. The maximum energy difference, ΔK_{max}, also known as the energy acceptance, is given by

$$\frac{\Delta K_{max}}{mc^2} = \sqrt{\frac{2qE_{acc}\gamma_s^3\beta_s^3\lambda}{\pi mc^2}(\phi_s\cos\phi_s - \sin\phi_s)}, \tag{5.214}$$

where E_{acc} is the accelerating gradient, and λ is the wavelength of the RF, and q is the particle charge, and the subscript s donates the synchronous particle's properties, with ϕ_s being the synchronous phase, γ_s being the Lorentz factor for the synchronous particle, and β_s being the relative velocity of the synchronous particle. This implies that a synchronous phase of $\phi_s = 0$ gives no energy acceptance and hence would not be suitable choice of synchronous phase. For the maximum allowable phase deviation (the phase acceptance), for any given synchronous phase, we solve the motion for the case where $\Delta K_{max} = 0$ and find two solutions, one for either side of the synchronous phase, ϕ_1 and ϕ_2, for the early and late particles respectively. We find one solution is $\phi_1 = -\phi_s$ while ϕ_2 is given by

$$\sin \phi_2 - \phi_2 \cos \phi_s = \sin \phi_s - \phi_s \cos \phi_s. \tag{5.215}$$

For small ϕ_s we find the phase acceptance is from $-\phi_s$ to $2\phi_s$. We find that both the energy and phase acceptance of the linac increases with increasing ϕ_s. However, the accelerating gradient decreases as $\cos \phi_s$ hence we do not want to have ϕ_s too large. Typically synchronous phases of around $20°$ are chosen as a good compromise. The electrons will now oscillate with simple harmonic motion with a frequency, ω_l equal to

$$\omega_l^2 = \omega_0^2 \frac{q E_{acc} \lambda \sin(-\phi_s)}{2\pi m c^2 \gamma_s^3 \beta_s}, \tag{5.216}$$

where ω_0 is the RF frequency and the amplitude is dependent on the particle's initial deviation from the synchronous particle in energy and phase. As can be seen, the frequency of the synchrotron motion in linacs is energy dependent, due to the increase in the Lorentz factor, with the frequency decreasing with increasing beam momentum. For sufficiently high energies the oscillation period will be longer than the length of time the bunch takes to traverse the linac and hence can be neglected.

Let's now think a little about transverse dynamics in linacs. We clearly need some kind of transverse stability, as we don't want the accelerating particles drifting off to larger transverse position as they accelerate. In Chapter 3 we discussed the accelerating action of RF resonant cavities and we see that when a particle in a cavity feels a longitudinal electric field, it also feels transverse fields. This means the particle receives transverse momentum kicks. Note the kicks felt as the particle enters and leaves the cavity are different as the field changes in time as the particle moves through the cavity and vary with radius. To get a feel for the necessity of transverse forces, imagine we transform to the rest frame of the particle in the cavity and hence only worry about electrostatic forces. These are described by Laplace's equation for the potential V in 2 dimensions,

$$\frac{\partial^2 V}{dx^2} + \frac{\partial^2 V}{dz^2} = 0, \tag{5.217}$$

meaning it is impossible for both the transverse (x) and the longitudinal (z) to be focusing (a minimum in $V(x, z)$) at the same time. For a full analysis of the transverse focusing from the changing fields in a cavity see [19, 18]. For longitudinal stability we need a synchronous phase off-crest between 0 and $\pi/2$. However, operating at synchronous phases other than $0°$ causes the beam to have a transverse voltage component. In a perfect pillbox cavity the longitudinal electric field is constant along the length of the cavity, however the introduction of the beam-pipes causes the longitudinal electric field to vary along the length. Gauss's law states that if the longitudinal electric field has a longitudinal variation then there must also be a radial electric field that varies radially. The radial electric field coupled with the azimuthal magnetic field gives rise to a transverse force that is zero on the beam axis ($r = 0$), but due to the finite bunch radius, the edges of the bunch experience the transverse force. If the beam is accelerated at an RF phase of $0°$ (i.e. the electric field is maximum when the beam reaches the halfway point of the cavity), then the transverse force at the first half of the cavity exactly cancels the force in the second half of the cavity for relativistic particles where the Lorentz factor doesn't change significantly over a single cell. If however the linac is designed with a non-zero synchronous phase, then there will be an RF focusing or defocusing term as the forces at the cavity entrance and exit no longer perfectly cancel as the beam reaches them at different phases of the RF. If the synchronous phase is chosen to be longitudinally stable where early particles receive less acceleration, then the RF is radially defocusing, and vice versa. The RF defocusing force, F_r, is a function of beam radial offset, r, and is given by [20]

$$F_r(r) = -\frac{er}{2}\left(\frac{dE_z}{dz} - \left(\frac{1}{\beta c} - \frac{\beta}{c}\right)\frac{\partial E_z}{\partial \Phi}\right), \tag{5.218}$$

where E_z is the longitudinal electric field applying the acceleration, which is a function of radius, longitudinal position and RF phase Φ. Integration along the beam path, offset from the central axis by a distance r, for a pillbox cavity yields

$$\Delta p(r) = -\frac{er\pi E_0 LT \sin\Phi}{\gamma^2 \beta^2 c^2 \lambda}, \tag{5.219}$$

where λ is the RF wavelength, and E_0LT is the cavity voltage as defined in Chapter 3. This shows the defocusing increases with increasing RF frequency and that, coupled with the large bunch lengths captured for a given synchronous phase, means that lower frequencies are preferred for proton linacs of low to intermediate energy. While a similar effect occurs in electron linacs below 0.5 MeV, this energy can be reached in a few cells hence higher frequencies are typically used. An example of this is the ESS linac which uses 352.2 MHz up to 201 MeV before transitioning to 704.4 MHz after 201 MeV, allowing higher gradient elliptical cavities to be used without using a very large radius structure. It starts with an RFQ, which allows the beam to have longitudinal bunching, radial (electrostatic) focusing and acceleration in the same structure, up to 3 MeV before being further accelerated in a DTL up to 79 MeV, which allows efficient acceleration for low particle velocities. In order to run at high duty cycles at higher gradient, the linac must then become superconducting so the beam is then injected into a 352.2 MHz superconducting spoke cavity up to 201 MeV. Above 201 MeV the beam is sufficiently relativistic to increase the frequency up to 704.4 MHz utilising elliptical cavities up to 2.5 GeV.

As well as the RF defocusing, there is also space-charge defocusing as will be discussed in Chapter 7. We can provide external transverse focusing to counteract this using magnets, but there is no way of providing an external longitudinal bunching force, hence the synchronous phase is usually chosen to be longitudinally focusing. Strong transverse focusing is required in low- to intermediate-energy proton and ion linacs in order to compensate for the space-charge effects and the RF defocusing. This means smaller phase advances per section are chosen in the lattice design than at higher beam velocities, typically increasing along the length of the linac.

5.10   Non-Linear Beam Dynamics

In this chapter we have studied beam dynamics of a single particle. The majority of this chapter has been the study of linear motion, and we used matrices to represent our transformations and we use the matrix M as the map that brings an initial state vector $X(s_0)$ to a final state vector $X(s_1)$, so that

$$X(s_1) = \mathrm{M} \cdot X(s_0). \tag{5.220}$$

We have also considered non-linear motion when we looked at chromaticity, as this involves a combination of x and δ. What does non-linear motion look like in general?

The extension to non-linear motion using the formalism of matrices is straightforward [21] if we use index notation. In this notation, we use indices to label rows and columns of our matrices and vectors. Now the linear transformation looks like

$$X_i = \sum_i \mathrm{M}_{ij} X_j$$
$$= \mathrm{M}_{ij} X_j. \tag{5.221}$$

In this formalism we make extensive use of the convention that repeated indices are summed over (in this case i). This can be extended to higher-order terms in the following way,

$$X_i = \mathrm{M}_{ij} X_j + \mathrm{T}_{ijk} X_j X_k, \tag{5.222}$$

where the additional term T_{ijk} generally describes the non-linear mapping. For example, we can think of T_{166} as describing the non-linear coupling between momentum deviation and position, with a term looking like

$$X_1 = \mathrm{M}_{16} X_6 + \mathrm{T}_{166} X_6^2. \tag{5.223}$$

Note that M_{16} is the linear dispersion and T_{166} is the first non-linear dispersion.

The majority of non-linear beam dynamics is beyond the scope of this book. However to motivate its use, consider a series of beam line elements consisting of an RF cavity, a drift and a four-dipole chicane. The RF cavity can change the beam's phase space by imparting a longitudinally dependent change in momentum, called a chirp. A linear chirp would increase or decrease momentum linearly when moving from the front of the bunch to the back. This chirp is useful when the beam enters the chicane, which essentially provides a structure where the particle's path length depends on the particle's momentum. This means we can adjust the longitudinal size of the bunch by rotating the longitudinal phase space. A short bunch is obtained at the expense of a large momentum spread, and vice versa. This linear rotation can be modelled by consideration of the M_{56} element of the map of the overall system, obtained from the composite map (all the matrices multiplied in the correct order) of the three elements. Similarly, the non-linear term T_{566} gives the non-linear distortion of the rotated longitudinal phase space through the compression system. In essence, it describes the non-linear chirp given to the beam by the non-linear compression system.

So how do we obtain the non-linear maps of our beam line elements? Well, that is a very big question and we refer the reader to the many good books available on the topics [1, 22]. Some non-linear maps are straightforward. Consider an RF cavity, which acts to boost or reduce the particle's momentum dependent on the arrival phase. This means the longitudinal position is unchanged, and the momentum deviation is changed by the voltage and the sine of the phase

$$\delta = \delta - \frac{qV}{E_0} \sin(\omega z/c), \tag{5.224}$$

where ω is the cavity frequency and we have followed the formalism of [1, 22]. This map is inherently non-linear, as the sine contains all odd powers of z. A series expansion of the right-hand side of this expression would generate the non-linear coefficients defined above.

For magnetic elements, there are many different ways to obtain the non-linear maps. We have already, in the section on the correction of chromaticity, obtained the map for the sextupole. At the time we did not use the term non-linear map, but that is what it is. This was done by using Newton's dynamics, and integrating the Lorentz force on the particle along the length of a magnet. The field of the sextupole is a quadratically-rising field,

$$B_x = Sxy, \quad B_y = \frac{S}{2}(x^2 - y^2), \tag{5.225}$$

where S defines the sextupole strength, $d^2 B_y/dx^2$, and hence the resulting kick to the transverse angle is quadratic in the transverse offset. We saw how this could be used for chromaticity correction. A more structured method to obtain non-linear maps is to use the Hamiltonian for our particle and Hamilton's equations to compute the dynamics. The resulting map can be expressed as a Taylor series in the dynamical variables (e.g. (x, p_x)) using these methods. More advanced and formal tools such as Lie analysis also provide methods to extract the maps. For an excellent discussion see [1, 22]. We also note that many good references tabulate the maps for many elements, e.g. [23]. When using these, be sure you understand the variables and the approximations used.

Exercises

---

1. Imagine a proton storage ring, with a beam momentum of 20 TeV/c and 17,000 proton bunches, with 1×10^{10} protons per bunch.

   (a) What is the stored energy of this machine's beam?

   (b) If the circumference is 83 km, and the field is 6.6 T, what fraction of the ring is filled with dipoles?

   (c) The LHC beam energy is 360 MJ. What problems might this cause?

2. Show, for forces that are constant in time, that the Hamiltonian is a conserved quantity.

3. Prove that magnetic fields bend the trajectory of a particle but do not do any work on the particle. This means the particle energy does not change. Try doing this two different ways.

4. Magnetic forces are generally transverse to the direction of motion. What does this mean about longitudinal control of a beam?

5. Substitute the Courant-Snyder ansatz into Hill's equation and derive the differential equation obeyed by the β-function. Comment on how this could be solved numerically.

6. We looked at a FODO cell where the focal lengths of the focusing and defocusing quadrupoles were the same. Find the focal length of two opposite-polarity quadrupoles of focal length f, separated by a distance d. Then, imagine they were different focal lengths; what would this mean for the phase advance in the x and the y plane? Assuming thin-lens optics, find expressions for the phase advance in each plane for a focusing quadrupole with focal length f_1 and a defocusing quadrupole with focal length f_2.

7. Describe the impact on a beam if a quadrupole of gradient $g=500$ T/m is displaced vertically by a millimetre. What is the impact of the displacement of a sextupole on the same beam?

8. Obtain the expression for β_D in the defocusing quadrupole of a FODO cell. Now explain using equations how to propagate these parameters from the quadrupoles and work out the β-function anywhere in the FODO cell.

9. For a FODO cell with parameters $K = \pm 0.54102$ m^{-2}, $l_q = 0.5$ m and $L_{\text{drift}} = 2.5$ m, show the phase advance per cell is $45°$.

10. Returning to the derivation of Hill's equation, derive the inhomogeneous Hill's equation in the presence of non-zero momentum deviation δ. You may need to consult the broader literature.

11. Show that the ellipse with a single-particle emittance of ϵ_{rms} corresponds to 68% of the particles in the beam.

12. A 3 GHz RF linac for accelerating protons at 100 MeV has an accelerating gradient of 50 MV/m. If the linac operates with a synchronous phase of $20°$, what is the energy acceptance of the linac?

13. (A little more open ended) Implement our linear transport equations for (x, x') in your favourite computer code. Transport some particles through a FODO cell by choosing some sensible initial particle coordinates and cell parameters. Now add calculation and evolution of the lattice functions (e.g. β-function).

References

1. A. Wolski. *Beam Dynamics in High Energy Particle Accelerators*. World Scientific, 2014.
2. H. Wiedemann. *Particle Accelerator Physics*. Springer, 2015.
3. S.Y. Lee. *Accelerator Physics, 3rd edn*. World Scientific, 2011.
4. Ernest D. Courant, M. Stanley Livingston, and Hartland S. Snyder. The strong-focusing synchrotron—a new high energy accelerator. *Phys. Rev.*, 88:1190–1196, Dec 1952.
5. The mad-x code, https://mad.web.cern.ch/mad/.
6. K. Floettmann. Astra: A space charge tracking algorithm. In *DESY, March 2017, Version 3.2*.
7. S. B. van der Geer and M. J. de Loos. *General Particle Tracer User Manual*. Pulsar Physics. Program Version 3.35.
8. Pulsar physics and general particle tracer code. http://www.pulsar.nl/gpt.
9. Alexander Wu Chao, Karl Hubert Mess, Maury Tigner, and Frank Zimmermann, editors. *Handbook of Accelerator Physics and Engineering 2nd Edition*. World Scientific, 2013.
10. H.A. Enge. Focusing of charged particles edited by A. Septier. *Academic Press, New York*, 2:203–264, 1967.
11. B. D. Muratori, J. K. Jones, and A. Wolski. Analytical expressions for fringe fields in multipole magnets. *Phys. Rev. ST Accel. Beams*, 18:064001, Jun 2015.
12. T. Persson, F. Carlier, J. Coello de Portugal, A. Garcia-Tabares Valdivieso, A. Langner, E. H. Maclean, L. Malina, P. Skowronski, B. Salvant, R. Toms, and A. C. Garca Bonilla. LHC optics commissioning: A journey towards 1% optics control. *Phys. Rev. Accel. Beams*, 20:061002, Jun 2017.
13. R B Appleby, L Thompson, B Holzer, M Fitterer, N Bernard, and P Kostka. The high luminosity interaction region for a ring-ring Large Hadron Electron Collider. *Journal of Physics G: Nuclear and Particle Physics*, 40(12):125004, Nov 2013.
14. J L Abelleira Fernandez et al. A large hadron electron collider at CERN report on the physics and design concepts for machine and detector. *Journal of Physics G: Nuclear and Particle Physics*, 39(7):075001, Jul 2012.
15. Edwin M. McMillan. The synchrotron—a proposed high energy particle accelerator. *Phys. Rev.*, 68:143–144, Sep 1945.
16. V. I. Veksler. A new method of acceleration of relativistic particles. *J. Phys.*, 9:153–158, 1945.
17. M. Jenkins, G Burt, AV Praveen Kumar, Y. Saveliev, P. Corlett, T. Hartnett, R Smith, A Wheelhouse, P McIntosh, and K Middleman. Prototype 1 MeV X-band linac for aviation cargo inspection. *Physical Review Accelerators and Beams*, 22(2):020101, 2019.
18. T. P. Wangler. *RF Linear Accelerators, Second Edition*. Wiley, 2008.

19. R.B. Appleby and D.T. Abell. Accurate dynamics in an azimuthally-symmetric accelerating cavity. *Journal of Instrumentation*, 10(02):P02005, Feb 2015.

20. RH Miller, H Deruyter, WR Fowkes, JM Potter, RG Schonberg, and JN Weaver. Rf phase focusing in portable x-band, linear accelerators. *IEEE Transactions on Nuclear Science*, 32(5):3231–3233, 1985.

21. K.L. Brown. A first- and second-order matrix theory for the design of beam transport systems and charged particle spectrometers. *Adv.Part.Phys. 1*, pages 471–1344, 1968.

22. A. Wolski. *Introduction to Beam Dynamics in High-Energy Electron Storage Rings*. Morgan & Claypool Publishers, 2018.

23. The mad physics guide, http://mad8.web.cern.ch/mad8/doc/phys˙guide.pdf.

6

Particles and Radiation

6.1   The Origin of Electromagnetic Radiation

6.1.1   The Fields around a Moving Charge

We begin by considering a *stationary* charge q at $\mathbf{r}'$, which has the associated field (for an observer at $\mathbf{r}$, and illustrated in Fig 6.1) of

$$\mathbf{E} = \frac{1}{4\pi\epsilon_0} \frac{q(\mathbf{r} - \mathbf{r}')}{|\mathbf{r} - \mathbf{r}'|^3}, \left(\mathbf{E} \propto \frac{1}{|\mathbf{r} - \mathbf{r}'|^2} \right),$$

$$\mathbf{B} = \mathbf{0}. \tag{6.1}$$

We can argue that a stationary charge does not radiate in two ways. Firstly, we know from our earlier discussion that electromagnetic radiation would have to have a component with a magnetic field, but we see for this stationary charge that $\mathbf{B} = \mathbf{0}$ everywhere. Secondly, energy flow from a charge radiating should vary as $S \propto 1/r^2$ to satisfy conservation of energy; in other words, $\mathbf{E}$ and $\mathbf{B}$ must vary as $\mathbf{E} \propto 1/r$, $\mathbf{B} \propto 1/r$ if there is electromagnetic radiation emanating from the point charge. We see from the diagram that there is no component of either field that has this variation, and hence there is no radiation emitted.

We next consider a charge moving at a constant velocity with respect to some observer. Both the charge and the observer are in inertial frames of reference, and so we may perform a Lorentz transformation from one frame to the other and still retain the observable which is the total amount of electric flux around the charge. A moving charge generates a magnetic

DOI: 10.1201/9781351007962-6

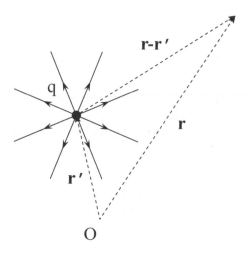

FIGURE 6.1   Field lines around a stationary charge, and a definition of the position of the charge, $\mathbf{r'}$, and where its electric field is experienced, $\mathbf{r}$.

field, but here we consider how the electric field of a moving charge appears to a stationary observer. We Lorentz transform the spherically-symmetric electric field and see that in the direction of motion, the field lines are compressed by a factor $1/\gamma$; the electric field is

$$\mathbf{E}(\mathbf{r}) = \frac{q}{4\pi\epsilon_0} \frac{1 - \beta^2}{(1 - \beta^2 \sin^2 \theta)^{3/2}} \frac{\hat{\mathbf{r}}}{|\mathbf{r}|^2} \tag{6.2}$$

where the polar angle $\theta = 0$ along the direction of the charge motion. It can be seen that the electric flux through a small area element dA still varies $\propto 1/r^2$ as it does around a stationary charge (Fig 6.2). One way to think of this is to remember that field lines from a uniformly-moving charge are still straight; hence their separation is proportional to the distance from the charge, regardless of the direction of the field lines. The area dA traced out by any four field lines varies as $dA \propto r^2$ regardless of the direction away from the charge. Hence the field strength still varies as $E \propto 1/r^2$ in any direction, and so again there is no emitted radiation. Another way to look at this is merely to transform to the rest frame of the charge, where of course it is stationary and therefore not emitting photons; the 'fact' that it is not emitting photons should still be true if we observe in a different inertial frame. A consequence of this 'compression' of the electric fields line around a rapidly-moving charge is that a detector that is sensitive to electric fields will see a pulse of electric field as a charge passes close by; this is the principle by which many diagnostic instruments work, such as beam position monitors (BPMs) where a capacitive pickup senses the voltage generated by a passing bunch of particles a few millimetres away.

Another way to consider a charge with constant velocity is to examine the magnetic field. The Biot-Savart law allows us to directly calculate the field created by a moving charge. According to this law, the field created by a moving charge is

$$\mathbf{B}(\mathbf{r}) = \frac{\mu_0}{4\pi} \frac{q\mathbf{v} \times \hat{\mathbf{r}}}{|\mathbf{r}|^2}. \tag{6.3}$$

Again, we see that the magnetic field in any direction only varies as $B \propto 1/r^2$, and hence there is no radiation. In summary, a charge moving at constant velocity emits no radiation. Therefore, radiation requires *acceleration* of the charge.

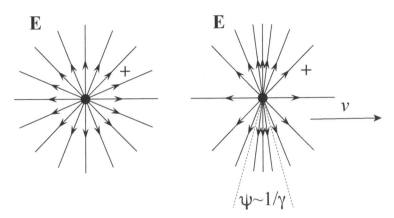

FIGURE 6.2 Electric field lines around a stationary charge (left), and around a charge moving at constant velocity $\beta = v/c$ (right). A moving charge has field lines compressed into a plane perpendicular to the charge's direction of motion, and the spread in field lines has typical width $\psi \sim 1/\gamma$ where $\gamma = E_{total}/E_0$.

6.1.2 Radiation from an Accelerated Charge

A Displaced Point Charge

It is possible to use a simple picture of a displaced point charge in order to derive the power and radiation pattern of an accelerated charge; this manner of describing the emission is generally attributed either to the physicist Edward Purcell, or perhaps earlier (according to Malcolm Longair) to J. J. Thomson.

We imagine a point charge initially at rest which is then subject to a short period of acceleration Δt, after which time it is moving at a constant velocity $u \ll c$. Hence $u = a\Delta t$ for an acceleration a. A short time later, we may observe two regions with different field configurations (Fig 6.3). Sufficiently close to the charge an observer sees the new location and speed of the charge, with field lines emanating radially away from it. Far from the charge an observer at a distance r still sees the field at the previous, stationary location; a time r/c has not yet elapsed to allow the observer to see the new motion of the charge. Between these two regions there must be a boundary where the field lines change from the old, stationary situation to the new, moving situation, and this boundary moves away from the charge's location at a speed c. We remember that in free space $\nabla \cdot \mathbf{E} = 0$, so there can be no discontinuities in electric field lines (this would imply extra charges at the boundary, which isn't true). The moving boundary must therefore appear as a kink in the electric field lines. This kink is the emitted radiation. We now calculate its properties.

By making the assumption that the final velocity $u \ll c$, we may say that the electric field lines are approximately parallel inside and outside the kink. We now imagine an observer looking at a time t at some angle θ to the final motion of the charge (Fig 6.4); the charge has a location ut which, viewed by the observer at θ, appears to be moving at $u_\perp t$ perpendicular to the line of observation and $u_\parallel t$ along it. We may then relate the perpendicular component of the electric field at the kink, $E_\perp$, to the radial component $E_\parallel$, in terms of the perpendicular motion and the radial motion. This is

$$\frac{E_\perp}{E_\parallel} = \frac{u_\perp t}{c\Delta t}. \tag{6.4}$$

We may see also that $u_\perp = a_\perp \Delta t$ (the component of the motion perpendicular to the line

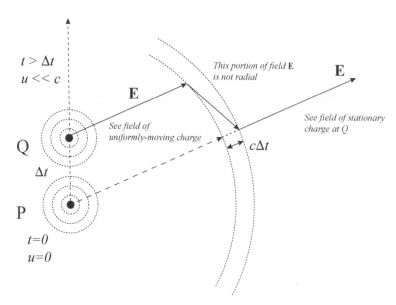

FIGURE 6.3 Illustration of how a small acceleration in a charge initially at P can generate radiation. After a uniform acceleration a for a time Δt, the charge is then moving uniformly at $u = a\Delta t$. When observed at a later time t there will be two portions of field. Within a radius $r = ct$ an observer sees the new, uniformly-moving charge; beyond $r = ct$ an observer sees the old field of the stationary charge at P. At the boundary $r = ct$ the field lines must still be continuous; hence there is a (small) kink in the electric field.

of observation), so that we may state

$$E_\perp = E_\parallel \frac{a_\perp t}{c} = E_\parallel \frac{a_\perp r}{c^2}, \tag{6.5}$$

since $r = ct$. But we also already know that

$$E_\parallel = \frac{q}{4\pi\epsilon_0 r^2}, \tag{6.6}$$

(the same field as if the charge were stationary), so we therefore obtain an expression for $E_\perp$ as

$$E_\perp = \frac{q a_\perp}{4\pi\epsilon_0 c^2 r}. \tag{6.7}$$

We see therefore that the magnitude of the kink electric field $E_\perp \propto 1/r$. However, we should also notice that the electric field at time t depends on the motion of the charge at an earlier time $\tau = t - r/c$ depending upon the distance of observation r.

Since we expect that at large distances there will be a plane wave emitted by the accelerated charge such that $\mathbf{B}$ is perpendicular to $\mathbf{E}$, we can from this obtain that

$$\mathbf{B} = \frac{1}{c}\hat{\mathbf{r}} \times \mathbf{E} \tag{6.8}$$

and that $B_\perp = E_\perp/c$. Alternatively, we can directly obtain a similar formula for $B_\perp$ to that for $E_\perp$ using Faraday's Law. Combining $\mathbf{E}$ and $\mathbf{B}$, we see immediately that the Poynting vector

$$\mathbf{S} = \frac{1}{\mu_0}(\mathbf{E} \times \mathbf{B}) \tag{6.9}$$

points radially outwards from the accelerated charge along $\hat{\mathbf{r}}$, and that its magnitude $S \propto EB \propto 1/r^2$.

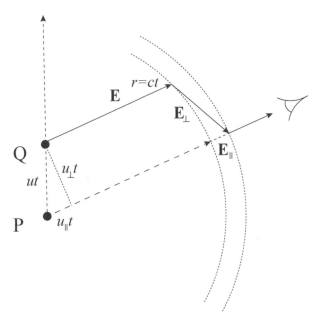

FIGURE 6.4 Illustration of how to obtain the magnitude of the perpendicular component $E_\perp$ of the electric field at some angle to the direction of charge acceleration and subsequent motion. The ratio of the parallel and perpendicular components of the electric field is just the ratio of the parallel and perpendicular components of the velocity.

Radiation Pattern from a Displaced Point Charge

We see from the previous section that

$$a_\perp = a \sin \theta, \tag{6.10}$$

and hence we may write the magnitude of the electric field $\mathbf{E}(\mathbf{r}, t)$ at some location $\mathbf{r}$ as

$$|\mathbf{E}(\mathbf{r}, t)| = \frac{q |\mathbf{a}(t - r/c)| \sin \theta}{4\pi\epsilon_0 c^2 r}. \tag{6.11}$$

This is illustrated in Fig 6.5. The Poynting flux (i.e. the power flow) is then

$$|\mathbf{S}(\mathbf{r}, t)| = \frac{q^2 |\mathbf{a}^2(t - r/c)| \sin^2 \theta}{16\pi^2 \epsilon_0 c^3 r^2}, \tag{6.12}$$

which has units of Wm^{-2} (see Fig 6.6). Note that $S \propto 1/r^2$ as it should.

Total Radiated Power

Since we now know the Poynting flux $\mathbf{S}(\mathbf{r}, t) \equiv \mathbf{S}(r, \theta, \phi)$ at a given distance r and polar angle (θ, ϕ) (noting that S has no dependence on azimuthal angle ϕ – see illustration in Fig 6.7), we may now integrate over the polar angle θ to obtain the total power $P(t)$ as

$$P(t) = \int_0^\pi S \mathrm{d}A \tag{6.13}$$

where

$$\mathrm{d}A = 2\pi r^2 \sin \theta \mathrm{d}\theta \tag{6.14}$$

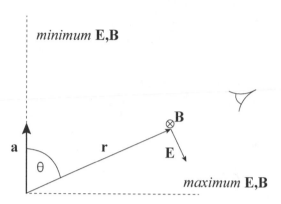

FIGURE 6.5   Illustration of how the magnitude of the emitted electric and magnetic fields vary with observation angle θ.

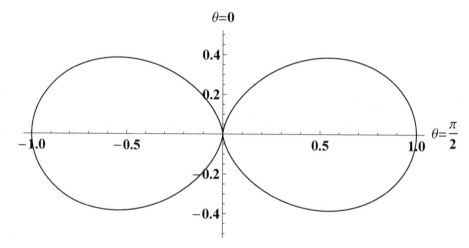

FIGURE 6.6   2D illustration of how the magnitude of the Poynting vector $\mathbf{S}$ (here shown as the distance of the solid from the origin, for any given angle θ) varies with observation angle θ.

is the area of a slice $d\theta$ at an angle θ of the overall sphere into which radiation is emitted (this is illustrated in Fig 6.8). Explicitly therefore,

$$P(t) = \int_0^\pi \left(S = \frac{q^2 a^2(t - r/c)}{16\pi^2 \epsilon_0 c^3 r^2} \sin^2 \theta\right)(\mathrm{d}A = 2\pi r^2 \sin\theta \mathrm{d}\theta) \tag{6.15}$$

or

$$P(t) = \frac{q^2 a^2(t - r/c)}{8\pi \epsilon_0 c^3} \int_0^\pi \sin^3 \theta \mathrm{d}\theta. \tag{6.16}$$

We may use a trigonometric identity to obtain the integral of $\sin^3 \theta$ as

$$\int_0^\pi \sin^3 \theta \mathrm{d}\theta = \int_0^\pi \sin\theta(1 - \cos^2 \theta)\mathrm{d}\theta = \left[-\cos\theta + \frac{1}{3}\cos^3 \theta\right]_0^\pi = \frac{2}{3} + \frac{2}{3} = \frac{4}{3}. \tag{6.17}$$

Hence we obtain an expression for the instantaneous total power emitted by an accelerated charge:

$$P(t) = \frac{q^2 a^2(t - r/c)}{6\pi \epsilon_0 c^3}. \tag{6.18}$$

This is Larmor's formula, and we used Edward Purcell's method to derive it.* Larmor's formula is the basis of all radiation calculations for charges.

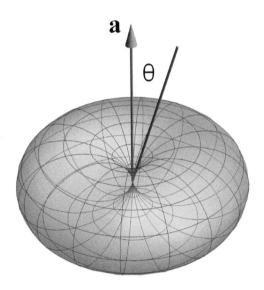

FIGURE 6.7 3D illustration of how the magnitude of the Poynting vector **S** varies with observation angle θ. $\theta = 0$ points up. There is no variation of emitted power with azimuthal angle ϕ.

6.1.3   The Hertzian Dipole

We may use the same basic argument that we used to obtain the Larmor formula to consider an oscillating current element and the resultant radiation that it emits. We will see that consideration of the current is equivalent to considering the motion of charge; the radiation pattern obtained is the same – it's just the method that is different. We will see below that the Hertzian dipole is the starting point for understanding the radiation emitted both from moving charges and from the radio-frequency sources that provide energy to them; both situations typically involve oscillatory motion that gives rise to Hertzian-like emission.

An Oscillating Current Element

We consider two locations (1) and (2) aligned along the **z** axis and separated by a small distance l that each have a variable amount of charge on them

$$q_1 = +q_0 e^{-i\omega t} \tag{6.19}$$

$$q_2 = -q_0 e^{-i\omega t} \tag{6.20}$$

---

*The reader is encouraged to be very careful here, since a number of textbooks will be encountered that use c.g.s. units, in which case the radiation formulae look quite different. We remind the reader that everything presented here uses the SI system of units.

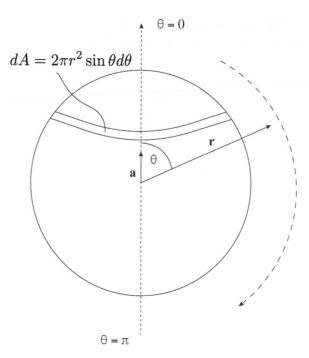

FIGURE 6.8   Illustration of how to calculate the total radiated power by considering slices of the sphere into which radiation is emitted, each slice having an area $dA = 2\pi r^2 \sin\theta d\theta$ for a particular polar angle θ.

such that the current flowing between the two points is

$$\mathbf{I} = \frac{dq_1}{dt}\hat{\mathbf{z}} = -i\omega q_0 e^{-i\omega t}\hat{\mathbf{z}}. \tag{6.21}$$

If the separation $l \to 0$, there is still a current $I_0 = -i\omega q_0$. We recall the formula for the magnetic vector potential

$$\mathbf{A}(\mathbf{r}, t) = \frac{\mu_0}{4\pi}\int_{V'}\frac{\mathbf{j}(\mathbf{r}', t')}{|\mathbf{r} - \mathbf{r}'|}dV' \tag{6.22}$$

for a current distribution $\mathbf{j}$ that exists within a volume V'. In the present case where the locations (1) and (2) are equidistant about the origin, we may write the vector potential as

$$\mathbf{A}(\mathbf{r}, t) = \frac{\mu_0}{4\pi}(I_0 l)\frac{e^{i(kr - \omega t)}}{r}\hat{\mathbf{z}}, \tag{6.23}$$

where we regard the product $I_o l$ as staying the same when $l \to 0$. We obtain the magnetic field in spherical polar coordinates as

$$\mathbf{B} = \nabla \times \mathbf{A} \equiv \frac{1}{r^2 \sin\theta}\begin{vmatrix} \hat{\mathbf{r}} & r\hat{\theta} & r\sin\theta\hat{\phi} \\ \frac{\partial}{\partial r} & \frac{\partial}{\partial \theta} & \frac{\partial}{\partial \phi} \\ A_r & rA_\theta & r\sin\theta A_\phi \end{vmatrix} \tag{6.24}$$

By looking at the terms in the cross product in sequence, we see that the components of $\mathbf{B}$ in polar coordinates are

$$B_r = 0,$$
$$B_\theta = 0,$$
$$B_\phi = \frac{\mu_0}{4\pi}(I_0 l)k\sin\theta\left(\frac{1}{kr} - i\right)\frac{e^{i(kr-\omega t)}}{r}. \tag{6.25}$$

We recall Ampere's Law in free space is

$$\nabla \times \mathbf{B} = \frac{1}{c^2}\frac{\partial \mathbf{E}}{\partial t} \tag{6.26}$$

so that we may then obtain the components of the electric field explicitly as

$$E_r = \frac{1}{4\pi\epsilon_0}\frac{2}{c}(I_0 l)\left(1 + \frac{i}{kr}\right)\frac{e^{i(kr-\omega t)}}{r^2},$$
$$E_\theta = \frac{1}{4\pi\epsilon_0}(I_0 l)\frac{k}{c}\sin\theta\left(\frac{i}{k^2 r^2} + \frac{1}{kr} - i\right)\frac{e^{i(kr-\omega t)}}{r},$$
$$E_\phi = 0. \tag{6.27}$$

These are quite complicated expressions, so it is just as well that we obtained them in the simplest way possible – via the vector potential. We may distinguish between two different regimes – near-field and far-field – where the far-field regime is the radiative part.

The near-field regime is when $kr \ll 1$; in other words, the distance of observation from the dipole is much less than the *wavelength* emitted. In this case, the dominant field components are

$$B_\phi \simeq \frac{\mu_0}{4\pi}(I_0 l)\sin\theta\frac{e^{i(kr-\omega t)}}{r^2},$$
$$E_r \simeq \frac{1}{4\pi\epsilon_0}\frac{2i}{c}(I_0 l)\frac{e^{i(kr-\omega t)}}{kr^3},$$
$$E_\theta \simeq \frac{1}{4\pi\epsilon_0}(I_0 l)\sin\theta\frac{e^{i(kr-\omega t)}}{kr^3}. \tag{6.28}$$

The components of E_r and E_θ look like the field around an electric dipole, and have a magnitude which falls as $\propto 1/r^3$ as one would expect.

The far-field regime is when $kr \gg 1$, in other words the distance of observation from the dipole is large compared to the wavelength emitted (this is similar to the Fraunhofer distance). Now, the dominant field components are

$$B_\phi(r,t) \simeq -i\frac{\mu_0}{4\pi}(I_0 l)k\sin\theta\frac{e^{i(kr-\omega t)}}{r},$$
$$E_\theta(r,t) \simeq -i\frac{1}{4\pi\epsilon_0}(I_0 l)\frac{k}{c}\sin\theta\frac{e^{i(kr-\omega t)}}{r} \tag{6.29}$$

(note that $k/c = \omega/c^2$). As we saw before, $\mathbf{r} \perp \mathbf{B} \perp \mathbf{E}$, and

$$\frac{|E_\theta|}{|B_\phi|} \simeq c. \tag{6.30}$$

We can see explicitly the direction of the electric and magnetic fields (they point in the $\hat{\theta}$ and $\hat{\phi}$ directions respectively). We also see that $\mathbf{E}$ and $\mathbf{B}$ oscillate *in phase* with each other, and that their magnitudes vary as $E \propto 1/r$ and $B \propto 1/r$ as they should. Combining the two components together, we may then obtain the power emitted by a Hertzian dipole as

$$P(t) = \frac{(l\omega)^2}{6\pi\epsilon_0 c^3} I_0^2 \sin^2(kr - \omega t). \tag{6.31}$$

There are several equivalent ways to write this formula, but this particular way shows explicitly that $P \propto I^2$; this is an important fact.

Radiation Resistance

We have seen that for a Hertzian dipole, $P \propto I^2$. This implies that there is an effective *resistance* for emitted radiation, which depends upon the length of the dipole l and upon the frequency ω that the current is oscillating at. We see from the formula for emitted power that we can define this *radiation resistance* as

$$R_{rad} = \frac{l^2 \omega^2}{6\pi\epsilon_0 c^3}. \tag{6.32}$$

Remembering that $\omega = 2\pi c/\lambda$ and that $c^2 = 1/\mu_0\epsilon_0$, we can re-write R_{rad} in a number of equivalent ways:

$$R_{rad} = \frac{2\pi}{3} \frac{1}{\epsilon_0 c} \left(\frac{l}{\lambda}\right)^2 = \frac{2\pi}{3} \mu_0 c \left(\frac{l}{\lambda}\right)^2 = \frac{2\pi}{3} Z_0 \left(\frac{l}{\lambda}\right)^2. \tag{6.33}$$

In the last of these expressions we have defined a quantity Z_0 as

$$Z_0 = \mu_0 c = \frac{1}{\epsilon_0 c} \simeq 377 \ \Omega \ (\text{ohm}). \tag{6.34}$$

Z_0 is known as the free-space impedance and, as you can see, it has the correct units. Note that Z_0 is often given in an approximate form

$$\frac{2\pi Z_0}{3} \simeq 80\pi^2, \tag{6.35}$$

so that that the radiation resistance may be variously written as

$$Z_0 \simeq 80\pi^2 \left(\frac{l}{\lambda}\right)^2 \simeq 800 \left(\frac{l}{\lambda}\right)^2 \simeq 790 \left(\frac{l}{\lambda}\right)^2. \tag{6.36}$$

As you can see, these equations allow a rough calculation of the radiation resistance (and hence the power emitted) for a given dipole emitter, as long as the length of the dipole and the emitted wavelength are known.

As an example, we set $l = \lambda/4$; this is a so-called quarter-wave antenna, also often called a monopole antenna. For example, a VHF antenna where $\lambda = 3$ m (frequency $f = 100$ MHz) could be made with a length $l = 0.75$ m. This gives a radiation resistance of $R_{rad} \simeq 50 \ \Omega$ (which you should recognise as a very common resistance found in electronic equipment). For a peak current $I_0 = 100$ A flowing through the antenna, the emitted power would be $P \simeq 500$ kW. This power is not unusual – it corresponds roughly to the kind of power emitted by the UK Winter Hill transmitter, which provides television signals for Manchester and the surrounding country.

An Oscillating Dipole

We derived our equations for a Hertzian dipole by considering current flowing back and forth between two points. We may equivalently view this situation as two points (separated by a small distance l in comparison to the emitted wavelength λ) upon each of which charge is deposited or removed; the charge on either end is equal and opposite. The charge on each end can be described as

$$q = \pm q_0 \sin \omega t \tag{6.37}$$

where $+$ is for one end of the dipole and $-$ is for the other end. The current flowing on/off the two ends of the dipole is then just

$$I = \frac{\mathrm{d}q}{\mathrm{d}t} = q_0 \omega \cos \omega t. \tag{6.38}$$

In other words, $I_0 = q_0 \omega$, and the oscillating current is equivalent to an oscillating dipole moment. The dipole moment of a given charge separation is just $p_0 = q_0 l$, so we may write

$$p = p_0 \sin \omega t = q_0 l \sin \omega t. \tag{6.39}$$

In other words, we may write

$$I_0 l = p_0 \omega,$$
$$I_0 l \omega = p_0 \omega^2. \tag{6.40}$$

Remembering that when we take an average over time $\left\langle \sin^2 \ldots \right\rangle = \frac{1}{2}$, we can re-write the Larmor formula for the time-averaged power as

$$\langle P \rangle = \frac{I_0^2 l^2 \omega^2}{12\pi \epsilon_0 c^3}, \qquad \text{('current picture')}$$

$$\langle P \rangle = \frac{p_0^2 \omega^4}{12\pi \epsilon_0 c^3}. \qquad \text{('dipole picture')} \tag{6.41}$$

The dipole picture is more interesting. It tells us what power will be radiated by a given amount of charge q_0 moving over a distance l. The power radiated varies very, *very* strongly with frequency. $P \propto \omega^4$: if the frequency is doubled, the power radiated goes up by a factor of sixteen! We will see below why that is of such importance.

To summarise: when we talk about a Hertzian dipole, we mean:

- non-relativistic charge motion;

- a source size $l \ll \lambda$ where λ is the emitted wavelength;

- an observation distance $r \gg \lambda$ (i.e. $kr \gg 1$);

- the emitted power $P \propto I_0^2$ and $P \propto \omega^4$.

6.1.4   Antennas

When talking about the Hertzian dipole, we briefly mentioned the idea of an antenna as a current-carrying object where the current oscillates with time. We will now formalise this concept by discussing different types of antenna. Antennas are hugely important in accelerator science, as they are a basic component used to couple electromagnetic power from one oscillating system (say, a waveguide) to another system (say, a cavity).

The Half-Wave Antenna

The half-wave antenna is also known as the dipole antenna, and is the conceptual basis of many sources of RF power described elsewhere in this textbook. It consists of a cable (usually coaxial) that connects to two aerials, one aerial to each of the cable conductors (this is illustrated in Fig 6.9). The aerials are not connected to each other, but see the same current from the cable. Since this real antenna has ends, obviously current can't flow out of those ends. Hence the current at the end of the antenna must be zero to be physical. Without knowing anything else, we can immediately say that the current in a real antenna must be maximal at its centre, and zero at the ends; if the antenna is fed with current at its centre, we can first assume that I linearly falls away from its centre. With this simple linear picture, we obtain that the *effective* current $I_{eff} = I_0/2$, and the *effective* antenna length $l_{eff} = l/2$. Hence the radiation resistance (also called the impedance) is

$$R_{rad} \simeq \frac{1}{4}\left(\frac{2\pi}{3}Z_0\right) \simeq 200 \ \Omega \left(\frac{l}{\lambda}\right)^2. \tag{6.42}$$

Comparing this equation to our previous equation for a 1/4-wave antenna, we find a radiation resistance $R_{rad} \simeq 12.5 \ \Omega$, a substantially smaller value. Hence for the same current in the drive cable we get *more* power. For example, a 100 A peak current would drive $P_{rad} = I_0^2 R_{rad} \simeq 125$ kW.

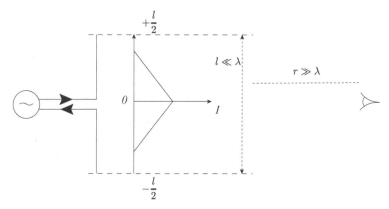

FIGURE 6.9   A first approximation to the current I flowing in a half-wave antenna is that the current fed into the middle by an AC source falls linearly towards the ends. The two leads that carry current into and out of the antenna generate radiation fields that cancel.

Let's now do the derivation more properly. We suspect in reality that a half-wave antenna doesn't really have a current along its length that falls linearly; we expect a standing wave to be set up, and the lowest mode of that standing wave is one where the current has the form of a half-sinewave. In other words, we assume that I is a maximum at the position the antenna is fed (i.e. at its centre), and $I = 0$ at the ends. The standing wave comes about because of the generation of transient voltages; the voltage can be momentarily different at different points on the antenna because it takes time for the currents to move from place to place. Hence we can describe the current at a different point z along the antenna as

$$I = I_0\left(e^{+ikz} + e^{-ikz}\right)e^{i\omega t} \tag{6.43}$$

(i.e. there are two waves moving up and down the antenna such that their currents cancel at the antenna ends). This is shown in Fig 6.10. We may therefore describe the real part of the current in the antenna as

$$I = I_0 \cos\frac{2\pi z}{\lambda}\cos\omega t \tag{6.44}$$

so that $I = 0$ at $z = \pm\lambda/4$, recalling that $\lambda = 2l$ where l is the (total) length of the antenna.

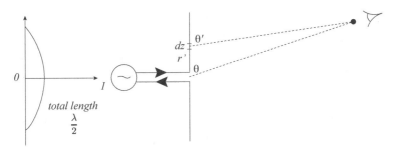

FIGURE 6.10 A better calculation of the electromagnetic radiation emitted by a half-wave antenna. As well as the variation of current along the length of the antenna, we must also account for the small phase difference in the emitted radiation, as seen by a distant (far-field) observer.

We may now sum up the contribution to the (far-field, radiative) component E_θ of each of the current elements passing at any moment through a short section of the antenna dz, viewed at some distance r at an observation angle θ. This is

$$E_\theta(r,t) = \int_{-l/2}^{+l/2} \frac{I(z')\omega}{4\pi\epsilon_0 rc^2} \sin\theta \sin(\omega t - k|\mathbf{r} - \mathbf{r}'|)dz'. \tag{6.45}$$

Each component dE_θ of the electric field seen at $\mathbf{r}$ contributes with a different phase such that the resulting E and B fields are

$$E_\theta(r,t) = \frac{2i}{4\pi\epsilon_0 c} I_0 \frac{\cos(\frac{\pi}{2}\cos\theta)}{\sin\theta} \frac{e^{i(\omega t - kr)}}{r},$$

$$B_\phi(r,t) = \frac{2i\mu_0}{4\pi} I_0 \frac{\cos(\frac{\pi}{2}\cos\theta)}{\sin\theta} \frac{e^{i(\omega t - kr)}}{r}, \tag{6.46}$$

noting once more that $E_\theta/B_\phi = c$ as it should. The total (time-averaged) Poynting flux is then just

$$\langle\mathbf{S}\rangle = \frac{I_0^2}{8\pi^2\epsilon_0 c} \frac{\cos^2(\frac{\pi}{2}\cos\theta)}{r^2 \sin^2\theta}. \tag{6.47}$$

The total power radiated by the antenna may be calculated straightforwardly by integrating over all angles θ and ϕ; it's just long-winded. The integration yields

$$\langle P\rangle = \int_0^{2\pi}\int_0^\pi \langle\mathbf{S}\rangle r^2 \sin\theta d\theta d\phi,$$

$$= \frac{I_0^2}{4\pi\epsilon_0 c} \underbrace{\left(\frac{1}{2\pi}\int_0^{2\pi} d\phi\right)}_{=1} \int_0^\pi \frac{\cos^2(\frac{\pi}{2}\cos\theta)}{r^2 \sin^2\theta} r^2 \sin\theta d\theta,$$

$$= \frac{I_0^2}{4\pi\epsilon_0 c} \int_0^\pi \frac{\cos^2(\frac{\pi}{2}\cos\theta)}{\sin\theta} d\theta. \tag{6.48}$$

Note that the integration over ϕ cancels with the factor $1/2\pi$, and one of the factors $\sin\theta$ cancels. The remaining integral can only be carried out numerically, and

$$\int_0^\pi \frac{\cos^2(\frac{\pi}{2}\cos\theta)}{\sin\theta} d\theta \simeq 1.22. \tag{6.49}$$

Hence the average emitted power is

$$\langle P \rangle \simeq 1.22 \frac{I_0^2}{4\pi\epsilon_0 c} \simeq 0.194 Z_0 I_{rms}^2, \tag{6.50}$$

remembering that $Z_0 = 1/\epsilon_0 c$ and $I_{rms}^2 = I_0^2/2$. The radiation resistance of a half-wave antenna is therefore just

$$R_{rad} \simeq 73.1 \ \Omega. \tag{6.51}$$

This impedance is close to that of a 'standard' 75 Ω coaxial cable. We summarise the half-wave antenna power as

$$\langle P \rangle \simeq 0.194 Z_0 \frac{I_0^2}{2} \simeq 73 \frac{I_0^2}{2}. \tag{6.52}$$

For example, to obtain 5 kW transmitted power from a single half-wave antenna, we require

$$73 \frac{I_0^2}{2} \simeq 5000, \tag{6.53}$$

and therefore $I_0 \simeq 11.7$ A.

Half-Wave Antenna Radiation Pattern

We saw earlier that a simple Hertzian dipole generates a radiation pattern with power distributed as

$$\langle S \rangle \propto \sin^2 \theta. \tag{6.54}$$

A realistic half-wave antenna has a radiation pattern distributed as

$$\langle S \rangle \propto \frac{\cos^2\left(\frac{\pi}{2}\cos\theta\right)}{\sin^2\theta}. \tag{6.55}$$

It's not immediately obvious how these compare, so let's plot them. One can see in the figure below that the half-wave antenna is more directional than a simple Hertzian dipole.

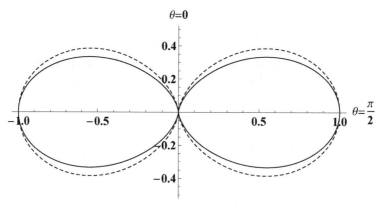

FIGURE 6.11   Comparison of the radiation pattern from a Hertzian dipole (dashed line) with that of a half-wave antenna (solid line). The half-wave antenna has a more directional output.

6.2 Radiation from Moving Charges

We now consider a most important aspect of particle accelerators; the phenomenon that the charges moving within them radiate. In this section, we will consider some simple cases of that radiation, although of course there are many more complex situations.

6.2.1 Cyclotron Radiation

We first consider a non-relativistic charge (in other words, $\gamma \simeq 1$) moving through a uniform magnetic field $\mathbf{B}$ which is oriented perpendicular to the velocity $\mathbf{v}$ of the charge, i.e. $\mathbf{v} \perp \mathbf{B}$; we assume to begin with that there is no electric field $\mathbf{E} = 0$. The usual Lorentz force $\mathbf{F} = q\left(\mathbf{E} + \mathbf{v} \times \mathbf{B}\right)$ reduces to the simpler form $F = qvB$ where $\mathbf{F}$ is perpendicular to both $\mathbf{v}$ and $\mathbf{B}$; it is *very* important to remind ourselves here that the magnetic field does no work upon the charge (and vice versa). No net energy is exchanged between the charge and the magnetic field (in this classical picture!). The charge will thus move in a circular path that remains at right angles to the field. Note that if there is a component of the motion parallel to $\mathbf{B}$ then the charge will move in a helical path around the field lines.*

Equating forces we have

$$\frac{mv^2}{\rho} = qvB \tag{6.56}$$

(where m is the mass of the charge) so therefore the radius of the circular path is just

$$\rho = \frac{mv}{qB}. \tag{6.57}$$

The acceleration of the charge is

$$a = \frac{qvB}{m} = \omega_c v, \tag{6.58}$$

where we have defined the cyclotron frequency (also known as the Larmor frequency)

$$\omega_c = \frac{v}{\rho} = \frac{qB}{m}. \tag{6.59}$$

This is of course the angular frequency of the cyclotron motion; the actual frequency (i.e. how many times the charge comes round past a fixed point) is just

$$f_c = \frac{1}{2\pi}\frac{qB}{m}. \tag{6.60}$$

The two most commonly used particles in accelerators are electrons and protons. Substituting their masses into this formula we obtain the electron cyclotron frequency as $\simeq 28$ GHz/T, and the proton cyclotron frequency as $\simeq 15.3$ MHz/T. In other words, low-energy electrons in a magnetic field of 1 T will gyrate in the field at 28 GHz; doubling the field will double that frequency. In a standard microwave oven magnetron, the electron gyration frequency is 2.45 GHz, and hence the magnetic field the electrons are immersed in must be about 0.09 T (made by ordinary permanent magnets).

---

*In plasma physics this is known as gyration.

Cyclotron Radiation: Power and Frequency

Substituting the acceleration a above into Larmor's formula directly gives us the power:

$$P = \frac{q^2 \omega_c^2 v^2}{6 \pi \epsilon_0 c^3}. \tag{6.61}$$

In contrast to an antenna, the acceleration of the charge in a magnetic field is constant and hence there is no factor $1/2$ in the average power. The output power $\propto v^2$, in other words $P \propto K$ where K is the kinetic energy of the charge. P is the total power emitted in all directions (i.e. over all angles ϕ). Also, note that P is the power emitted by *each* charge; if we have N charges then the power simply adds up.*

If we observe side-on a charge moving in a magnetic field, it looks at a sufficient distance like a Hertzian dipole.* This is of course why we examined the case of the Hertzian dipole earlier. We therefore expect the emitted radiation (in the far field) to be just the same: the frequency of the emitted radiation is *the same* as the cyclotron frequency. Also, the polarisation of the emitted radiation is parallel to the plane of the circular motion.

In a real cyclotron there may be many, many protons moving together.* For example, a typical modern cyclotron might have protons circulating with a kinetic energy of 10 MeV (corresponding to a velocity $v = 44 \times 10^6$ ms^{-1} in a magnetic field of $B = 1$ T; $\gamma \simeq 1$ and the protons are not relativistic). The cyclotron frequency is $\omega_c = 96 \times 10^6$ s^{-1} or $f_c = 15$ MHz; proton cyclotrons have cyclotron frequencies which are tens of MHz. The power emitted per proton is $P \sim 10^{-22}$ W, or $P \sim 10^{-16}$ W/pC. This is a *very* small value. What does it tell us? It tells us that protons don't radiate very much, and so they don't lose a significant amount of their energy when circulating in a magnetic field; hence our original assumption of circular motion is valid. We will see below that in some circumstances the radiation given out by a charge can be significant with respect to its initial kinetic energy.

In this first derivation of the cyclotron frequency we obtained

$$f_c = \frac{1}{2\pi} \frac{qB}{m} \tag{6.62}$$

by equating the centripetal force to the Lorentz force on the moving charge. However, we should remember that at a sufficient velocity the charge will gain mass. We know of course that a charge increases in mass according to $m = \gamma m_0$ where $\gamma = E/E_0$, so our derivation for the revolution frequency should really be

$$\frac{mv^2}{\rho} = qvB \tag{6.63}$$

where $v = \beta c$ and $m = \gamma m_0$. Hence

$$m\beta c = qB\rho,$$
$$\rho = \frac{\beta \gamma m_0 c}{qB}, \tag{6.64}$$

---

*We must be careful about this point; the radiation only adds up if it is emitted by the charges *incoherently*, i.e. they act as separate emitters. Look ahead at the discussion of coherent synchrotron radiation in Section 7.4

*'Sufficient' here means $\gg r$, where r is the cyclotron radius; this means the side-to-side motion looks completely sinusoidal.

*Each *bunch* might be around 1 pC.

so that the revolution frequency f_r is

$$f_r = \frac{v}{2\pi\rho} = \frac{\beta c}{2\pi\rho} = \frac{1}{2\pi}\frac{qB}{\gamma m_0} = \frac{f_c}{\gamma}. \tag{6.65}$$

This means that relativistically-moving charges emit cyclotron radiation at this modified frequency rather than at the classically-obtained value. Below, we will see that with a sufficiently-large γ there are a number of other important differences.

Whilst the radiation emitted *per charge* might not be very much and therefore doesn't change the kinetic energy of the charges, if we have a very, very large number of charges then the output power can be quite significant. For example, a plasma* immersed in a magnetic field can give out a significant amount of radiation; at any given temperature the electrons will have much greater typical (thermal) velocity v than the ions, and so only the electrons will be significantly radiating. If we have a density N_e electrons per unit volume, then the radiation from the plasma can be written as

$$P = \frac{N_e q^2 \omega_c^2 v^2}{6\pi\epsilon_0 c^3} \tag{6.66}$$

(per unit volume). Re-writing v in terms of the kinetic energy K, we have $P \simeq 6.2 \times 10^{-20} N_e B_0^2 K$ Wm^{-3} where K is the kinetic energy in eV [1].

If the plasma lies within a uniform (i.e. constant) magnetic field, then the emitted radiation has a well-defined single frequency equal to the cyclotron frequency f_c (for the electrons). If the magnetic field is not constant (say, it varies by some small amount across the region occupied by the plasma), then there will be other frequencies also emitted which are harmonics of f_c, i.e. at $2f_c$, $3f_c$ and so on. We can calculate the intensity of these other frequency components by calculating the Fourier transform of the magnetic field variation. An example is that of the typical domestic/commercial fluorescent lamp*; these contain a plasma of ionised mercury vapour caused by an 'arc' ignited with a sufficient voltage; the voltage causes the gas molecules to break down (ionise), and the free electrons then move and cause further ionisations. This is known as a gas discharge lamp; mercury vapour discharges give out blue/ultraviolet wavelengths, and a phosphor coating on the inside of the glass envelope of the tube converts that into a decent spectrum of white light. A typical kinetic energy of the moving charges inside the fluorescent tube might be $K \sim 1$ eV, and the charge density might be $N_e \sim 10^{17}$ m^{-3}. If we take a (switched-on) fluorescent tube and place a magnet near it (giving a field in the tube of – say – 0.1 T), then the emitted power from the electrons in the plasma is $P \simeq 6 \times 10^{-5}$ Wm^{-3}. Note that the volume of plasma inside a fluorescent tube is a small fraction of 1 m^3, so the emitted power is quite small; but it is quite detectable. Also, in contrast to the visible light from the phosphor (which emits frequencies around 10^{14} Hz), the cyclotron radiation from the plasma electrons has an emitted frequency here of $f_c \sim 2.8$ GHz (the cyclotron frequency of electrons is much higher than for protons or ions in the same magnetic field).

We here mention a paradox. Elsewhere in this textbook we frequently encounter current-carrying coils which – for example – are used to generate magnetic fields (usually using iron poles and yokes). Since there are electrons circulating in those coils we might expect that they should radiate, since they obviously must be accelerating inwards in order to go around the coils. However, they do not. One way to see that they do not is to remark (in our other

---

*A plasma here is defined as a volume of ionised atoms in which the numbers of positive and negative charges add up to give a quasi-neutral overall charge.

*The long 'tubes' you often see above your head in lecture theatres and labs.

argument for whether radiation occurs) that in a coil of wire carrying a constant current I there are no time-varying electric or magnetic fields; hence there should be no radiation. How do we resolve this apparent paradox? Clearly, for a sufficiently-smooth distribution of charges (that gives a constant current), there must be cancellation of the radiative fields from each charge. Conversely, we therefore would expect that a non-uniformity of the electron distribution will give rise to net radiation; the frequency spectrum of the emitted radiation should be the Fourier transform of the time variation of the electron density. This is what we see, but we shall not derive it here. Another phenomenon is that nearby charges can give an enhancement of the radiation; this is the phenomenon of coherent radiation, which will be discussed in Chapter 7.

6.2.2   Synchrotron Radiation

We saw that cyclotron radiation is the electromagnetic radiation emitted by non-relativistic charges deflected by moving through a magnetic field – often in a circular path. Synchrotron radiation is the equivalent process, but for when the charges are moving relativistically ($\gamma \gg 1$). In the previous section we saw that relativity modifies the formula for the cyclotron frequency; it also greatly changes the pattern and strength of the emitted radiation.

We saw earlier that a Lorentz transformation of the electric field around a moving charge compresses the electric field lines into a 'pancake' with characteristic width $\sim 1/\gamma$ transverse to the direction of charge motion (as seen by an observer in a different frame of reference). A similar Lorentz transformation of the classical Hertzian dipole radiation pattern results in the pattern of radiation emitted by a relativistically-moving charge. The difference here is that the radiation is compressed into a typical width $\sim 1/\gamma$ in the direction of charge motion. Also, the compression is not symmetric: there is much more radiation emitted in the forward direction than in the backward direction (see Fig 6.12). It is possible to obtain the radiation pattern directly by considering the Liènard-Wiechert potentials [2].

The fact that the opening angle of the radiation is compressed to $\theta \sim 1/\gamma$ by a Lorentz transformation has some important consequences for the nature of the radiation emitted. We can explain those using some simple arguments. Firstly, we picture the effect of the Lorentz transformation on the apparent acceleration experienced by the charge. The (transverse) acceleration in the magnetic field is $a = \mathrm{d}^2 x/\mathrm{d}t^2$ in the charge's frame; as seen by a (stationary) observer, the apparent distance $\mathrm{d}x^* = \mathrm{d}x$ is unchanged by the Lorentz transformation, however the apparent time $\mathrm{d}t = \gamma \mathrm{d}t^*$ to give $a^* = \gamma^2 a$. The acceleration appears to the charge to be occurring over a longer time, and hence there is more radiation emitted. Hence the power emitted is

$$P = \frac{q^2 (a^*)^2}{6\pi\epsilon_0 c^3} = \frac{q^2 a^2 \gamma^4}{6\pi\epsilon_0 c^3}. \tag{6.67}$$

The radiated power is increased by a factor γ^4, which can be enormous if γ is significant.

As an example, we consider a proton and electron with the same kinetic energy $K = 250$ MeV. The proton has $\gamma \simeq 1.3$ – the cyclotron radiation power is increased by a factor $1.3^4 \sim 3$ – and the very small radiated power per proton is still very small. In contrast, the electron has $\gamma \simeq 500$; the radiated power is increased by a factor $500^4 \sim 10^{11}$, which is huge. The limit of this radiated power upon the ultimate energy achievable by electrons was first realised in 1946 by John Blewett [3], and later described by Julian Schwinger in 1949 [4].

The comparison between protons and electrons is hugely technologically significant. As we have seen, using electrons means two things: the radiated power is much higher than it would be for protons; also the radiation is much more forward-directed, which means it is easier to utilise in experiments. Conversely, when colliding particles together in storage

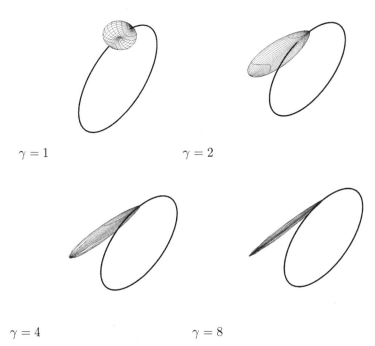

$\gamma = 1$ $\gamma = 2$

$\gamma = 4$ $\gamma = 8$

FIGURE 6.12 Variation in the radiation pattern of charge orbiting (anticlockwise) in a magnetic field as γ rises. When $\gamma = 1$ (non-relativistic motion) the emitted radiation follows the Larmor formula. A dramatic reduction in the opening angle is already apparent for moderate values of γ.

rings* to do particle physics experiments, protons are advantageous because they emit far less electromagnetic radiation; e^+-e^- colliders have the limitation that each doubling of the collision energy gives rise to *sixteen* times the amount of energy lost to radiation, which eventually becomes too costly to replace. Some people think the LEP-2 collider, in which the stored electron/positron energies were in excess of 100 GeV, is the largest energy one can store electrons – even in a relatively large circumference of 27 km.

The Spectrum of Synchrotron Radiation

We saw above that cyclotron radiation is mostly emitted at the same frequency with which the charges are oscillating (either back and forth or when orbiting in a field). Synchrotron radiation is completely different. We may understand what's going on by imagining an observer viewing an orbiting, relativistically-moving charge; rather than seeing continuous cyclotron radiation, the observer only sees synchrotron radiation for a short time per orbit. The observed pulse length is shortened because of the $1/\gamma$ factor of the radiation opening angle, and shortened by another factor $1/\gamma$ because of Lorentz contraction; Fig 6.13 illustrates this. Hence the typical emitted frequency of the synchrotron radiation is related to the cyclotron frequency as

$$f_s \sim f_c \gamma^2. \tag{6.68}$$

---

*We saw in Chapter 2 that a storage ring is a particle accelerator in which the particles are stored for long times by making them orbit repeatedly using dipole magnets.

Another way to explain it is to take the emitted cyclotron frequency in the charge's rest frame, and apply a relativistic Doppler shift into the observer's frame of reference. Hence the observed frequency of the synchrotron radiation is

$$\frac{f_s}{f_c} = \sqrt{\frac{1+\beta}{1-\beta}} \simeq \frac{2}{1-\beta^2} = 2\gamma^2, \tag{6.69}$$

where we can make the approximation given for the Doppler formula because $\beta \simeq 1$.

Radiation pulse shortened by factor γ

Back of pulse catches up with
front of pulse by factor γ

FIGURE 6.13   Illustration of why the frequency spectrum of synchrotron radiation is pulsed with typical frequency $\gamma^2 f_c$.

Both these arguments lead us to conclude that synchrotron radiation has a typical emitted frequency which is *many* times higher than the cyclotron frequency. However, whilst cyclotron radiation is emitted at a single frequency (i.e. at the cyclotron frequency), synchrotron radiation is emitted over a wide range of frequencies; this is because of the pulse length shortening we just mentioned.

Synchrotron radiation – even from a single electron – is pulsed because the narrow angle of emission has the effect that it is only observed fleetingly. It is this pulsed nature that means that it is composed of a wide range of frequency components.[*] We can quantify this by comparing two quantities. Let's consider a charge of rest mass m_0 moving relativistically in a circle due to a uniform magnetic field B. The pulse period (how long it takes the charge to orbit once in the field) is

$$t_r = \frac{1}{f_r} = \frac{2\pi\gamma m_0}{eB} = \frac{\gamma}{f_c} \tag{6.70}$$

where $f_c = eB/2\pi m_e$. The pulse *duration* is

$$dt = \frac{1}{f_s} = \frac{1}{\gamma^2 f_c}. \tag{6.71}$$

Hence the pulse duration is *much* shorter than the period:

$$dt = \frac{t_r}{\gamma^3}. \tag{6.72}$$

So-called synchrotron radiation facilities (or 'sources') utilise these various properties of the radiation emitted by relativistically-moving charges. Because the overall power from electrons far exceeds that from protons, all synchrotron radiation sources utilise electrons.[*]

---

[*]A shorter time duration means a wider frequency spread because one is the Fourier transform of the other.
[*]Actually, some facilities have used positrons instead; positrons have the same rest mass of 0.511 MeV$/c^2$ as electrons giving the same power output for the same stored energy and beam current.

As an example, suppose a synchrotron radiation facility has electrons of $K = 1$ GeV circulating in a constant magnetic field of 1 T. Hence $\gamma \simeq 2000$ and $f_c = 28$ GHz. The orbital period of the electrons is 71 ns, but the pulse duration is 0.9 attoseconds; the synchrotron radiation typical frequency is $f_s \simeq 10^{17}$ Hz. In other words, photons of typical energy $E = hf_s \simeq 460$ eV are emitted; these are so-called soft X-rays. This sets a scale for synchrotron radiation sources; to make X-rays in typical dipole fields of 1 T, we need to use electrons with energies ~ 1 GeV or more.

Our stored electrons above emit light pulses that are basically periodic δ-functions with period t_r. Taking the Fourier transform of this, we can see that the *frequency spectrum* of the synchrotron radiation extends up to $\sim f_s$, with frequency components spaced at $1/t_r$, in other words, spaced apart in frequency by f_c/γ; this is shown schematically in Fig 6.14. $f_s \gg f_c/\gamma$, so the frequency spectrum of synchrotron radiation is basically continuous up to f_s.

In summary: relativistically-moving charges emit light which appears pulsed in time to an observer. The pulsed nature of the light means that it must be composed of many different frequencies from zero up to the typical frequency f_s. Hence, we can see that synchrotron radiation is composed of photons from zero energy up to $\epsilon_s \sim hf_s$. However, each emitted photon is still polarised in the same direction as the electron motion; hence, synchrotron radiation observed *in the plane of the orbit* is linearly polarised, whilst when observed out of the orbit plane (at some angle ψ, say) the radiation will be elliptically polarised.

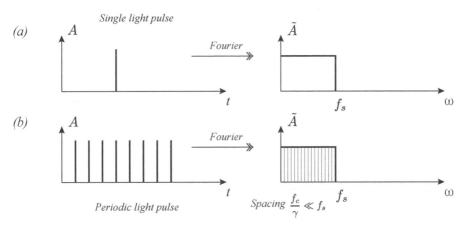

FIGURE 6.14 (a) A single light pulse of duration dt gives a frequency spectrum which is continuous up to a frequency $f_s = 1/dt$. (b) A train of light pulses each of duration dt and separated by a period $T = \gamma/f_c$ gives a frequency spectrum that still extends up to f_s, but is now composed of a set of discrete lines separated in frequency by $1/T = f_c/\gamma = f_s/\gamma^3$. An observer of the radiation from a relativistic electron moving in a circle will see periodic pulses of light of this nature; γ^3 may be very large indeed since typical values of γ encountered are $\sim 10^3$.

Another way to write f_s is as follows. We recall once more that

$$f_s = \gamma^2 f_c = \gamma^2 \frac{qB}{2\pi m_0}. \tag{6.73}$$

But we know that the radius of curvature of a charge moving in a magnetic field B is just

$$\rho = \frac{\beta\gamma m_0 c}{qB}. \tag{6.74}$$

We see that

$$\frac{qB}{m_0} = \frac{\beta\gamma c}{\rho}, \tag{6.75}$$

so that we can express f_s as

$$f_s = \frac{c\beta\gamma^3}{2\pi\rho}. \tag{6.76}$$

Hence the typical emitted photon energy is

$$\epsilon_s = hf_s = \frac{hc\beta\gamma^3}{2\pi\rho}. \tag{6.77}$$

Using our example synchrotron radiation source, our 1 GeV electron moving in a 1 T magnetic field has a bending radius of $\rho = 3.336$ m. We again obtain $f_s \simeq 10^{17}$ Hz, and a typical emitted photon energy of $\epsilon_s = hf_s \simeq 460$ eV. Note that for very relativistic electrons ($\gamma \gg 1$), a very useful formula relating the electron energy $E = \gamma m_e c^2$ to the magnetic field and bending radius is

$$E \text{ [GeV]} \simeq 0.3B\rho \text{ [Tm]}, \tag{6.78}$$

where the units to use are given in square brackets (some books will give the slightly more accurate formula E [GeV] $\simeq 0.2998B\rho$ [Tm]) [5]. This formula doesn't work at all for protons, of course.

Critical Photon Energy and the Emitted Photon Number

In the preceding discussion we have calculated the *typical* photon frequency f_s, where

$$f_s = \frac{c\beta\gamma^3}{2\pi\rho}. \tag{6.79}$$

A fuller calculation can be done than the one we have done here, where a critical frequency can be defined such that, half the radiation power is emitted in photons above the critical frequency, and half the radiation power is emitted in photons below the critical frequency (this is shown later in Section 6.2.3). Hence the critical frequency is also known as the half power point. Since the energy of photons below the critical frequency is obviously lower than that of the energy of photons above the critical frequency, and also the frequency can extend all the way down to zero, synchrotron radiation is composed of very, very many low-frequency photons and rather fewer high-energy photons. The derivation for the critical frequency gives

$$f_{crit} = \frac{3}{2}\frac{c\gamma^3}{2\pi\rho}, \tag{6.80}$$

(since we always use high-energy electrons, we have here set $\beta = 1$ so that there is a corresponding *critical energy* of

$$\epsilon_c = \frac{3}{2}\frac{\hbar c\gamma^3}{\rho}. \tag{6.81}$$

We can write the critical energy in convenient units as ϵ_c [keV] $\simeq 2.218E^3/\rho$ or ϵ_c [keV] $\simeq 0.665E^2B$, where E is given in GeV and B, ρ are in SI units. Alternatively we may calculate the corresponding critical wavelength [6], which is λ_{crit}[Å] $\simeq 18.64/E^2B$.

Since there are many, many more low-energy photons than high-energy photons, the average photon energy is lower than the critical energy. The mean energy of the photons (see below) is

$$\langle\epsilon\rangle = \frac{8\sqrt{3}}{45}\epsilon_c. \tag{6.82}$$

It is worth remarking here about the effect of the quantised nature of the photon emission; an electron experiences a small recoil that lowers the emitted photon energy. Schwinger in

1954 calculated the approximate effect on the overall emitted power [7]; the corrected power (including the quantum effect) is

$$P_q \simeq P \left(1 - \frac{55}{16\sqrt{3}} \frac{\epsilon_c}{\gamma m_0 c^2} \right), \tag{6.83}$$

where P is the power calculated without that correction. For a 3 GeV electron orbiting in a 1.4 T magnetic field (critical energy $\epsilon_c = 8.3$ keV), the correction is around 5.5×10^{-6}, which is small enough to ignore entirely.

It's instructive to consider how many photons are emitted per orbit of the charge. In a uniform field B, a charge emits electromagnetic radiation with average power

$$P = \frac{q^2 a^2 \gamma^4}{6\pi \epsilon_0 c^3}, \tag{6.84}$$

The charge orbits in a circle of radius $\rho = \gamma m_0 c / eB$, so that the acceleration $a = v^2/\rho$ where $v = \beta c$. Hence the power P may be written as

$$\begin{aligned} P &= \frac{q^2 \beta^4 c^4 \gamma^4}{6\pi \epsilon_0 c^3 \rho^2} \\ &= \frac{q^2 c \beta^4 \gamma^4}{6\pi \epsilon_0 \rho^2}. \end{aligned} \tag{6.85}$$

The radiation energy U_0 emitted during one orbit (which takes t_r to happen) is

$$\begin{aligned} U_0 = P t_r &= \frac{e^2 \beta^4 c^4 \gamma^4}{6\pi \epsilon_0 c^3 \rho^2} \frac{2\pi\rho}{\beta c} \\ &= \frac{q^2 \beta^3 \gamma^4}{3\epsilon_0 \rho}. \end{aligned} \tag{6.86}$$

U_0 is known as the energy loss per turn.

So far we have not considered a specific particle type. However, in nearly all practical cases we are dealing with electrons that have a large kinetic energy (say, 10 MeV or higher – usually much higher). Hence $q = e$, $m_0 = m_e$, and $\beta = 1$ to a very good approximation. Therefore we have

$$P = \frac{e^2 c \gamma^4}{6\pi \epsilon_0 \rho^2}, \langle \epsilon \rangle = \frac{8\sqrt{3}}{45} \frac{\hbar c \gamma^3}{\rho}, U_0 = \frac{e^2 \gamma^4}{3\epsilon_0 \rho}. \tag{6.87}$$

We can then estimate the number of photons emitted per orbit as

$$N_\gamma = \frac{U_0}{\langle \epsilon \rangle} = \frac{45}{8\sqrt{3}} \frac{2}{3} \frac{\rho}{\hbar c \gamma^3} \frac{e^2 \gamma^4}{3\epsilon_0 \rho}. \tag{6.88}$$

We note here that the fine-structure constant, α, can be written as

$$4\pi\alpha = \frac{e^2}{\epsilon_0 \hbar c}. \tag{6.89}$$

We can therefore write N_γ more simply as

$$N_\gamma = \frac{5\pi}{\sqrt{3}} \alpha \gamma \simeq 0.0662\gamma. \tag{6.90}$$

This is a very pleasing formula, since it contains only dimensionless constants. Note that N_γ is independent of the circumference, and therefore independent of the bending field B;

a larger bending field leads to a greater rate of photon production but a smaller orbit time. The overall *rate* of photon production is therefore just

$$\dot{N} = \frac{5\pi}{\sqrt{3}} \frac{\alpha\gamma}{\tau_r}. \tag{6.91}$$

A typical storage ring will have $\gamma \sim 10^3$ and $\tau_r \sim 10^{-6}$ s, so that each electron emits $\sim 10^8$ photons per second.

Calculating Synchrotron Radiation Output

We will now carry out some example calculations of photon output for the Diamond storage ring in Oxfordshire (UK); this is a typical storage ring source used to generate X-rays for a variety of scientific experiments.[*] The electrons in Diamond are maintained at a kinetic energy $K = 3$ GeV, and pass through dipole magnets that give a field of 1.4 T, which corresponds to a bending radius $\rho = 7.1$ m; note that the circumference L of the storage ring is not $L = 2\pi\rho$, since not all of the path taken by the electrons has a bending field B applied.[*] In Diamond, the circumference $L = 561.6$ m, so that the revolution period is $\tau_r = L/c \simeq 1.87$ μs. Hence the critical energy of the photons is $\epsilon_c = 8.3$ keV and critical wavelength $\lambda_{crit} = 1.48$ Å, and the mean photon energy is $\langle \epsilon \rangle = 2.6$ keV.

Of course, there isn't just one electron orbiting in Diamond. Knowing that an ammeter placed at any point in the storage ring measures a typical passing current of 300 mA and that obviously $I \equiv \Delta Q/\Delta t$, the total charge in the storage ring ΔQ is

$$\Delta Q = I\Delta t = \frac{IL}{c} \tag{6.92}$$

(see Fig 6.15) where the circumference is $L = 561.6$ m, and $\Delta t = \tau_r$. The number of electrons is then just $N_e = \Delta Q/e \simeq 3.5 \times 10^{12}$ (the stored charge $\Delta Q \simeq 560$ nC) for a current of 300 mA.

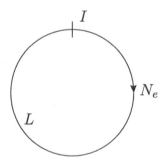

FIGURE 6.15 Relating the current I (observed at some point along the circumference L) for a total number of electrons N_e.

---

[*]There are over fifty such third-generation facilities in the world today.

[*]In fact, in most storage rings only a small fraction of the particle path has dipole field; in Diamond only about 8% of the circumference is dipole magnets. The word 'circumference' when used for storage rings is therefore a bit of a misnomer; by 'circumference' we mean the total distance travelled by the particle in one orbital period.

By comparing the synchrotron radiation power to the revolution period, we can straightforwardly obtain that the energy loss per turn is

$$U_0 = \frac{e^2 \gamma^4}{3\epsilon_0 \rho} \simeq 1.0 \text{ MeV}. \tag{6.93}$$

The total power radiated by each electron is $P_e = 86$ nW, but since there are $\sim 10^{12}$ electrons, the total power emitted is $P_{total} = N_e P_e \simeq 300$ kW. This is a simply enormous power. Synchrotron radiation facilities such as Diamond are the only known method of producing such a large quantity of X-ray photons; they are one of the brightest artificial sources of photons. The number of photons emitted by each electron as it executes a single orbit is $N_\gamma = 5\pi\alpha\gamma/\sqrt{3} \simeq 380$ photons, or $\sim 2 \times 10^8$ photons per second. Hence the radiation may be treated as 'quasi-continuous'.

You might be wondering what the effect is of splitting up the dipoles into pieces, rather than having a continuous bending B field as we had in our original derivation for N_γ. The way to understand it is that photons are only being emitted when the electrons are passing through the dipole magnets; they're not being emitted when there is no B field accelerating them.* Hence, N_γ is just the same regardless of whether there are 'gaps' in the B field. $t_r\gamma/f_c$ is still calculated the same way, but now it is not the same as τ_r, the time it takes for the electrons to go around the storage ring; τ_r is only needed to convert beam *current* into a number of electrons.

In our example of the Diamond storage ring we note that the radiation from N_e electrons has a power $P \propto N_e$; this radiation is incoherent synchrotron radiation (ISR). For a beam current I_b the total emitted power is

$$P_{\text{total}} = \frac{e\gamma^4 I_b}{3\epsilon_0 \rho}; \tag{6.94}$$

this may be expressed in practical units as

$$P_{\text{total}} [\text{kW}] = 88.4 \frac{E \text{ [GeV]}^4 I_b \text{ [A]}}{\rho \text{ [m]}}. \tag{6.95}$$

Another way to express the emitted power is simply as $P_{\text{total}} [\text{kW}] = U_0 [\text{keV}] I_b [\text{A}]$.

6.2.3 The Spectrum of Emitted Synchrotron Radiation

In this section we consider only the emission of radiation by ultra-relativistic electrons, since these are the only particles used practically in synchrotron radiation sources. As we saw above, an electron circulating in a uniform magnetic field B has an effective angular frequency ω_0 that may be written as

$$\omega_0 = \frac{\beta c}{\rho} \tag{6.96}$$

where ρ is the bending radius of the electron. An (angular) critical frequency may also be defined as

$$\omega_c = 3c\gamma^3/2\rho. \tag{6.97}$$

---

*Actually, in real storage rings there are extra 'insertion' devices (see later in this chapter) that can produce additional photons, but this fact doesn't change the basic argument about N_γ.

It can be shown that [8] the horizontal and vertical electric field components of the far-field radiation at a given frequency ω – as seen by an observer looking at the electron at some angle ψ to the plane of the electron orbit – are

$$
\begin{aligned}
E_x(\omega) &= \frac{\sqrt{3}e\gamma}{4\pi\sqrt{2\pi}c\epsilon_0 R}\left(\frac{\omega}{\omega_c}\right)\left(1+\gamma^2\psi^2\right)K_{2/3}(G), \\
E_y(\omega) &= \frac{i\sqrt{3}e\psi\gamma^2}{4\pi\sqrt{2\pi}c\epsilon_0 R}\left(\frac{\omega}{\omega_c}\right)\left(1+\gamma^2\psi^2\right)^{1/2}K_{1/3}(G).
\end{aligned}
\tag{6.98}
$$

R is the distance of the observer from the electron, $K_{2/3}(G)$ and $K_{1/3}(G)$ are the standard modified Bessel functions, and

$$
G = \left(\frac{\omega}{2\omega_c}\right)\left(1+\gamma^2\psi^2\right)^{3/2}.
\tag{6.99}
$$

In the far field, the radiation seen by the observer must be an electric field $\mathbf{E}$ perpendicular to the observation direction $\mathbf{n}$, so that the Poynting vector is

$$
\mathbf{S} = \epsilon_0 cE^2\mathbf{n}.
\tag{6.100}
$$

If the observer sees the radiation over a solid angle $\Delta\Omega$, the total energy passing through this area ($R^2\Delta\Omega$) in some time Δt is

$$
W = (\boldsymbol{n}\cdot\boldsymbol{S})\Delta tR^2\Delta\Omega = \epsilon_0 cE^2R^2\Delta t\Delta\Omega.
\tag{6.101}
$$

We can relate this to the total power emitted $P(t)$ over time as

$$
P(t) = \frac{\mathrm{d}W}{\mathrm{d}t} = \int_0^{4\pi}\epsilon_0 cE(t)^2R^2\mathrm{d}\Omega.
\tag{6.102}
$$

Writing $E(t)$ in terms of its Fourier transform

$$
E(t) = \frac{1}{\sqrt{2\pi}}\int_{-\infty}^{\infty}E(\omega)e^{-i\omega t}\mathrm{d}\omega
\tag{6.103}
$$

we can obtain the energy passing through a solid angle as

$$
\frac{\mathrm{d}W}{\mathrm{d}\Omega} = 2\epsilon_0 cR^2\int_0^{\infty}|E(\omega)|^2\mathrm{d}\omega.
\tag{6.104}
$$

We may thus define the *spectral* angular distribution (i.e. into a bandwidth $\mathrm{d}\omega$ around a given frequency ω) as

$$
\frac{\mathrm{d}^2W}{\mathrm{d}\Omega\mathrm{d}\omega} = 2\epsilon_0 cR^2|E(\omega)|^2.
\tag{6.105}
$$

Since the electron executes $c/2\pi\rho$ revolutions per second in the magnetic field B, we may write the spectral power density as

$$
\frac{\mathrm{d}^2P}{\mathrm{d}\Omega\mathrm{d}\omega} = \frac{c}{2\pi\rho}\frac{\mathrm{d}^2W}{\mathrm{d}\Omega\mathrm{d}\omega} = \frac{R^2}{\pi\mu_0\rho}|E(\omega)|^2.
\tag{6.106}
$$

Since $|E(\omega)|^2 = E_x^2(\omega)+E_y^2(\omega)$, we may finally obtain the spectral power density of bending magnet (dipole) radiation as

$$
\frac{\mathrm{d}^2P}{\mathrm{d}\Omega\mathrm{d}\omega} = \frac{3e^2\gamma^2}{32\pi^4\epsilon_0\rho}\left(\frac{\omega}{\omega_c}\right)^2\left(1+\gamma^2\psi^2\right)^2\left[K_{2/3}^2(G)+\frac{\gamma^2\psi^2}{(1+\gamma^2\psi^2)}K_{1/3}^2(G)\right].
\tag{6.107}
$$

Integrating over all angles gives the spectral power

$$\frac{\mathrm{d}P}{\mathrm{d}\omega} = \int \frac{\mathrm{d}^2 P}{\mathrm{d}\Omega \mathrm{d}\omega} \mathrm{d}\Omega = 2\pi \int \frac{\mathrm{d}^2 P}{\mathrm{d}\psi \mathrm{d}\omega} \mathrm{d}\psi$$
$$= \frac{P_0}{\omega_c} S\left(\frac{\omega}{\omega_c}\right) = \frac{P_0}{\omega_c}\left(S_x\left(\frac{\omega}{\omega_c}\right) + S_y\left(\frac{\omega}{\omega_c}\right)\right) \tag{6.108}$$

where we have now written the total power as

$$P_0 = \frac{ce^2\gamma^4}{6\pi\epsilon_0\rho^2} \tag{6.109}$$

and the relative spectral powers of the horizontal and vertical polarised components S_x and S_y (and their sum S) are

$$S_x\left(\frac{\omega}{\omega_c}\right) = \frac{9\sqrt{3}\omega}{16\pi\omega_c}\left[\int_{\omega/\omega_c}^\infty K_{5/3}(u)\mathrm{d}u + K_{2/3}\left(\frac{\omega}{\omega_c}\right)\right],$$
$$S_y\left(\frac{\omega}{\omega_c}\right) = \frac{9\sqrt{3}\omega}{16\pi\omega_c}\left[\int_{\omega/\omega_c}^\infty K_{5/3}(u)\mathrm{d}u - K_{2/3}\left(\frac{\omega}{\omega_c}\right)\right],$$
$$S\left(\frac{\omega}{\omega_c}\right) = \frac{9\sqrt{3}\omega}{8\pi\omega_c}\int_{\omega/\omega_c}^\infty K_{5/3}(u)\mathrm{d}u. \tag{6.110}$$

The function $S(\omega/\omega_c)$ is universal (common to any ω_c) and is shown in Fig 6.16 [8, 6, 5]. Integrating over all frequencies we obtain

$$\int_0^\infty S_x\left(\frac{\omega}{\omega_c}\right)\mathrm{d}\left(\omega/\omega_c\right) = \frac{7}{8},$$
$$\int_0^\infty S_y\left(\frac{\omega}{\omega_c}\right)\mathrm{d}\left(\omega/\omega_c\right) = \frac{1}{8},$$
$$\int_0^\infty S\left(\frac{\omega}{\omega_c}\right)\mathrm{d}\left(\omega/\omega_c\right) = 1. \tag{6.111}$$

We see that, of all the power radiated, exactly $7/8$ is horizontally-polarised whilst $1/8$ is vertically-polarised, in other words $P_x/P_y = 7$. This is in contrast to cyclotron radiation, where $P_x/P_y = 3$. Integrating frequencies up to $\omega = \omega_c$ only, we see

$$\int_0^1 S\left(\frac{\omega}{\omega_c}\right)\mathrm{d}\left(\omega/\omega_c\right) = \frac{1}{2}. \tag{6.112}$$

This demonstrates what we said earlier, which is that half the total radiation power is emitted at frequencies below the critical frequency ω_c. Finally, we may similarly obtain how the emitted power varies with observation angle ψ as

$$\frac{\mathrm{d}P}{\mathrm{d}\psi} = \frac{21P_0\gamma}{32\left(1 + \gamma^2\psi^2\right)^{5/2}}\left[1 + \frac{5\gamma^2\psi^2}{7\left(1 + \gamma^2\psi^2\right)}\right]. \tag{6.113}$$

Photon Flux

We have so far derived the emitted synchrotron radiation power as a function of emission frequency. It is straightforward to re-express this in terms of the number of photons. Each emitted photon has an energy $\epsilon = h\omega/2\pi = \hbar\omega$, so the critical energy of emission is $\epsilon_c = \hbar\omega_c$.

If the number of photons emitted per second with energy ϵ is $\dot{N}(\epsilon)$, then the power emitted at that energy is just $\epsilon\dot{N}(\epsilon)$. We may use this to determine the number of photons emitted into a bandwidth $\delta\epsilon/\epsilon$ as

$$
\begin{aligned}
\frac{\mathrm{d}^2\dot{N}}{\mathrm{d}\Omega\mathrm{d}\epsilon/\epsilon} &= \frac{\epsilon\mathrm{d}^2\dot{N}}{\mathrm{d}\Omega\mathrm{d}\epsilon} \\
&= \frac{\mathrm{d}^2P}{\mathrm{d}\Omega\mathrm{d}\epsilon} = \frac{\mathrm{d}^2P}{\hbar\mathrm{d}\Omega\mathrm{d}\omega} \\
&= \frac{3e^2\gamma^2}{32\pi^4\hbar\epsilon_0\rho}\left(\frac{\omega}{\omega_c}\right)^2(1+\gamma^2\psi^2)^2\left[K_{2/3}^2(G) + \frac{\gamma^2\psi^2}{(1+\gamma^2\psi^2)}K_{1/3}^2(G)\right].
\end{aligned}
\tag{6.114}
$$

Using the definition of the fine-structure constant

$$
\alpha = \frac{e^2}{2ch\epsilon_0} \approx \frac{1}{137}
\tag{6.115}
$$

we may obtain

$$
\begin{aligned}
\frac{\mathrm{d}\dot{N}}{\mathrm{d}\Omega} &= \frac{3\alpha\gamma^2}{4\pi^2}\frac{c}{2\pi\rho}\left(\frac{\delta\epsilon}{\epsilon}\right)\left(\frac{\omega}{\omega_c}\right)^2 \\
&\quad\times(1+\gamma^2\psi^2)^2\left[K_{2/3}^2(G) + \frac{\gamma^2\psi^2}{(1+\gamma^2\psi^2)}K_{1/3}^2(G)\right].
\end{aligned}
\tag{6.116}
$$

This quantity $\mathrm{d}\dot{N}/\mathrm{d}\Omega$ is the spectral intensity. If we have N_e electrons, we may re-cast this expression in terms of the beam current $I_b = N_e ec/2\pi\rho$ to obtain

$$
\begin{aligned}
\frac{\mathrm{d}\dot{N}}{\mathrm{d}\Omega} &= \frac{3\alpha\gamma^2}{4\pi^2}\frac{I_b}{e}\left(\frac{\delta\epsilon}{\epsilon}\right)\left(\frac{\omega}{\omega_c}\right)^2(1+\gamma^2\psi^2)^2 \\
&\quad\times\left[K_{2/3}^2(G) + \frac{\gamma^2\psi^2}{(1+\gamma^2\psi^2)}K_{1/3}^2(G)\right].
\end{aligned}
\tag{6.117}
$$

This expression can be given for the on-axis ($\psi = 0$) emission from a beam current I_b (in Amperes) of electrons of energy E (given in GeV) as

$$
\left.\frac{\mathrm{d}\dot{N}}{\mathrm{d}\Omega}\right|_{\psi=0} = 1.33 \times 10^{13} E^2 I_b \left(\frac{\omega}{\omega_c}\right)^2 K_{2/3}^2\left(\frac{\omega}{2\omega_c}\right)
\tag{6.118}
$$

which has been given in the usual units of *photons per second per milliradian squared per 0.1% bandwidth*. From this we can obtain the total rate of photon emission into a given energy bandwidth $\delta\epsilon/\epsilon = \delta\omega/\omega$ as

$$
\dot{N} = \sqrt{3}\alpha\gamma\frac{I_b}{e}\left(\frac{\delta\epsilon}{\epsilon}\right)\left(\frac{\epsilon}{\epsilon_c}\right)\int_{u=\epsilon/\epsilon_c}^{\infty}K_{5/3}(u)\mathrm{d}u.
\tag{6.119}
$$

This quantity – known as the spectral photon flux – is consistent with the earlier expression for photons emitted in an orbit, given in terms of the typical photon energy, $N_\gamma = 5\pi\alpha\gamma/\sqrt{3}$. We may write this as

$$
\dot{N} = 2.46 \times 10^{13} E I_b \left(\frac{\epsilon}{\epsilon_c}\right)\int_{u=\epsilon/\epsilon_c}^{\infty}K_{5/3}(u)\mathrm{d}u.
\tag{6.120}
$$

Here, $\dot{N}$ has been given in units of photons per second, per horizontal milliradian, per 0.1% bandwidth. This is a very useful expression because it follows the universal curve $\epsilon/\epsilon_c\int_{u=\epsilon/\epsilon_c}^{\infty}K_{5/3}(u)\mathrm{d}u$. The spectral flux peaks at around $\epsilon/\epsilon_c \simeq 0.25$, and falls off sharply above $\epsilon/\epsilon_c \simeq 5$; when $\epsilon/\epsilon_c = 10$ the emitted power has fallen by about $3400\times$.

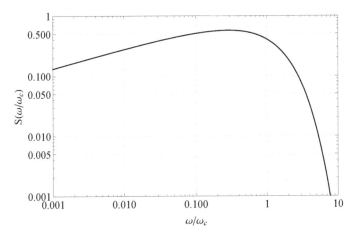

FIGURE 6.16 The universal curve $S(\omega/\omega_c)$ for synchrotron radiation.

Synchrotron Radiation Sources

We have so far discussed dipole radiation by relativistic electrons – also known as bending magnet radiation – which arises naturally in any particle accelerator where those electrons have sufficient kinetic energy. One important application of this is in so-called synchrotron radiation sources, where this radiation is deliberately generated at specific wavelengths for use in a wide variety of scientific research. For example, X-ray diffraction experiments utilise photons whose wavelength is comparable to the inter-atomic separation in solids – i.e. about 1 Å; a crystal formed from many atoms has a regular, periodic arrangement of its constituent atoms that will give rise to interference between the radiation scattered from each atom – a diffraction pattern is formed that can be used to elucidate the arrangement of the atoms. Whether a simple crystal such as NaCl or one made of more complex molecules such as proteins, the formation of clear diffraction spots depends both upon the spread of wavelengths incident upon the crystal and upon how parallel the X-rays are. A good X-ray beam brightness depends upon a small electron beam source size at the location of synchrotron radiation emission, and the natural broad spectrum of synchrotron emissions must have suitable wavelengths selected from it using a monochromator. To quickly form a distinct diffraction pattern requires the highest possible X-ray intensity, and is limited by the ability of the monochromator and other X-ray optics to handle the heat load; typical limits are several hundred W/mm^2.

Synchrotron radiation sources are dedicated facilities – usually large (e.g. covering areas exceeding 100 m × 100 m) – providing a variety of experimental beamlines with radiation tailored differently depending upon the use [9, 6, 10]. The predominant sources in use today are so-called third-generation sources, which are electron storage rings within which electrons of several GeV in energy circulate for many hours at a time; third-generation sources, which are designed by definition to incorporate insertion devices (see below), have largely supplanted the earlier first-generation sources that parasitically used electron synchrotrons (SURF, Tantalus-I, NINA), and second-generation sources that relied mostly on dipole radiation (such as the Daresbury SRS, and NSLS in the United States).

One simple way to vary the output wavelength in a storage ring is to use a wavelength shifter, which is essentially a high-field dipole inserted amongst the other dipoles which are needed to form the overall storage ring. Hence a wavelength shifter is a type of *insertion device*, whose magnetic field can be turned on or off without significantly affecting the operation of the rest of the storage ring. The first wavelength shifters attempted to extend the output radiation emission to shorter wavelengths, and hence a higher field was required. Often, superconducting magnets are used to achieve fields as high as 6 T or more; an example is shown in Fig 6.17.

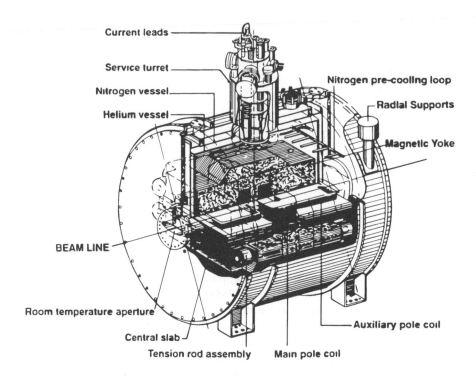

FIGURE 6.17 An example of a high-field wavelength shifter, as used at the 2 GeV Daresbury Synchrotron Radiation Source in the 1990s and up to 2008. A central 6 T field created by superconducting coils is used to generate a larger critical photon energy than was obtained using the main 1.2 T dipoles. Two lower-field ancillary poles lie either side of the main pole to create a localised orbit 'bump' within the device so that it can be turned on and off without changing the overall geometry of the storage ring [11]; some small beam-optical corrections are however still required. It is hence known as an insertion device. (Diagram adapted from original © STFC.)

6.2.4 Wiggler Radiation

An extension to the idea of the wavelength shifter is the multipole wiggler; a multipole wiggler comprises an alternating field arranged along an electron's path, provided by poles of alternating polarity [8]. An example of a multipole wiggler is shown in Fig 6.18. Assuming to begin with that there is only a vertical magnetic field B_y (as in an ordinary dipole magnet), an alternating magnetic field may be approximately described as sinusoidal with a spatial period λ_u equal to the distance between neighbouring north poles. Thus

$$B_y(s) = -B_0 \sin\left(\frac{2\pi s}{\lambda_u}\right) \tag{6.121}$$

where B_0 is the peak field in the wiggler. The resultant acceleration on the electron is $\ddot{x} = \mathrm{d}^2x/\mathrm{d}s^2 = eB_y/\gamma m_0 c$ and only in the horizontal plane. The electron deflection angle $\dot{x}$

is

$$\dot{x}(s) = \frac{K}{\gamma} \cos\left(\frac{2\pi s}{\lambda_u}\right), \tag{6.122}$$

where the so-called K-parameter is

$$K = \frac{B_0 e}{m_0 c} \frac{\lambda_u}{2\pi} = 0.9336 \, B_0 \lambda_u \tag{6.123}$$

and where in the right-hand expression, B_0 is expressed in T and λ_u in centimetres. K is dimensionless since $\dot{x} = dx/ds$. Integrating, we obtain the path through the wiggler as

$$x(s) = \frac{K}{\gamma} \frac{\lambda_u}{2\pi} \sin\left(\frac{2\pi s}{\lambda_u}\right). \tag{6.124}$$

The usefulness of K is that the maximum angular deflection is K/γ. Since the opening angle of the emitted radiation is $\sim 1/\gamma$, then if $K < 1$ the radiation from each pair of poles overlaps – giving rise to interference of the radiation – whilst if $K \gg 1$ then there is little overlap and the radiation from each pole pair is effectively independent. The condition $K \gg 1$ defines a multipole wiggler, and $K < 1$ defines an undulator; otherwise, they are much the same. Obviously, undulators typically utilise magnetic fields lower than those in wigglers – say, less than ~ 1 T. In practice there is a regime between $K \sim 1$ and $K \sim 5$ where there is some interference between the poles.

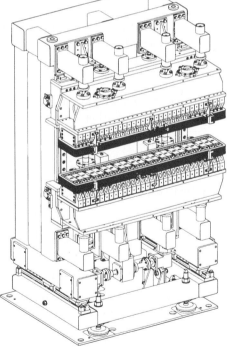

FIGURE 6.18 An example of a multipole wiggler, here generating an on-axis field with a maximum 2.4 T using a hybrid arrangement of permanent-magnet pieces and steel poles. Half-poles at each end compensate the overall orbit shift, and an adjustable gap allows variation of the field (and hence the output photon energy). (Photograph © STFC, diagram adapted from original © STFC.)

Wigglers emit radiation with a critical energy

$$\epsilon_c = \frac{3hc\gamma^3}{4\pi\rho} \tag{6.125}$$

that depends on the instantaneous bending radius ρ. As with dipole radiation we may state this in practical units as

$$\epsilon_c \; [\text{eV}] = 665.025 \, E[\text{GeV}]^2 B \; [\text{T}]. \tag{6.126}$$

Since the wiggler is an insertion device, it can be turned on and off, but more importantly its field may be adjusted more or less at will between those values.* Electromagnetic wigglers (EMWs) may adjust their field simply by varying the current that drives the field through the wiggler poles; permanent-magnet wigglers (PMWs) can vary their field by varying the gap between the poles (with some limitations).

As discussed previously in Chapter 4, PMWs typically use poles made from either SmCo (remanent field 0.9–1.1 T) or NdFeB (remanent field 1.1–1.4 T), either in a pure permanent magnet (PPM) arrangement using only permanent magnetic material [12, 9, 8], or with the addition of steel pole pieces in a *hybrid* arrangement to augment the on-axis field (see for example Fig 4.23 in Section 4.5). In a PPM configuration the maximum on-axis field attainable is around

$$B_y \simeq 1.72 B_r e^{-\pi g/\lambda_u}, \tag{6.127}$$

where g is the gap between the poles (i.e. the available height for beam and vacuum vessel); in practice $g/\lambda_u < 0.1$ is a realistic limit, so that PPM wigglers are limited to fields no more than around 1.5 T. A hybrid wiggler might augment the on-axis field by perhaps 30%.

Multipole wigglers (MPWs) are different from ordinary dipoles in that the effective critical energy seen depends upon the horizontal observation angle with respect to the axis of the wiggler. Viewed by an observer looking along the wiggler axis, the maximum critical energy (denoted ϵ_{c0}) is

$$\epsilon_{c0} \; [\text{eV}] = 665.025 \, E[\text{GeV}]^2 B \; [\text{T}]. \tag{6.128}$$

Radiation into other observation angles θ is determined by the critical energy when the electron is pointing at θ; this varies with s and is

$$\epsilon_c = \epsilon_{c0} \sin\left(\frac{2\pi s}{\lambda_u}\right). \tag{6.129}$$

Knowing that $\cos(2\pi s/\lambda_u) = \theta\gamma/K$, we find the critical energy at angle θ is

$$\epsilon_c = \epsilon_{c0} \sqrt{1 - \left(\frac{\theta\gamma}{K}\right)^2} = \epsilon_{c0} \sqrt{1 - \left(\frac{\theta}{\theta_{max}}\right)^2}. \tag{6.130}$$

Multipole Wiggler Flux and Tuning

The previous discussion allows us to summarise the emission advantages of a multipole wiggler over a dipole magnet:

- The (on-axis) critical energy may be conveniently adjusted, independently of the electron energy and of other devices in the synchrotron radiation facility.

---

*Usually some small but manageable adjustments are made to the beam focusing when the wiggler field is changed.

- The critical energy varies with observation angle, which allows some additional tuning.

- The total wiggler flux is increased by a factor N_u, where N_u is the number of wiggler poles.

To give a sense of the advantages, a typical EMW might comprise 50 pole pairs ($N_u = 50$), each with a peak magnetic field of perhaps 1.6 T if normal conducting, or 4 to 5 T if superconducting, which is variable down to zero. The total power emitted from *any* insertion device with a sinusoidally-varying field may be simply obtained as

$$P_{total} = \frac{1}{3} r_e m_e c^2 \gamma^2 K^2 \frac{4\pi^2}{\lambda_u} = 632.8 E^2 B_0^2 L I_b, \qquad (6.131)$$

where E is the electron energy in GeV. Similar to dipole radiation, radiation from a multipole wiggler is linearly-polarised when viewed in the plane of the electron oscillations.

6.2.5 Undulators

An undulator is defined as a multipole device where the output is dominated by interference effects [13]. We already know that when $K < 1$ there is significant overlap of the emitted radiation from each pole pair; we therefore expect interference will occur at certain wavelengths, enhancing the output intensity; this idea was first proposed by Vitaly Ginzberg in 1947 [14, 4] and verified experimentally by Hans Motz in 1953 [15]. To determine the wavelengths for which constructive interference occurs, we again assume the electron motion is sinusoidal; the combination of its finite electron velocity $\beta < 1$ and its periodic transverse velocity causes the electron to fall behind the photons it has emitted. The average velocity in the forward direction is

$$\langle \beta_s \rangle \simeq \beta - \frac{K^2}{4\beta\gamma^2} \simeq 1 - \frac{1}{2\gamma^2} - \frac{K^2}{4\beta\gamma^2}. \qquad (6.132)$$

The condition for interference to occur is that each electron should fall behind its emitted wavefront by a whole number of wavelengths per period of undulator passed through. Observed at a horizontal angle θ, the condition for constructive interference at emitted wavelength λ is

$$n\lambda = \frac{\lambda_u}{\langle \beta_s \rangle} - \lambda_u \cos\theta. \qquad (6.133)$$

Substituting the value for $\langle \beta_s \rangle$ and for small angles of θ, we can hence obtain the so-called undulator equation

$$\lambda = \frac{\lambda_u}{2n\gamma^2} \left(1 + \frac{K^2}{2} + \theta^2 \gamma^2 \right). \qquad (6.134)$$

Each value of n is known as the harmonic number of the emission (not to be confused with the storing ring harmonic number h, which is the number of circulating bunches). For example, a 3 GeV electron passing through an undulator with a period of 50 mm and $K = 1$ emits on-axis photons ($\theta = 0$) with a wavelength of 1.1 nm, or an energy of 1.1 keV. The most important thing to note about the undulator equation is the γ^2 factor between the undulator wavelength and the emission wavelength; this arises because of Lorentz contraction acting to shrink the period of the undulator as observed by the electron and a Doppler shift of the electron emission into the observation (laboratory) frame. γ is typically a few thousand and undulators have a typical period λ_u of a few centimetres, so we immediately see that emission wavelengths will typically be $\sim 10^{-9}$ m; again, this is useful for typical X-ray experiments [8]. Also, note that as an undulator gap is closed, K increases which makes the output wavelength λ longer.

Since interference is occurring from all the N_u poles of the undulator, the emitted radiation will be confined within a certain bandwidth

$$\frac{\Delta\lambda}{\lambda} \simeq \frac{1}{N_u n}. \tag{6.135}$$

For example, a 100-period undulator will emit photons in the first harmonic with a wavelength spread of about 1%, which is around ten times larger than the typical spread of the electron energies in a storage ring (see later); the bandwidth is determined by the undulator and not by the electron energy spread. The opening angle of the radiation is limited also by interference effects to

$$\Delta\theta \simeq \sqrt{\frac{2\lambda}{N_u \lambda_u}}. \tag{6.136}$$

For example, a 50 mm-period undulator with 100 periods has $\Delta\theta = 40$ μrad, which is less than the radiation opening angle $1/\gamma \sim 170$ μrad at 3 GeV. It should also be noted that the on-axis radiation contains only the odd harmonics $n = 1, 3, 5, 7...$.

One of the key advantages of undulators is that they greatly enhance the radiation output at desired wavelengths whilst suppressing it at unwanted wavelengths. For a given photon flux at an experimental sample this means that much less unwanted X-ray power is dissipated on the monochromators.* The angular flux density of the emission is given in practical units as

$$\left. \frac{\mathrm{d}\dot{N}}{\mathrm{d}\Omega} \right|_{\theta=0} = 1.74 \times 10^{14} N_u^2 E^2 I_b F_n(K), \tag{6.137}$$

where

$$F_n(K) = \frac{n^2 K^2}{(1 + K^2/2)^2} \left(J_{(n+1)/2}(Y) - J_{(n-1)/2}(Y) \right)^2 \tag{6.138}$$

and

$$Y = \frac{nK^2}{4(1 + K^2/2)}. \tag{6.139}$$

When K is very small (<0.5), $F_n(K)$ is only significantly greater than zero in the first harmonic; in other cases undulators can be utilised routinely up to harmonic number 15 or so. The angular flux density $\mathrm{d}\dot{N}/\mathrm{d}\Omega \propto N_u^2$, so for example an undulator with $N_u = 100$ periods gives a photon flux density in the first harmonic which is nearly $N_u^2 = 10,000$ times larger than from a simple dipole magnet. It is possible to show that the photon output in the fundamental harmonic from each electron passing through a magnet period is

$$N_\gamma = \frac{2\pi}{3}\alpha K^2, \tag{6.140}$$

from which the total photon output can be readily estimated.

We have in the above discussion only considered ordinary undulators that deflect the electrons in a single plane; in this case the emitted photons are still linearly polarised when observed in the plane of electron oscillation, as they are from dipoles and multipole wigglers. There also exist a wide variety of more complicated magnetic arrangements in which the electrons may execute both horizontal and vertical motion, and devices may be constructed to give radiation with both tuneable wavelength and polarisation.

---

*A monochromator is a device used to select one X-ray wavelength from a broadband source, and is usually made from a large single silicon crystal in conjunction with collimation

6.3   Scattering of Electromagnetic Radiation

6.3.1   The Scattering Cross Section

An electromagnetic wave passing over atoms causes the charges in those atoms to accelerate. Hence those charges radiate; this idea is shown schematically in Fig 6.19. The process of absorption of electromagnetic energy by atoms and then re-radiation of that energy is *scattering*.

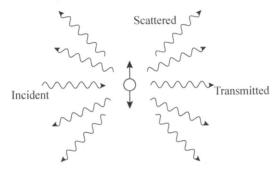

FIGURE 6.19   Scattering may be explained in a classical description by considering that incident radiation provides an electric field that accelerates the charges in a medium, removing energy from the incident radiation field. The accelerated charges emit radiation in many directions; the fraction in the original direction may be considered – along with the part of the incident radiation field that was not absorbed – the 'transmitted' radiation field. The radiation emitted in other directions is the 'scattered' radiation field. Note that in this classical picture there is not such a thing as an individual photon which is both incident and then scattered; rather, the incident photon is absorbed and then re-emitted in the radiated field.

Atoms contain bound electrons, which will move to a position $\mathbf{z}$ due to the force imparted by a passing electromagnetic wave (the nuclei are more massive and effectively stay still). The displaced electrons give rise to an oscillating dipole moment in the atom of

$$\mathbf{p}(t) = -e\mathbf{z} = \frac{\frac{e^2}{m_e} E_0 \cos \omega t}{(\omega_0^2 - \omega^2) + i\omega\gamma}. \tag{6.141}$$

Note that this expression holds for a single resonant frequency ω_0 of the electrons, but we can extend this analysis for multiple frequencies if we wish. This oscillating dipole will radiate quasi-isotropically, i.e. like a Hertzian dipole, with total emitted power

$$\mathbf{P}(t) = \frac{\omega^4 |\mathbf{p}(\tau)|^2}{6\pi\epsilon_0 c^3}, \tag{6.142}$$

where $\tau = t - r/c$ is the usual retarded time for an observer at a distance r from the moved electron. We can immediately combine those equations to obtain the average power emitted from each atom as

$$\langle P \rangle = \frac{e^4}{12\pi\epsilon_0 m_e^2 c^3} \frac{E_0^2 \omega^4}{(\omega_0^2 - \omega^2)^2 + \omega^2 \gamma^2}. \tag{6.143}$$

A question arises: can we relate the incident power to the radiated power? To do this, we use the idea of the cross section, which is the effective area (in this case) of the scattering objects. The cross section for an atom can be defined as

$$\sigma = \frac{P}{\langle S \rangle}, \tag{6.144}$$

i.e. it is the ratio of the emitted power to the incident power. This is the *total* cross section, in other words it describes the rate at which the incident power is converted to radiated (scattered) power for any direction of the radiated power. We can subdivide this total cross section into the rate emitted at different angles θ and ϕ; this is a type of differential cross section.

The cross section σ is an effective area presented by the atom (really, by the electrons in the atom) to the incoming radiation. Thinking of the incoming radiation as being made of discrete photons, some of those photons will strike the area σ and those photons will be scattered; other photons that do not strike this effective area will not be scattered. σ thus describes the proportion of photons that are scattered. We can understand how a cross section operates by counting the number of photons in the incident and scattered radiation as

$$\underbrace{P}_{\#/s} = \underbrace{\sigma}_{m^2} \underbrace{S}_{\#/m^2 s}. \tag{6.145}$$

Since we can write the Poynting vector in terms of the energy density as

$$\langle S \rangle = c \langle U \rangle = \frac{1}{2} c \epsilon_0 E_0^2, \tag{6.146}$$

we can re-write the scattered power (collecting separately the constants, frequency dependence and Poynting vector) as

$$\begin{aligned}
\langle P \rangle &= \frac{e^4}{12 \pi \epsilon_0 m_e^2 c^3} \frac{E_0^2 \omega^4}{(\omega_0^2 - \omega^2)^2 + \omega^2 \gamma^2} \\
&= \frac{e^4}{6 \pi \epsilon_0^2 m^2 c^4} \frac{\omega^4}{(\omega_0^2 - \omega^2)^2 + \omega^2 \gamma^2} \frac{1}{2} c \epsilon_0 E_0^2 \\
&= \frac{e^4}{6 \pi \epsilon_0^2 m^2 c^4} \frac{\omega^4}{(\omega_0^2 - \omega^2)^2 + \omega^2 \gamma^2} \langle S \rangle.
\end{aligned} \tag{6.147}$$

Hence, the total scattering cross section is

$$\sigma = \frac{e^4}{6 \pi \epsilon_0^2 m^2 c^4} \frac{\omega^4}{(\omega_0^2 - \omega^2)^2 + \omega^2 \gamma^2}. \tag{6.148}$$

We can re-write this more simply by defining a constant (that you have probably seen before): we define the classical electron radius – which obviously has dimensions of length – as

$$r_e = \frac{e^2}{4 \pi \epsilon_0 m c^2} \simeq 2.818 \times 10^{-15} \text{ m}. \tag{6.149}$$

Substituting into our expression for σ, we obtain

$$\sigma = \frac{8 \pi r_e^2}{3} \frac{\omega^4}{(\omega_0^2 - \omega^2)^2 + \omega^2 \gamma^2}. \tag{6.150}$$

We note that σ has the correct dimensions (m^2) for a cross section. This is a general form for the scattering cross section of atoms.

Let's look at some special cases of the scattering cross section. The first is for low-frequency photons for which $\omega \ll \omega_0$. Hence this situation describes the scattering of light of wavelength much longer than the wavelengths at which absorption will be taking place in those atoms. We obtain a scattering cross section of

$$\sigma_R \simeq \frac{8 \pi r_e^2}{3} \frac{\omega^4}{\omega_0^4} \propto \frac{1}{\lambda^4}, \tag{6.151}$$

where λ is the wavelength of the incident/scattered radiation. We use the subscript R since this cross section (and the phenomenon that goes with it) is known as Rayleigh scattering. We see that shorter wavelengths are scattered *much* more than longer wavelengths. Consider the scattering of visible photons in air, which is an example of Rayleigh scattering. The relative rate of scattering of e.g. red and blue photons is given by

$$\left(\frac{\lambda_{red}}{\lambda_{blue}}\right)^4 \simeq \left(\frac{780 \text{ nm}}{390 \text{ nm}}\right)^4 = 16. \tag{6.152}$$

Despite the two wavelengths being comparatively close, the rate of scattering is dramatically different. This is the explanation for why the daytime sky is blue, and why sunsets are red. It is very important to note here: in the Rayleigh scattering process we have described, there is not *net* transfer of energy from the light to the electrons. This is an elastic scattering process, and we see two important facts: the energy in the scattered radiation is equal to that lost in the incident radiation; the scattered wavelength is equal to the incident wavelength.

Near the resonant frequency we have $\omega \simeq \omega_0$. We obtain

$$\sigma_{res} \simeq \frac{8\pi r_e^2}{3}\frac{\omega^2}{\gamma^2}. \tag{6.153}$$

This cross section describes so-called resonant scattering, and the cross section for this is large because γ is typically small. Most interesting however is the high-frequency case where $\omega \gg \omega_0 \gg \gamma$. We now obtain a very simple form for the cross section, which is

$$\sigma_T = \frac{8\pi r_e^2}{3}. \tag{6.154}$$

This cross section may also be written equivalently in terms of the other fundamental constants as

$$\sigma_T = \frac{e^4}{6\pi\epsilon_0^2 m^2 c^4} = 6.74 \times 10^{-30} \text{ m}^2 = 0.0674 \text{ barns}. \tag{6.155}$$

The *barn* unit is convenient for scattering calculations; 1 barn= 10^{-28}m^2. There is no frequency dependence at all in this expression; the likelihood of scattering does not depend on the incident photon frequency as long as it's high enough. The cross section is given the subscript T because this regime is known as Thomson scattering; the behaviour of the scattering cross section with incident frequency is shown for atoms in Fig 6.20. We recall that since ω is very large, incident radiation does not see that the electrons are bound, so that we have free electrons; therefore, as well as describing the scattering of high-frequency radiation by atoms, this cross section also describes the scattering of radiation from free electrons. It is therefore important when considering the mutual passage of photons over electrons in certain laser-plasma interactions. We saw above that $\langle S \rangle = c \langle U \rangle$, so therefore the power emitted by an electron due to Thomson scattering is

$$P = \sigma_T c \langle U \rangle. \tag{6.156}$$

The emitted power is due to the electric field energy passing over the electrons at speed c.

6.3.2 Synchrotron Radiation and the Field Energy

One reason for deriving the scattering cross section of photons from electrons is to look once more at the synchrotron radiation emitted power. We saw earlier that the power emitted by an electron moving in a magnetic field with radius ρ is

$$P = \frac{e^2 c \beta^4 \gamma^4}{6\pi\epsilon_0 \rho^2}. \tag{6.157}$$

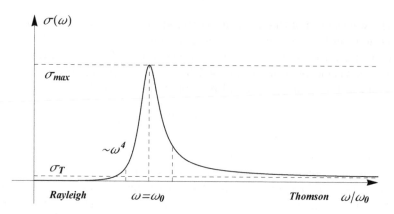

FIGURE 6.20 Different regimes for scattering. At low frequencies (large wavelengths) there is Rayleigh scattering whose cross section varies as ω^4. At high frequencies the cross section tends to the Thomson cross section σ_T. In between there is a resonant region of width γ.

Let's write the synchrotron radiation power in terms of the Thomson cross section, which is

$$P = \sigma_T \frac{m^2 \epsilon_0 c^5 \beta^4 \gamma^4}{e^2 \rho^2}. \tag{6.158}$$

We can then write the emitted power in terms of the magnetic field B by remembering that $\rho = \beta \gamma m c / e B$, so that

$$P = \sigma_T B^2 \epsilon_0 c^3 \beta^2 \gamma^2. \tag{6.159}$$

We saw for Thomson scattering that $P = \sigma_T c \langle U \rangle$. Let's write the synchrotron radiation power also in terms of a field energy density – this time, the energy density of the magnetic field B. The energy density in the magnetic field is $U_B = B^2 / 2\mu_0$, which gives an expression for the synchrotron radiation power as

$$P = 2\sigma_T U_B \epsilon_0 \mu_0 c^3 \beta^2 \gamma^2. \tag{6.160}$$

We recognise that $\epsilon_0 \mu_0 = 1/c^2$, so that finally

$$P = 2\sigma_T c U_B \beta^2 \gamma^2. \tag{6.161}$$

What does this expression mean? We can regard the electron as taking energy from the magnetic field at some rate σ_T, where the magnetic field has an effective Poynting flux $S_B = c U_B$; the difference is the extra factor γ^2 from the motion of the electron.

6.3.3   Thomson and Compton Scattering

In our scattering derivation above, we calculated the rate of scattering for high-frequency radiation; this was the Thomson scattering cross section. This is an elastic process in which the incident and scattered wavelengths are the same. However, we also know that individual photons carry momentum, and therefore should transfer some of that if they interact with an electron; this is the process that we call Compton scattering. Clearly, there must be some way of reconciling these two phenomena; we realise that Thomson scattering applies as long as the energy of the photon is much less than the rest energy of an electron, in other words $hf \ll m_e c^2$. At higher frequencies the momentum transfer starts to become important and we have Compton scattering. In ordinary Compton scattering, a high-energy photon (with energy ϵ_i) is incident upon a stationary electron; the photon is scattered by an angle β

causing a recoil of the electron. The scattered photon therefore has a lower energy ϵ_f and a longer wavelength λ_f, given by the standard Compton formula

$$\lambda_f - \lambda_i = \lambda_c(1 - \cos \beta) \tag{6.162}$$

where

$$\lambda_c = \frac{h}{m_e c} \simeq 0.002 \text{ nm} \tag{6.163}$$

is the Compton wavelength; this process is shown schematically in Fig 6.21. We may measure the electron mass by measuring the energy change of photons at a specific scattering angle β, and for practical values of this angle we have the requirement that $\lambda_i \sim \lambda_c$; in other words, the energy ϵ_i of the incident gamma ray photon should be comparable to the rest energy $m_e c^2$ of an electron.

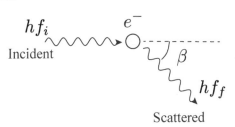

FIGURE 6.21 In the quantum picture of scattering, an incoming photon of frequency f_i is scattered through an angle β into a scattered photon of frequency f_f. In the case of Thomson scattering, where $hf_i \ll m_e c^2$, there is no appreciable transfer of momentum from the photon to the electron, and therefore $f_f = f_i$; this is an elastic scattering process. When there is an appreciable transfer of momentum, $f_f < f_i$ and we call it Compton scattering.

It can be shown that the total Compton cross section tends to the Thomson cross section for low-frequency photons. Defining the so-called recoil parameter as

$$X = \frac{4\gamma\epsilon_i}{m_e c^2}, \tag{6.164}$$

a quantum electrodynamics analysis [16, 17] yields the following exact expression for the Compton cross section:

$$\sigma_c = \sigma_T \frac{3}{4X} \left[\left(1 - \frac{4}{X} - \frac{8}{X} \right) \log(1 + X) + \frac{1}{2} + \frac{8}{X} - \frac{1}{2(1 + X)^2} \right]. \tag{6.165}$$

The scattering of long-wavelength light from electrons implies $X \ll 1$, for which $\sigma_c \simeq \sigma_T(1 - X)$; at long enough wavelengths $(X \to 0)$, the Compton cross section tends to the Thomson value as it should. A significant recoil parameter can be obtained if gamma rays are incident upon a stationary electron; for example, the gamma rays from the decay of cobalt-60 have energies $\epsilon_i \simeq 1$ MeV, which yields a recoil parameter of $X \simeq 7.82$. In this case the Compton cross section is $\sigma_c \simeq 0.22\sigma_T$, a substantial reduction. At these larger values of X the Compton cross section tends to

$$\sigma_C \simeq \sigma_T \left(\frac{3}{4X} \left[\log X + \frac{1}{2} \right] \right), \tag{6.166}$$

which is accurate to about 10% when $X > 2$.

We should also compare the difference in the angular distributions of the Thomson- and Compton-scattered radiation fields. Thomson scattering has an intensity distribution like a Hertzian dipole (for incoming photons that are polarised), i.e. $I \propto \cos^2 \beta$ (where β is the angle of observation compared to the direction of incident radiation); the scattered radiation has the same wavelength as the incident wavelength, and is polarised in the same direction. The proportion of scattered radiation does not change with incident wavelength. In contrast, the Compton scattering intensity distribution is peaked in the forward direction (this is compared to the Thomson rate in Fig 6.22), and scattered photons have wavelengths that are generally larger because of the momentum transfer; the wavelength varies with angle to conserve momentum. As the wavelength reduces, so does the rate of scattering.

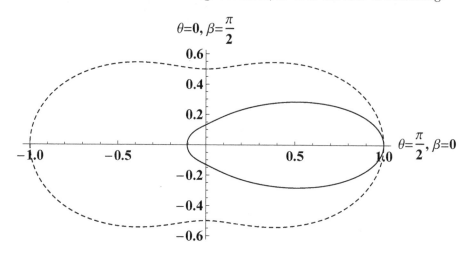

FIGURE 6.22   Rate of Thomson (classical) scattering and Compton scattering (for 1 MeV photons) as a function of scattering angle β, obtained from the Klein-Nishina formula. The Compton-scattered photons are forward-peaked due to the conservation of momentum. The dotted 'peanut' shape for Thomson scattering differs from the Larmor formula; the Larmor formula is the scattering rate for polarised photons, whilst the Thomson scattering rate shown here is the rate for unpolarised photons.

6.3.4   Inverse Compton Scattering (ICS)

We have seen that in ordinary Compton scattering – where the electron is initially stationary – that the scattered photon always reduces in energy. *Inverse* Compton scattering is the situation where the electron is moving sufficiently fast that a collision may cause the photon to *increase* in energy. For this to occur, the electron typically must be moving relativistically with $\gamma \gg 1$. We will see that an incident photon can be scattered to a *much* larger outgoing energy.

Energy Change from the Inverse Compton Process

We consider an electron moving with velocity **v** where $\gamma \gg 1$, and an electromagnetic wave is incident upon it at some angle θ to the direction of the electron. $\theta = 0$ corresponds to the photons being incident head-on with the electron.

Since the electron is moving relativistically, we must perform a Lorentz transformation of each photon frequency f in the laboratory frame to the frequency f' in the electron's rest frame. For a head-on approach of the electron and the photon, we may write the ordinary relativistic Doppler formula for the frequency change of the photon as

$$\frac{f'}{f} = \gamma(1 + \beta) = \frac{1 + \beta}{\sqrt{1 - \beta^2}} = \sqrt{\frac{(1 + \beta)^2}{(1 + \beta)(1 - \beta)}} \tag{6.167}$$

$$= \sqrt{\frac{1 + \beta}{1 - \beta}}. \tag{6.168}$$

For a photon approaching the electron at an angle θ to the head-on direction, it can be shown that the frequency change is given by

$$\frac{f'}{f} = \gamma(1 + \beta \cos \theta) \tag{6.169}$$

or equivalently that the photon energy changes as

$$\epsilon' = \epsilon_i \gamma(1 + \beta \cos \theta) \tag{6.170}$$

Also, the apparent angle of incidence θ' of the photon upon the electron (in the electron's frame of reference) is related to the laboratory-frame angle of incidence by

$$\sin \theta' = \frac{\sin \theta}{\gamma(1 + \beta \cos \theta)}, \tag{6.171}$$

$$\cos \theta' = \frac{\cos \theta + \beta}{\gamma(1 + \beta \cos \theta)}. \tag{6.172}$$

We see that when θ is small and β is large (i.e. the electron velocity $v \simeq c$), the apparent change in frequency is

$$f' \simeq 2\gamma f. \tag{6.173}$$

As an example, we consider an electron with kinetic energy $T = 1000$ MeV, so that $\gamma \simeq 1957$ and $\beta = 1$ to a *very* good approximation. Visible photons of wavelength 500 nm are incident upon the electron, so that $\epsilon_i = hc/\lambda = 2.48$ eV. The Doppler shift into the electron's rest frame changes the photon energy to $\epsilon_i' = \gamma \epsilon_i = 9707$ eV. Hence we see that if the electron energy is less than ~ 1 GeV and the incident photons are near-visible ($\epsilon_i \sim$eV), in the rest frame of the electrons $\epsilon_i' = hf' \ll m_e c^2$ is still small in comparison to the electron rest energy. There is therefore no significant transfer of momentum to the electron, and we have ordinary – essentially isotropic – Thomson scattering where the scattered energy (in the electron rest frame) is $\epsilon_f' = \epsilon_i'$. When we transform ϵ_f' back into the laboratory frame another Doppler shift is performed. For the head-on case ($\theta = 0$) we see that the outgoing photon energy is

$$\epsilon_f = \gamma^2(1 + \beta)^2 \epsilon_i \simeq 4\gamma^2 \epsilon_i. \tag{6.174}$$

This is a very important result. A typically-used incident laser wavelength is 1064 nm (near-infrared, corresponding to $\epsilon_i \simeq 1$ eV); Compton scattering from 50 MeV electrons means that the scattered photon energy is around 10 keV (suitable for X-ray scattering experiments), from 500 MeV electrons we obtain ~ 1 MeV photons (suitable for exciting resonances in atomic nuclei), and 5000 MeV electrons deliver ~ 1 GeV photons! In other words, 'ordinary' electron energies (up to $\sim$GeV) can be used to generate Compton-scattered photons with energies extending far above those available from other sources. Moreover, the generated photon energies are tuneable, and this is mostly achieved by varying the energy of the electrons rather than by varying the energy of the photons [18].

A more careful analysis of the scattering process yields the better formula

$$\epsilon_f \simeq \frac{4\gamma^2 \epsilon_i}{1 + (\gamma\theta)^2 + 4\gamma\epsilon_i/(m_e c^2)}, \tag{6.175}$$

where θ is the observation angle; the second term in the denominator therefore tells us that the produced photons are monochromatic to within some bandwidth as long as the angular spread of photons seen by the observer is restricted – we may collimate the scattered photons to select a desired bandwidth of photon energies. Above energies of around 100 keV there is no alternative source of near-monochromatic photons, and therefore inverse Compton scattering is an important method. The third term describes the degree of electron recoil, which as earlier, results in a reduction in the scattered photon energy; for 1 eV incident photons, this recoil parameter is small even for large electron energies of ~ 1 GeV, but can be significant if the incident photons have keV or higher energies [19].

Inverse Compton Scattering Cross Section and Output Power

We recall that the scattered radiation power in Thomson scattering for an incident power $\langle S \rangle$ is just

$$P' = \sigma_T \langle S' \rangle = \sigma_T c \langle U' \rangle \tag{6.176}$$

where $\langle U' \rangle$ is the average energy density in the incident electromagnetic wave (in the electron's frame of reference). The instantaneous radiated/scattered power from the Thomson scattering is quasi-isotropic in the electron's rest frame with the usual Hertzian dipole pattern

$$|\mathbf{S}'(\mathbf{r}, t)| = \frac{q^2 |\mathbf{a}^2(t - r/c)| \cos^2 \xi'}{16\pi^2 \epsilon_0 c^3 r^2}, \tag{6.177}$$

where ξ' is the angle of the emitted radiation from the electron with respect to the incident photon direction in the electron rest frame. In the laboratory (observation) frame the angular and energy distribution of the scattered photons is therefore effectively completely determined by the relativistic Doppler transformation, known as a kinematic restriction.

We note that the Thomson-scattered power is defined as a rate of energy emission, and hence the total scattered power is invariant under a Lorentz transformation. Hence the total emitted power P' in the electron rest frame is the same as the emitted power P in the observer's frame. To calculate P we need only to calculate P' and therefore to calculate the energy density U' of the photons in the rest frame of the electron. To do this, we first note that photons of a given frequency f and volumetric number density n have an energy density

$$U = nhf \tag{6.178}$$

so that the incident flux is

$$S = Uc = nhfc. \tag{6.179}$$

The interval in the arrival time of these photons at the electron is reduced in the electron's rest frame by the Doppler shift, and so in the head-on case the effective number density of the photons increases to

$$n' = n\gamma(1 + \beta). \tag{6.180}$$

We earlier showed that the Doppler shift increases the photons' apparent frequency to $f' = f\gamma(1 + \beta)$, so that

$$U' = U\gamma^2(1 + \beta)^2. \tag{6.181}$$

The Thomson-scattered power in the rest frame of the electron is now just

$$P' = \sigma_T c U' = \sigma_T c U \gamma^2 (1 + \beta)^2, \tag{6.182}$$

which is the same as the power in the laboratory frame, i.e. $P = P'$; this is the total power contained in the scattered photons, which now have higher energies than they did before. However, those photons already had an initial power

$$P_{initial} = \sigma_T cU \tag{6.183}$$

before they interacted. So, the net power given to those interacting photons is

$$
\begin{aligned}
P_{\text{ICS}} = P' - P_{initial} &= \sigma_T cU\gamma^2(1+\beta)^2 - \sigma_T cU \\
&= \sigma_T cU[\gamma^2(1+\beta)^2 - 1] \\
&= \sigma_T cU\gamma^2 \left[(1+\beta)^2 - \frac{1}{\gamma^2}\right] \\
&= \sigma_T cU\gamma^2[(1+\beta)^2 - (1-\beta^2)] \\
&= 2\sigma_T cU\gamma^2\beta(\beta+1).
\end{aligned} \tag{6.184}
$$

Inverse Compton Scattering Flux

We know that the energy density U in the (incoming) photon beam is given by the photon density, which in most practical situations has a Gaussian transverse profile of some size σ_L. Assuming the photons are scattered directly backwards, it is possible to show that the number of Compton-scattered photons is

$$N_f = \sigma_T \frac{N_e N_L}{2\pi(\sigma_e^2 + \sigma_L^2)}, \tag{6.185}$$

where N_e, N_L are the numbers of electrons and photons, respectively, focused into circular spots of r.m.s. size σ_e and σ_L. We see the same basic scaling with photon number N_L.

We may alternatively express the laser power in terms of its normalised vector potential

$$\mathbf{a} = \frac{e\mathbf{A}}{m_e c^2}, \tag{6.186}$$

where the associated field strength parameter is

$$a_0 = \frac{eE_{0i}\lambda_i}{2\pi m_e c^2} \simeq 0.855 \times 10^{-9}\sqrt{I}\lambda_i. \tag{6.187}$$

E_{0i} is the maximum strength of the incident laser electric field, and a_0 has been expressed in convenient units in which the incident laser intensity I is given in W/cm^2 and the wavelength λ_i is given in μm. One may obtain the number of Compton-scattered photons as

$$N_f = \frac{2}{3}\pi\alpha N_\lambda a_0^2 N_e, \tag{6.188}$$

where N_λ is the number of wavelengths in the incident laser pulse; again we see the scaling with number of incident photons as expected. This expression is only valid for the linear regime where $a_0 \ll 1$. Multi-photon scattering occurs as the intensity approaches $a_0 \sim 1$, such that higher-energy scattered photons can be obtained. A number of methods and codes are available to estimate the ICS flux, and a good summary is given by Krafft and Priebe [18] that includes a number of useful approximations.

6.4   Radiation Damping

In general, radiation damping is the phenomenon of the reduction of some charge's oscillation amplitude due to the emission of radiation. For example, in the classical description of an electron orbiting an atomic nucleus, the (classical) emission of radiation by the charge would cause it to spiral inwards into the nucleus in some time $t \sim 10^{-8}$ s. In particle accelerators, the term is used to describe the effect of what is basically a quantum phenomenon; it arises particularly in the context of electron storage rings, an important case we describe here.

An orbiting electron in a storage ring emits photons continuously with the spectrum derived in Section 6.2.3; the photon emission is quantised, which means the energy change of the electron is discrete (rather than smoothly changing). At the moment of photon emission the electron's energy changes by a finite amount and the electron experiences a (small) recoil; more importantly, the electron has a new, lower, energy and – if the dispersion function η in either plane of motion x or y is non-zero – the electron will now start to oscillate with respect to a new closed orbit. This is the phenomenon of quantum excitation; obviously, to limit the amount of excitation a good storage ring design should limit the typical values of η. Storage rings are usually planar (i.e. bending only in the x direction) so that the photon emission is mainly in the x plane, and also η_y is essentially zero; quantum excitation essentially only occurs in a storage ring in the x plane and gives an effective horizontal momentum. η_x is limited by making use of achromats (double bend achromats, triple bend achromats, etc.) to limit the size of η_x that is generated in the dipole magnets, and hence limit the excitation.

Radiation damping competes with this excitation process; in a storage ring we are implicitly saying that energy loss from radiation is replaced by re-acceleration of the electrons by means of RF cavities; the re-acceleration is only in the beam direction, so any prior transverse momentum will be steadily damped. For example, in the Diamond storage ring, each electron loses about 1 MeV per turn; the actual voltage supplied is somewhat larger, firstly because the quantised emission gives rise to a typical energy spread and also because scattering (mainly Touschek scattering) gives rise to electron energy changes of $\sim 1\%$ that must be tolerated. Without re-acceleration, the electron lifetime would be $\tau \sim E/U_0 t_r \sim 5$ ms; *with* re-acceleration the typical time for electrons to damp to some equilibrium value is the same – this is the idea of the damping time. Quantum excitation competes with radiation damping to give an equilibrium oscillation amplitude – this is just the equilibrium emittance; in the absence of quantised emission, the equilibrium emittance would be zero.

6.4.1   The Radiation Integrals

Matthew Sands derived a useful formalism for describing the effect of quantised radiation not only for the equilibrium emittance but also for other associated electron beam parameters [20]. The quantities he obtained are known as the radiation integrals, and Sands derived five expressions but today many use a sixth [5]; the original integrals also assumed horizontal bending only. The complete set of radiation integrals (now allowing for vertical bending as well) are

$$\mathcal{I}_1[\text{m}] = \oint \left(\frac{\eta_x}{\rho_x} + \frac{\eta_y}{\rho_y} \right) ds,$$

$$\mathcal{I}_2[\text{m}^{-1}] = \oint \frac{1}{\rho_x^2} ds,$$

$$\mathcal{I}_3[\text{m}^{-1}] = \oint \frac{1}{|\rho_x^3|} ds,$$

$$\mathcal{I}_{4x}[\text{m}^{-1}] = \oint \left[\frac{\eta_x}{\rho^2 \rho_x} + \frac{2}{\rho_x} (k\eta_x + k'\eta_y) \right] ds,$$

$$\mathcal{I}_{4y}[\text{m}^{-1}] = \oint \left[\frac{\eta_y}{\rho^2 \rho_y} + \frac{2}{\rho_y} (k'\eta_x - k\eta_y) \right] ds,$$

$$\mathcal{I}_{5x}[\text{m}^{-1}] = \oint \frac{\mathcal{H}_x}{|\rho_x^3|} ds,$$

$$\mathcal{I}_{5y}[\text{m}^{-1}] = \oint \frac{\mathcal{H}_y}{|\rho_y^3|} ds,$$

$$\mathcal{I}_{6x}[\text{m}^{-1}] = \oint (k\eta_x + k'\eta_y)^2 ds,$$

$$\mathcal{I}_{6y}[\text{m}^{-1}] = \oint (k'\eta_x - k\eta_y)^2 ds, \tag{6.189}$$

$$\tag{6.190}$$

where the integrals are each evaluated over a single turn of the storage ring; k is the quadrupole strength, k' the skew quadrupole strength, and $1/\rho^2 = 1/\rho_x^2 + 1/\rho_y^2$. We see that the $\mathcal{I}_{5x}$ and $\mathcal{I}_{5y}$ integrals – which describe the quantum excitation – are dependent on the functions $\mathcal{H}_x(s)$ and $\mathcal{H}_y(s)$, which are determined by the Twiss functions and dispersion as

$$\mathcal{H}_x = \beta_x \eta_x'^2 + 2\alpha_x \eta_x \eta_x' + \gamma_x \eta_x^2,$$
$$\mathcal{H}_y = \beta_y \eta_y'^2 + 2\alpha_y \eta_y \eta_y' + \gamma_y \eta_y^2. \tag{6.191}$$

These are fairly involved general expressions, but they simplify considerably in the (very typical) case where there is only horizontal bending and no focusing in the bending magnets. Then $\mathcal{I}_{4x} = \oint (\eta_x/\rho_x^3) ds$, $\mathcal{I}_{4y} = 0$, $\mathcal{I}_{5y} = 0$, and $\mathcal{I}_{6x} = \mathcal{I}_{6y} = 0$. With this formalism, the synchrotron radiation power (per electron) may be expressed as

$$P = C_\gamma \frac{E^4 \mathcal{I}_2}{2\pi t_r} \tag{6.192}$$

where the so-called quantum constant is

$$C_\gamma = \frac{55\hbar}{32\sqrt{3}m_e c} \simeq 3.8319 \times 10^{-13} \text{ m}. \tag{6.193}$$

Hence the energy loss per turn

$$U_0 = C_\gamma \frac{E^4 \mathcal{I}_2}{2\pi} = \frac{2r_e E^4 \mathcal{I}_2}{3m_e^3 c^6}. \tag{6.194}$$

There are three damping times – one for each direction of electron motion x, y, s with respect to the moving electron bunch centre – which are

$$\tau_{x,y,s} = \frac{3m_e c^5 L \rho_x}{2\pi r_e J_{x,y,s} E^3} \tag{6.195}$$

where L is the storage ring circumference. J_x, J_y, J_s are the damping partition numbers obtained as

$$J_x = 1 - \frac{\mathcal{I}_{4x}}{\mathcal{I}_2},$$

$$J_y = 1 - \frac{\mathcal{I}_{4y}}{\mathcal{I}_2},$$

$$J_s = 2 + \frac{\mathcal{I}_{4x} + \mathcal{I}_{4y}}{\mathcal{I}_2}, \tag{6.196}$$

leading to the Robinson Sum Rule

$$J_x + J_y + J_s = 4. \tag{6.197}$$

In our planar ring situation we have $J_y = 1$, and usually $\mathcal{I}_{4x} \ll \mathcal{I}_2$ so that $J_x \simeq 1$ so that longitudinal oscillations (of the energy) have twice the damping time of the lateral amplitudes.[*] For example, in the 96 m circumference Daresbury SRS at its injection energy of 600 MeV and with a dipole bending radius of 5.56 m ($B = 0.36$ T), the damping time $\tau_s = 93$ ms; after ramping of the dipole field to 1.2 T to circulate 2 GeV electrons, the damping time reduces to 2.5 ms. Another way to state the damping times when $J_x \simeq 1$ is

$$\tau_x = \tau_y = \frac{3t_r}{r_e \gamma^2 \mathcal{I}_2}, \qquad \tau_s = \frac{\tau_x}{2} = \frac{3t_r}{2r_e \gamma^2 \mathcal{I}_2}. \tag{6.198}$$

The electrons emit photons at a rate $\dot{N} = 5\pi\alpha\gamma/\sqrt{3}t_r$ as derived earlier. The RMS photon energy $\langle \epsilon^2 \rangle = 11\epsilon_c^2/27$, so that in a planar ring the rate of photon production is

$$\dot{N}\langle \epsilon^2 \rangle = \frac{55}{24\sqrt{3}} \hbar c^2 r_e m_e c^2 \frac{\gamma^7}{|\rho_x^3|}. \tag{6.199}$$

The induced energy spread also therefore scales $\propto \gamma^7$, and over some distance L (such as the ring circumference) is

$$\Delta\sigma_E^2 = \frac{55\alpha\hbar^2 c^2 \gamma^7}{48\sqrt{3}} \int_0^L \frac{1}{|\rho_x^3|} \mathrm{d}s. \tag{6.200}$$

This in turn gives an emittance growth

$$\Delta\epsilon_x = \frac{55 r_e \hbar \gamma^5}{24\sqrt{3} m_e c} \int_0^L \frac{1}{|\rho_x^3| \mathcal{H}_x} \mathrm{d}s. \tag{6.201}$$

6.4.2 Equilibrium Properties

The above expressions can be used in any electron accelerator to determine the emittance and energy spread growth, whether it's a ring or not. In a storage ring, however, an equilibrium is formed when the excitation rate equals the damping rate, such that the energy spread becomes

$$\frac{\sigma_E^2}{E^2} = C_\gamma \gamma^2 \frac{\mathcal{I}_3}{2\mathcal{I}_2 + \mathcal{I}_{4x}} \tag{6.202}$$

---

[*]To show that $\mathcal{I}_{4x} \ll \mathcal{I}_2$, consider a typical storage ring where $\oint ds = L \sim 100$ m, $\eta_x \sim 0.1$ m and $\rho_x \sim 10$ m.

(assuming here again that $\mathcal{I}_{4y} = 0$). Hence the relative energy spread in a planar ring is

$$\frac{\sigma_E}{E} \simeq \sqrt{\frac{C_\gamma \gamma^2}{2\rho_x}}. \tag{6.203}$$

Note that this is the energy spread of *one* electron, i.e. the typical range of energies of that electron over time. We saw that in a typical storage ring, there are $\sim 10^{12}$ electrons, each of which independently has an energy spread σ_E/E so that the whole beam has that energy spread. The corresponding bunch length depends upon the momentum compaction factor α_c that couples energy to longitudinal position (Equation 5.159 in Section 5.7.3). Thus we have

$$\sigma_s = \frac{c \, |\eta_c|}{\omega_s} \frac{\sigma_E}{E}, \tag{6.204}$$

where the phase-slip factor $\eta_c = \alpha_c - 1/\gamma^2$ and ω_s is the synchrotron frequency; in electron rings $\eta_c \simeq \alpha_c$ since γ is typically a few thousand. The synchrotron frequency – the rate at which electrons oscillate back and forth within the electron bunch – is

$$\omega_s = \omega_r \sqrt{\frac{\alpha_c h \cos(\phi_s) e V_{\rm rf}}{2\pi E}}, \quad \sin(\phi_s) = \frac{U_0}{e V_{\rm rf}}, \tag{6.205}$$

where ϕ_s is called the synchronous phase and $q = eV_{\rm rf}/U_0$ is the overvoltage; $\omega_r = 2\pi/t_r$ is the electron angular revolution frequency. The overvoltage gives an energy acceptance – here called the RF acceptance – of

$$\epsilon_{\rm RF} = \pm \sqrt{\frac{2U_0}{\pi \alpha_c h E} \left[\sqrt{q^2 - 1} - \cos^{-1}\left(\frac{1}{q}\right) \right]}. \tag{6.206}$$

$\epsilon_{\rm RF}$ is typically several percent to accommodate Touschek scattering between the electrons in a bunch (see Chapter 7). Using the radiation integrals, we have simply that

$$\alpha_c = \frac{\mathcal{I}_1}{L}. \tag{6.207}$$

We find in many electron storage rings that $\alpha_c \sim 10^{-4}$, although it can take different values (including negative ones) depending upon the average value of η_x in the dipole magnets. The *equilibrium* emittance is obtained when there is a balance between the quantum excitation rate and the emittance damping, such that $d\epsilon_x/dt = -2\epsilon_x/\tau_x$. From this we can obtain the natural emittance value (in a planar ring) as

$$\epsilon_{x0} = C_\gamma \gamma^2 \frac{\mathcal{I}_{5x}}{J_x \mathcal{I}_2}, \tag{6.208}$$

showing the contribution of the $\mathcal{I}_{5x}$ excitation and $\mathcal{I}_2$ terms. We label this ϵ_{x0} the 'natural' emittance, and in the absence of field errors we have $\epsilon_{y0} = 0$ (there is actually a lower limit on the vertical emittance due to the photon emission, which is rather small [5]). Typical values of the natural emittance in modern synchrotron light sources are from 0.1 nm-rad to 10 nm-rad, for electron energies around 1 to 8 GeV. A central activity in the design of new electron storage rings is to generate as small a value of $\mathcal{I}_{5x}$ as possible; it has proven advantageous to split the bending dipoles into as many pieces as possible to allow quadrupoles to be interleaved to minimise the average dispersion $\langle \eta_x \rangle$ that drives $\mathcal{I}_{5x}$, creating what are known as multi-bend achromats (MBAs). The consequence is the strong non-linear limitation to the dynamic aperture brought about by having so much strong focusing to correct the dispersion; a number of beam-optical cancellation schemes have been proposed, primarily based upon setting an appropriate phase advance (for example $\mu = \pi$) between the non-linear kicks.

In real storage rings, small vertical B fields from dipole and quadrupole misalignments weakly couple the horizontal and vertical planes of motion; hence the vertical emittance is not zero. The horizontal and vertical emittances may be written as $\epsilon_x + \epsilon_y = \epsilon_{x0}$; for a weak coupling factor κ we can write approximately

$$\epsilon_x \simeq \epsilon_{x0}, \quad \epsilon_y \simeq \kappa \epsilon_{x0}. \tag{6.209}$$

Today, position alignments of 10s of μm and roll accuracies of 10s of μrad allow $\kappa \simeq 10^{-3}$ to give very small vertical emittances in the picometre regime. Unintended vertical bends give small residual vertical dispersion of typically a few millimetres, which also contributes to 'effective' coupling, and so instead, the term emittance ratio is used interchangeably to describe ϵ_y/ϵ_x.

It is common also in electron storage ring design to separate the emittance contributions from the dipoles and insertion devices, in particular the multipole wigglers (MPWs) where both $\mathcal{I}_{5x}$ and $\mathcal{I}_2$ can potentially be significant. Strong MPWs (i.e. many poles and high field) are used to maximise $\mathcal{I}_2$ and hence minimise the emittance; this *damping wiggler* technique may be used either in storage rings (where an equilibrium is obtained) or in damping rings where the MPWs help an initially-injected large emittance to be damped to a small desired value as quickly as possible. Separating out the contributions of the wigglers (labelled 'w') from the rest of the ring (labelled '0') we can simply state

$$\epsilon_w = C_\gamma \frac{\gamma^2}{J_x} \frac{\mathcal{I}_{5x}^0 + \mathcal{I}_{5x}^w}{\mathcal{I}_2^0 + \mathcal{I}_2^w}, \tag{6.210}$$

where ϵ_w indicates that this is the emittance *with* MPWs. The benefit of the MPWs can be written as

$$\frac{\epsilon_w}{\epsilon_{x0}} = \frac{1 + \mathcal{I}_{5x}^w/\mathcal{I}_{5x}^0}{\mathcal{I}_2^w + \mathcal{I}_2^0}. \tag{6.211}$$

We may re-state this ratio in terms of the MPW properties as

$$\frac{\epsilon_w}{\epsilon_{x0}} = \frac{1 + \frac{4C_\gamma}{15\pi J_x} N_p \frac{\langle \beta_x \rangle}{\epsilon_0 \rho_w} \gamma_r^2 \frac{\rho_0}{\rho_w} \theta_w^3}{1 + \frac{1}{2} N_p \frac{\rho_0}{\rho_w} \theta_w}. \tag{6.212}$$

N_p is the total number of MPW poles, $\langle \beta_x \rangle$ the average horizontal beta function in the MPW, ρ_w the minimum MPW bend radius at the peak field B_w, and $\theta_w = \lambda_w/2\pi\rho_w$ is the peak deflection angle in the MPW, where λ_w is the MPW period. The MPW bend angle must be limited to avoid too great a 'self-dispersion'; there is a maximum field for which the MPW does not reduce the emittance and for which $\epsilon_w/\epsilon_{x0} > 1$. Damping wigglers are ideally long, and with lower field.

6.5   Bremsstrahlung Radiation

A quite different practical phenomenon from synchrotron radiation, but one that is ultimately derived from the same basic physics, is bremsstrahlung. The word bremsstrahlung is German, and was derived from the words 'bremsen' (to brake) and 'strahlung' (radiation). Bremsstrahlung is therefore 'braking radiation'. Bremsstrahlung is the name given to the phenomenon whereby a charged particle is caused to radiate (and therefore lose kinetic energy and slow down) due to it passing close to an atomic nucleus, and so feeling a very strong electric field. A strong electric field at right angles to the particle's motion causes much the same thing as a strong magnetic field: electromagnetic radiation is emitted [1].

A common example of bremsstrahlung is in radiotherapy, in which a patient's cancer is treated with X-rays. These X-rays are generated using bremsstrahlung: electrons from a suitable accelerator are directed into a metal target (usually something like tungsten that has a large atomic number Z, or some other refractory metal*); this is shown schematically in Fig 6.23. The electrons may have an initial kinetic energy of, say, 10 MeV, and when some of those electrons pass close to one of the atomic nuclei, they experience a strong (transverse) force from the electric field of that nucleus. This causes the electron to radiate photons. Obviously, the largest photon energy that is emitted cannot be greater than the initial kinetic energy of the electron; most of the time, the electrons pass somewhat further away from the nuclei and emit lower-energy ('softer') photons. Overall, a broad spectrum of radiation is emitted, with a larger number of softer photon energies.

A high Z obviously gives a stronger nuclear field, and a high density ρ means there are more nuclei per unit volume to hit.* As an electron passes by a nucleus, its distance from the nucleus obviously changes, and hence so does the electric field seen by the electron (Fig 6.24). The closest distance is called the *impact parameter*, which we label b. The varying electric field seen by the electron gives a varying acceleration, although the overall effect is that the electron is deflected by some angle between its initial velocity $\mathbf{v}$ and its final velocity $\mathbf{v}'$. The power emitted by the electron as a function of time is just the same as with any other acceleration:

$$P(t) = \frac{e^2}{6\pi\epsilon_0 c^3} a^2(t').$$
(6.213)

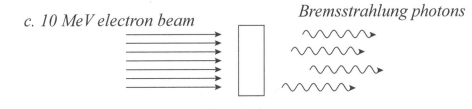

FIGURE 6.23 Radiotherapy is a use of bremsstrahlung radiation; electrons generate X-rays as they are accelerated by the nuclei of the metal atoms in the target. A broad spectrum of X-ray emission is observed – a large number of low-energy photons and a small number of high-energy photons; the maximum energy of the emitted photons is very nearly the initial kinetic energy of the electrons.

The Electron-Ion Collision

Whilst the electron does not strictly collide with the atomic nucleus, we nevertheless still call it an electron-ion collision (the ion being the positive nucleus bit). We may calculate features of the output spectrum of the emitted photons as follows:

---

*The so-called refractory metals are those metals with extremely high melting points; these are tungsten (W), tantalum (Ta), rhenium (Re), molybdenum (Mo) and niobium (Nb). They are also physically quite robust.

*Although you should take account that high-Z atoms also have a larger mass number A; a good exercise is to compare the atomic number density of some common metals like aluminium, tungsten, and lead.

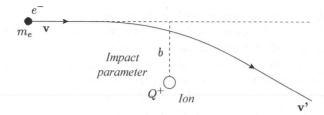

FIGURE 6.24 Definition of the impact parameter b for an interaction of an electron with an ion.

- For a large enough impact factor b (which is basically always true), we can make the statement that the nuclear electric field $\mathbf{E} \perp \mathbf{v}$; therefore the nuclear electric field does no work on the passing electron.

- The acceleration a given to the electron varies over the course of a nuclear collision; the typical time over which the collision takes place is $\Delta t \sim 2b/v$, and so the frequency of the photons emitted is spread over values from zero to $f_{max} \sim v/2b$.

- The maximum photon energy must be less than the initial kinetic energy E_k of the electron.

One very simple law (the Duane-Hunt Law) is that the cut-off frequency above which no photons are emitted is

$$\nu_c = \frac{E_k}{h}. \tag{6.214}$$

Next, let's look at the case of large-enough impact factor b such that $|\mathbf{v}| \simeq |\mathbf{v}'|$; the electron is only slightly deflected by the nucleus and doesn't change much in energy due to the collision. We may then obtain the distance from the nucleus as a function of time as

$$r(t) = \sqrt{b^2 + v^2 t^2} \tag{6.215}$$

where $r = b$ at $t = 0$ (Fig 6.25). Obviously, for a nuclear charge Q, the acceleration experienced by the electron is

$$a = \frac{1}{m_e} \frac{eQ}{4\pi\epsilon_0 r^2}, \tag{6.216}$$

so that the emitted power as a function of time is

$$P(t) = \frac{e^4 Q^2}{96\pi^3 \epsilon_0^3 m_e^2 c^3} \frac{1}{(v^2 t^2 + b^2)^2}. \tag{6.217}$$

The total energy released as photons is

$$W = \int_{t=-\infty}^{\infty} P(t)\mathrm{d}t = \frac{e^4 Q^2}{192\pi^2 \epsilon_0^3 m_e^2 c^3} \frac{1}{vb^3}, \tag{6.218}$$

since

$$\int_{t=-\infty}^{\infty} \frac{1}{(v^2 t^2 + b^2)^2}\mathrm{d}t = \frac{\pi}{2vb^3}. \tag{6.219}$$

W is measured in joules per electron.

If we have a beam of electrons,* then each electron will have a different closest distance b from a nucleus. Obviously, larger values of b are more probable, with that probability

---

*And there are a *lot* of electrons passing through a radiotherapy target!

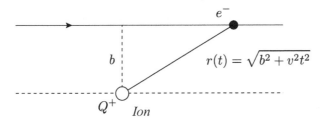

FIGURE 6.25 In a shallow collision there is a small deflection of the electron, and we can write the distance of the electron from the ion as $r(t) = \sqrt{b^2 + v^2 t^2}$.

$\propto 2\pi b db$ (Fig 6.26). Let's therefore try to calculate the overall energy emitted per unit length traversed by electrons in a target, by integrating over the likelihood of having a certain b value. For N_e electrons in the beam, and a number density per unit volume of nuclei in the target N_i, the energy loss per unit length of target the electrons move through is

$$\frac{\mathrm{d}E}{\mathrm{d}l} = N_i \int_{b_{min}}^{b_{max}} N_e W 2\pi b db$$

$$= \frac{N_i N_e e^4 Q^2}{96\pi\epsilon_0^3 v m_e^2 c^3} \left[\frac{1}{b}\right]_{b_{min}}^{b_{max}}. \tag{6.220}$$

If we know the numbers of electrons and target nuclei, we can work out the bremsstrahlung power; except, what values of b_{min} and b_{max} should we use? We have to pick some. Looking carefully at our expression for $\mathrm{d}E/\mathrm{d}l$, we see that placing an upper limit $b_{max} = \infty$ is fine; electrons can, in principle, travel very far from the nucleus. But what about b_{min}? If we were to allow a very close approach $b_{min} \to 0$, this would lead our expression to diverge and predict an infinite emitted power; clearly this is not okay and must be unphysical. We recall that our original assumption that $|\mathbf{v}| \simeq |\mathbf{v}'|$ must break down at small values of b; another way of saying this is that the total energy emitted must be less than the initial kinetic energy E_k of the electron. If we choose a value for b_{min}, we may obtain the radiated power $P = \mathrm{d}E/\mathrm{d}t$ for electrons of velocity v as

$$P_{brem} = \frac{N_i N_e e^4 Q^2}{96\pi\epsilon_0^3 m_e^2 c^3} \frac{1}{b_{min}}. \tag{6.221}$$

Notice how v has disappeared from the denominator when we went from writing dE/dl to writing P_{brem}.

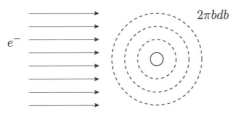

FIGURE 6.26 The likelihood of an impact factor b is $\propto 2\pi b db$.

However, our calculation of the power is still deficient: we don't know what b_{min} to use. We can't easily improve on it unless we do a full quantum calculation (which we won't do here), but we can get an idea by placing a couple of limits on b_{min}; this is equivalent to calculating the emitted power up to some maximum photon energy, and is still useful. One limit is to cut off our calculation by saying our integral becomes invalid when $\Delta v \sim v$. We can calculate what this change in v is as

$$\Delta v = \frac{Qe}{m_e} \int_{t=-\infty}^{\infty} \frac{b}{(v^2 t^2 + b^2)^{3/2}} \mathrm{d}t = \frac{2Qe}{m_e b v}. \tag{6.222}$$

Hence the limit to apply is

$$b_{min} = \frac{4Qe}{\pi m_e v}. \tag{6.223}$$

Another limit one could apply is when quantum effects become significant, in other words, that

$$b_{min} \sim \frac{\hbar}{m_e v}. \tag{6.224}$$

In summary, we have tried to do a classical calculation of what is really a quantum process. The method is deficient because of the question about what b_{min} we can use; in practice people typically set $b_{min} = \hbar/mv$ (the quantum limit). This gives

$$P_{brem} \simeq \frac{N_e N_i e^4 Q^2 v}{48 \epsilon_0^3 m_e c^3 h} \ \mathrm{Wm}^{-3}. \tag{6.225}$$

Notice something important here: the radiated power strongly depends on the charge of the nucleus. We can re-write the bremsstrahlung power in a more convenient form for singly-charged ions as

$$P_{brem} \simeq 1.85 \times 10^{-38} N_e N_i \sqrt{E_k} \ \mathrm{Wm}^{-3}, \tag{6.226}$$

where E_k is the kinetic energy of the electrons in eV [1].

We have found in this discussion that a classical derivation of bremsstrahlung radiation power gives us some idea of what is going on – we can obtain the right spectrum and we can place limits on the classical integral that give about the right power. In practice a so-called Gaunt Factor is used to describe the numerical factor difference between the classical calculation we have just done, and the proper quantum calculation.

Examples of Bremsstrahlung

Electron bremsstrahlung from X-ray tubes is an instructive example. The photon spectrum from an X-ray tube contains contributions from the electron bremsstrahlung of accelerated electrons impinging upon the thick (in most cases) anode target, and also lines of X-ray transitions of the atoms of the anode material. Here we ignore the X-ray transitions, and discuss the bremsstrahlung part of the tube emission spectrum. We saw above in our classical calculation that the rate of energy loss into the target $\mathrm{d}E/\mathrm{d}l \propto N_i Q^2$ (where Q was the charge of the ion). The probability of bremsstrahlung emission from a moving charge q of energy E is proportional to $q^2 Z^2 E/m^2$. Again we see that electrons impacting upon high-Z materials will generate more radiation, and electrons generate about $(m_p/m_e)^2 \sim 3 \times 10^6$ times more bremsstrahlung than protons do, because they have much less mass. The energy loss in a target due to bremsstrahlung is, according to the quantum Thomas-Fermi model,

$$\frac{\mathrm{d}E}{\mathrm{d}l} \simeq -4\alpha r_e^2 N_0 E Z^2 \ln \frac{183}{Z^{1/3}}. \tag{6.227}$$

Hence the bremsstrahlung loss is proportional to energy, and we can write

$$\frac{\mathrm{d}E}{\mathrm{d}l} = -\frac{1}{X_0} E \tag{6.228}$$

where the *radiation length* X_0 is given by

$$\frac{1}{X_0} = 4\alpha r_e^2 N_0 Z^2 \ln \frac{183}{Z^{1/3}}. \tag{6.229}$$

The *radiation thickness* is defined as

$$x_r = \rho X_0 = \frac{A}{4\alpha r_e^2 N_0 Z^2 \ln \left[\,\right] 183/Z^{1/3}\right]}, \tag{6.230}$$

where ρ is the target density and A is its atomic mass.

We already saw that the maximum photon energy was limited by the energy E of the incoming electrons. $E \simeq U$ where U is the tube voltage, and so the Duane-Hunt law can be stated as

$$\lambda_{thresh} = \frac{hc}{eU}, \tag{6.231}$$

where λ_{thresh} is the minimum possible wavelength emitted. The higher the voltage, the smaller λ_{thresh} is; we can write this in convenient units as

$$\lambda_{thresh}[\text{nm}] = 1.239 \times 10^9 \frac{1}{U[\text{kV}]}. \tag{6.232}$$

Another example is so-called free-free emission in a plasma, so called because the electrons are free to move both before their encounter with an ion and after that encounter; there is no capture of the electrons by the ions. Both the ions and electrons in a finite-temperature plasma can see ('encounter') both ions and electrons and thereby see accelerations and radiate. However, we know of course that $m_e \ll m_{ion}$, and so the only significant radiation from an encounter is when the electrons are accelerated by encounters with the ions – the electron acceleration is much larger than the ion acceleration. Hence in a plasma the electrons radiate bremsstrahlung and the ions do not; the ions cause acceleration and the electrons don't. Knowing it's the electrons doing most of the radiating, plasmas behave basically the same as electrons passing through a target and we may straightforwardly calculate the bremsstrahlung power for the free-free radiation in the same way as we did previously, using

$$P_{brem} \simeq \frac{N_e N_i e^4 Q^2 v^2}{48 \epsilon_0^3 m_e c^3 h} \text{ Wm}^{-3}. \tag{6.233}$$

Here, N_e is the density of the free electrons, and N_i is the density of the ions. There can be more electrons than ions as long as the plasma is neutral overall. Usually, $N_e = N_i$ as might be expected.

Our last example involve the tokamaks proposed for fusion power, which typically contain a large volume of plasma; we wish to maintain a high plasma temperature, but unfortunately the plasma is cooled by bremsstrahlung that occurs as electrons pass close by nuclei. An interesting feature is that the bremsstrahlung power $P_{brem} \propto Q^2$, where Q is the ionisation state of the plasma ions. One example of a tokamak is JET, the Joint European Torus. This has a plasma volume of around 100 m^3. A typical temperature during JET operation might be 100 million kelvin (10^8 K), which corresponds to a kinetic energy of the electrons of around 13 keV. The number density of the electrons/ions in the plasma during fusion would be around 10^{20} m$^{-3} \equiv 10^{14}$ cm^{-3}. This density should be compared to the density of a typical solid material, water, which has $\rho N_A / M \sim 10^{22}$ molecules in a cubic centimetre. A 'high-density', 'high-temperature' plasma is therefore still rather tenuous, and contains low-energy electrons. Using these values for the JET tokamak, we predict that the lost bremsstrahlung power is quite high: 2 MW. Therefore, it is hard to keep a tokamak plasma hot because free-free radiation will cause it to cool itself down.

Electron Bremsstrahlung Spectrum

An important practical situation is that of the bremsstrahlung generated by electrons in high-Z targets, for example the generation of X-rays in the head of a radiotherapy machine. Already in 1959 Koch and Motz tabulated convenient formulae [21] to estimate the photon output as a function of energy and emission angle, later augmented by Berger and Selzer for specific metal targets such as tungsten [22]. An example is shown in Fig 6.27. Zschornack's handbook of X-ray data contains a wealth of useful information [23]. In addition, a number of codes – including earlier ones such as `EGS` and more modern ones such as `GEANT4` – enable the calculation of photon output, normally using Monte-Carlo sampling methods; care should be taken when using such codes that sufficient accuracy is used for the cross section, geometry, number of primaries simulated and simulation 'cuts' that reliable results are generated. Alex Bielajew's numerous publications should be consulted [24]. Nordell and Brahme [25] augment the earlier results of Stearns [26], and in particular they give an estimate of the angular spread of the photons, θ_p, due to the bremsstrahlung process, whose RMS value is

$$\theta_{p\text{RMS}} = k \frac{m_e c^2}{T_e} \ln \frac{T_e}{m_e c^2} \tag{6.234}$$

where T_e is the electron energy and $k \simeq 0.26$ is an empirical factor derived from measurements. To the spread in photon angle must be added the angular spread due to electron scattering in the target and the initial angular spread of the incident electrons.

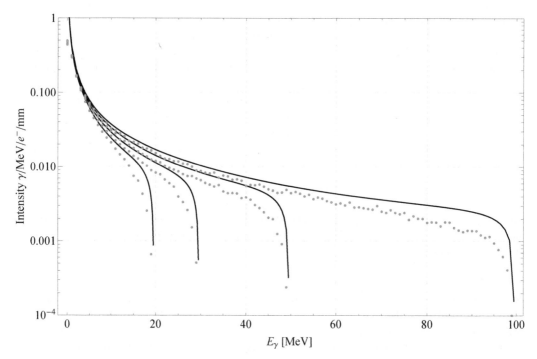

FIGURE 6.27 Comparison of the energy-resolved photon production from a (thin) 1 mm tungsten target, for incident electron energies of 20, 30, 50, and 100 MeV. The solid line shows the Bethe-Heitler formula as given in Koch and Motz is used [21], and is compared to Monte-Carlo simulated production made with `GEANT4`.

6.6 Čerenkov Radiation

All the way through this chapter, we have made the statement that an *acceleration* is required to give rise to a radiative component to the electromagnetic field; this is certainly true in most situations. However, a *uniformly*-moving charge may radiate in certain circumstances; one of those is Čerenkov radiation. Čerenkov radiation is the process in which radiation is emitted when a uniformly-moving charge is moving in some medium with some velocity v_c that is greater than the speed of light v_p in that medium. We can explain Čerenkov radiation* pictorially by considering the electric field exerted by the moving charge [27].

We first consider a uniformly-moving charge with $v_c < v_p$; even though the charge is moving, the field lines from that charge still point in straight lines away from it, and are still symmetric in strength ahead of and behind the charge. The converse situation – corresponding to Čerenkov radiation – is when $v_c > v_p$. Now, there is a contradiction as the charge is moving faster than the electric field itself can propagate; the charge outruns its field lines. We may think of the charge at each point in time as a different source of the electric field, and the retarded time as a wavefront that propagates away from the charge. We may then apply Huygens' principle to deduce the behaviour of these wavefronts. We see that, for the case here where $v_c > v_p$, an overall wavefront is formed that propagates at an angle to the direction of charge motion (Fig 6.28); the individual point sources (from the charges 'emitting' field at different times) overlap at an angle θ. We can obtain θ by comparing the distance travelled by the charge in a time Δt, which is $v_c \Delta t = \beta c \Delta t$, to the distance travelled by the individual wavefronts, which is $v_p \Delta t = (c/n) \Delta t$. The overall (linear) wavefront – obtained by overlapping the infinitesimal wavefronts from each emitting point – is obviously perpendicular to the direction it's moving, so we can then obtain θ as

$$\cos \theta = \frac{\frac{c}{n} \Delta t}{v \Delta t} = \frac{c}{\beta c n} = \frac{1}{\beta n}, \tag{6.235}$$

where βc is the velocity of the charge and n is the refractive index of the medium through which the charge is moving (Fig 6.29); notice that n may vary with wavelength, so that different wavelengths may be emitted at different angles.

If we assume that the charge is moving with a large kinetic energy and therefore large velocity (and that is very often the case), we may state $\beta \simeq 1$ and our Čerenkov angle takes a very simple form:

$$\cos \theta \simeq \frac{1}{n}. \tag{6.236}$$

For example, water has a refractive index of about 1.33 for visible wavelengths. Hence the Čerenkov angle is $\theta \simeq 41°$. Notice that Čerenkov radiation is emitted when any charge is moving through a material with $v_c > v_p$, but we will only see that radiation if the material itself is transparent to it. Also, the Čerenkov radiation is emitted at 41° at any *azimuth* around the direction of the charge – the radiation is emitted as a *cone*. Slower-moving particles ($\beta < 1$) give rise to a radiation cone which is *narrower*, and obviously the minimum velocity where Čerenkov radiation is produced is when $v_c = v_p$, in other words when $\beta = 1/n$. In water, charged particles have to move faster than $\beta = 0.75$ to generate

---

*Pavel Čerenkov (the surname is pronounced 'Cherenkov') was a most interesting Russian scientist, and also said to be the inspiration for the Star Trek character Pavel Chekov. As a doctoral student Čerenkov studied under Sergey Vavilov, another notable Russian physicist known in radiation physics for his description of energy loss of charged particles, and also as the co-discoverer with Čerenkov of what is now known as Čerenkov radiation.

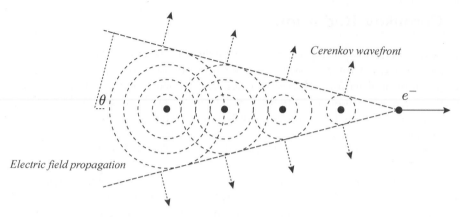

FIGURE 6.28 An electron moving to the right at a velocity $v_c > v_p$, pictured at successive time intervals; the spherical fronts corresponding to different retarded times are shown. The combination of wavefronts as the charge moves gives rise to a conical Cerenkov wavefront whose normal is an angle θ to the direction of charge motion.

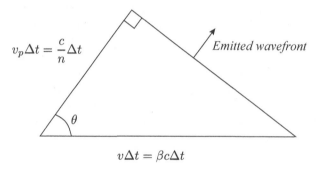

FIGURE 6.29 Geometry of emitted wavefront in Cerenkov radiation.

Čerenkov radiation (Fig 6.30). Note that we haven't said what *kind* of charged particle can do this – any charge can generate Čerenkov radiation. However, usually it's electrons that we talk about since they are the most common situation.

If we can measure the angle θ of the Čerenkov cone and we know the refractive index n, we can measure the velocity v_c of the charge. Since a charged particle slows down in a transparent material (such as water) by means of ionisation slowing, there is thus a 'ring' of Čerenkov light emitted for the time the charge's velocity remains larger than v_p. A so-called ring-imaging detector (for example the RICH (Ring-Imaging CHerenkov) detector) can then determine θ and combined with a separate measurement of the momentum p_c of the charge (by measuring the deflection angle caused by a dipole field B), we can determine the mass of the charge using $p_c = m_c v_c$ – a useful process in particle physics experiments.

Pavel Čerenkov first observed the radiation named after him when observing the blue glow in a bottle of water caused by radioactive particles travelling rapidly through it. The interesting part of that statement is the glow was *blue*, not white. Why is Čerenkov radiation blue? The reason is that, whilst photons of many different frequencies are generated, *more* blue photons than other visible frequencies are generated. Ilya Frank and Igor Tamm obtained a description of this behaviour.* The basic Frank-Tamm formula describing the

---

*Pavel Čerenkov, Ilya Frank and Igor Tamm were jointly awarded a Nobel prize for the discovery and

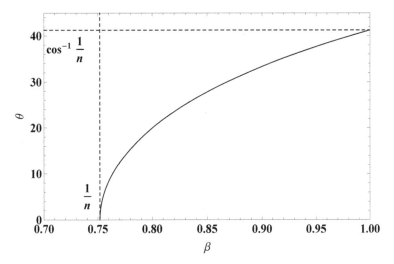

FIGURE 6.30 Water has a refractive index of about $n = 1.33$ for visible light. The minimum velocity where Cerenkov radiation can be generated is $\beta = 1/n$. Ultra-relativistic particles with $\beta \simeq 1$ have a Cerenkov angle of $\theta = \cos^{-1}(1/n) \simeq 41°$.

number of photons $\mathrm{d}N$ liberated over a given frequency range $\mathrm{d}\omega$ is given by

$$\frac{\mathrm{d}^2 N}{\mathrm{d}\omega \mathrm{d}x} = q^2 \frac{\mu_0}{4\pi} \sin^2 \theta \qquad (6.237)$$

over a distance dx, where θ is the Čerenkov angle. Converting the number of photons into their energies, we may obtain the energy lost at different frequencies as

$$\frac{\mathrm{d}E}{\mathrm{d}x} = q^2 \frac{\mu_0}{4\pi} \omega \left(1 - \frac{1}{\beta^2 n^2}\right) \mathrm{d}\omega. \qquad (6.238)$$

This explains why more intensity is produced at blue wavelengths than red wavelengths – Čerenkov light is blue. However, consider the simple case of a constant refractive index n (i.e. constant for all frequencies ω). The total energy emitted into Čerenkov light becomes

$$\left(\frac{\mathrm{d}E}{\mathrm{d}x}\right)_{total} = \frac{q^2 \mu}{4\pi} \left(1 - \frac{1}{\beta^2 n^2}\right) \int_0^\infty \omega \mathrm{d}\omega. \qquad (6.239)$$

The integral $\int_0^\infty \omega \mathrm{d}\omega$ diverges. Hence there must be a maximum frequency of emission when n becomes less than 1. High-frequency (short wavelength) radiation above the UV range typically has $n < 1$ (n later rises to around $n = 1$ for X-rays and gamma rays), so the finite amount of energy emitted into Čerenkov radiation is basically limited by the fact that the refractive index varies (this is illustrated in Fig 6.31).

description of Čerenkov radiation.

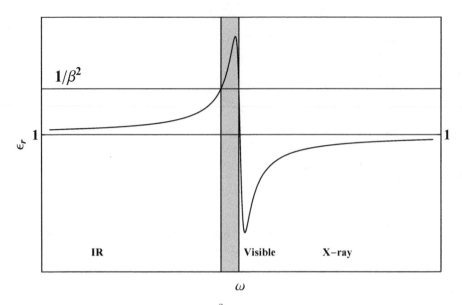

FIGURE 6.31   Variation of permittivity $\epsilon_r = n^2$ with frequency ω in a dielectric material such as water. If the Cerenkov condition were satisfied at all radiation frequencies then an infinite amount of Cerenkov radiation light would be emitted. However, the Cerenkov condition is only satisfied for a narrow range of frequencies so a finite amount of radiation is emitted. The condition is typically satisfied in the visible part of the spectrum, and the light output is more blue than white because of the frequency dependence of the Cerenkov intensity; this is why Cerenkov light in a nuclear reactor appears to be blue.

Exercises

1. The United Kingdom JET tokamak utilises a toroidal field system in which the toroidal coils have an aperture around 5.5 m in height and 4 m in width; the outer diameter of the toroid is 10 m. Estimate the stored energy in the toroid if it generates a magnetic field of 3.45 T.

2. For a plane electromagnetic wave, show that the real part of the time-averaged Poynting vector is
$$\langle S \rangle = \frac{E_{rms}^2}{Z_0}.$$

3. A focused laser pulse generates a power of 27 TW over a circular area of radius 0.1 mm. Calculate the RMS electric and magnetic fields at the focus.

4. A Hertzian dipole antenna with length 1 cm radiates with a power of 100 W at 100 MHz. Find the amplitude of the alternating current fed to the antenna. Determine the magnitudes of the electric and magnetic fields at points a distance of 100 M away, (i) along that antenna axis and (ii) perpendicular to the antenna axis.

5. A proton is accelerated by a potential difference of 700 kV in a static electric field, over a distance of 3 m. Obtain an expression for the ratio of radiated energy to the final kinetic energy, and hence show that radiation losses are negligible.

6. Consider an isochronous cyclotron that produces protons at its extraction point with an energy of 20 MeV and an average current of 1 mA; the average field at the extraction radius is 1.8 T. What is the emitted cyclotron radiation power for the outermost turn,

and what frequency does it have? Describe qualitatively the pattern of emitted radiation and its polarisation.

7. For the same cyclotron as the previous problem, now calculate the total cyclotron power, assuming the voltage gain per turn is 20 kV.

8. A non-relativistic charged particle orbits in a uniform magnetic field. Defining the energy decay time as

$$\tau = U \left(\frac{dU}{dt} \right)^{-1}$$

(where U is the charge's energy), show that τ is given by

$$\tau = \frac{3\pi\epsilon_0 c^3 m^3}{e^4 B^2}.$$

If the magnetic field strength is $B = 2$ T, calculate the energy decay time due to cyclotron radiation both for a proton and for an electron.

9. Consider synchrotron radiation from a highly-relativistic electron gyrating with radius ρ in a magnetic field B. Let $\Delta\theta$ be the angular width of the emitted radiation as seen by a (stationary) observer and τ its duration. Obtain an expression for τ', the time interval of the radiation in the reference frame of the electron, and hence show that

$$\tau \simeq \frac{R\Delta\theta}{c} \frac{1}{\gamma^2}.$$

Writing $\Delta\theta = 1/\gamma$, show that

$$\tau = \frac{m_e}{eB} \left(\frac{m_e c^2}{E} \right)^2,$$

where E is the electron energy. This pulse length determines the maximum frequency of emitted synchrotron radiation. For electrons of 3 GeV energy moving perpendicular to the magnetic field of 0.8 T, estimate the associated maximum emitted photon energy.

10. Consider an electron-positron collider with an energy in each beam of 2000 GeV per particle. For a 100 km tunnel length and a dipole field of 0.1 T, estimate the fractional energy loss per turn an electron undergoes. Relate this to the likely momentum acceptance in such a ring and thereby estimate the minimum number of cavities needed; assume that the ring lattice is tuned for 2000 Gev at all positions in the ring.

11. The Daresbury Synchrotron Radiation Source generated photons from a circulating electron current of 200 mA and an average electron energy of 2 GeV. Given that the ring dipole bending radius was 5.56 m, show that the on-axis emitted power was 20.8 W/mrad^2. A 6 T wavelength shifter (WS) was used to increase the short-wavelength flux of photons. Calculate the ordinary dipole and WS critical energies, and show that the spectral photon flux at 30 keV is increased in the WS by about a factor of 100.

12. Consider a 5 GeV storage ring with a circulating electron current of 100 mA. A 4 metre-long undulator with $K = 1$ is installed in one of the ring straight sections; calculate its bandwidth and the number of photons emitted in the first harmonic. If six such undulators are installed, estimate the fractional increase in energy loss per turn U_0.

13. A 1 kW beam of monochromatic, 10 keV X-rays is incident upon the end of a column of fully-ionised hydrogen plasma in which the number density is 10^{16} cm^{-3}; the plasma column has a cross-sectional area of 1 cm^2 and is 10 cm long. Estimate the X-ray power scattered by the plasma column, and the wavelength of the scattered radiation.

14. A laser of wavelength 1 μm is scattered from an electron beam to generate 50 keV photons. Assuming the 'head-on' geometry in which the scattered photon direction is at 180 degrees to the incident photon direction, what is the required mean electron energy? What is the bandwidth of the 180-degree scattered photons if the electron energy spread is 0.1%? For a laser pulse energy of 1 μJ, an electron bunch charge of 100 pC, and an interaction repetition rate of 10 Hz, estimate the rate of 50 keV photon production if both the laser and electron beams are focused to a size of 10 μm at their interaction point.

15. A 4 GeV storage ring is designed to have a so-called 'theoretical minimum emittance' (TME) lattice such that $\int_{dipole} \mathcal{H}_x(s)ds$ has its minimum possible value in each dipole; there are 40 dipoles in the ring. By minimising the value of $\mathcal{I}_{5x}$ in each dipole (by varying the values of η_x, η'_x, β_x and α_x at the dipole entrance, and assuming that the dipoles are all the same), show that the natural emittance may be given by

$$\epsilon_{x0} = \frac{1}{12\sqrt{15}}C_\gamma\gamma^2\theta^3,$$

where θ is the bend angle in the electrons created by each dipole magnet. From this, estimate the natural emittance and (assuming that $\mathcal{I}_{4x} \simeq 0$) the equilibrium energy spread.

References

1. G. Bekefi and A. H. Barrett. *Electromagnetic Vibrations, Waves, and Radiation.* Massachusetts Institute of Technology, 1977.
2. J. D. Jackson. *Classical Electrodynamics (3rd edition).* Wiley, 1998.
3. John P. Blewett. Radiation losses in the induction electron accelerator. *Phys. Rev.*, 69:87–95, Feb 1946.
4. Julian Schwinger. On the classical radiation of accelerated electrons. *Phys. Rev.*, 75:1912–1925, Jun 1949.
5. Alexander Wu Chao, Karl Hubert Mess, Maury Tigner, and Frank Zimmermann, editors. *Handbook of Accelerator Physics and Engineering 2nd Edition.* World Scientific, 2013.
6. H. Wiedemann. *Particle Accelerator Physics.* Springer, 2015.
7. Julian Schwinger. The quantum correction in the radiation by energetic accelerated electrons. *Proceedings of the National Academy of Sciences*, 40(2):132–136, 1954.
8. J. A. Clarke. *The Science and Technology of Undulators and Wigglers.* Oxford University Press, 2004.
9. S.Y. Lee. *Accelerator Physics, 3rd edn.* World Scientific, 2011.
10. A. Wolski. *Introduction to Beam Dynamics in High-Energy Electron Storage Rings.* Morgan & Claypool Publishers, 2018.
11. V.P. Suller. The interaction of wigglers and undulators with stored electron beams. *Nuclear Instruments and Methods*, 172(1):39–44, 1980.
12. Klaus Halbach. Application of permanent magnets in accelerators and electron storage rings (invited). *Journal of Applied Physics*, 57(8):3605–3608, 1985.
13. A. Hofmann. Quasi-monochromatic synchrotron radiation from undulators. *Nuclear Instruments and Methods*, 152(1):17–21, 1978.
14. Vitaly Ginzburg. *Lizv. Akad. Nauk SSSR Ser. Fiz.*, 11:165, 1947.
15. H. Motz, W. Thon, and R. N. Whitehurst. Experiments on radiation by fast electron beams. *Journal of Applied Physics*, 24(7):826–833, 1953.
16. V. B. Berestetskii, E. M. Lifshitz, and L. P. Pitaevskii. *Quantum Electrodynamics (Volume 4*

of Course of Theoretical Physics); translated by J. B. Sykes and J. S. Bell. Pergamon Press, 1982.

17. C. Curatolo, I. Drebot, V. Petrillo, and L. Serafini. Analytical description of photon beam phase spaces in inverse compton scattering sources. *Phys. Rev. Accel. Beams*, 20:080701, Aug 2017.

18. Geoffrey A. Krafft and Gerd Priebe. Compton sources of electromagnetic radiation. *Reviews of Accelerator Science and Technology*, 03(01):147–163, 2010.

19. F. V. Hartemann, W. J. Brown, D. J. Gibson, S. G. Anderson, A. M. Tremaine, P. T. Springer, A. J. Wootton, E. P. Hartouni, and C. P. J. Barty. High-energy scaling of compton scattering light sources. *Phys. Rev. ST Accel. Beams*, 8:100702, Oct 2005.

20. Matthew Sands. The physics of electron storage rings: an introduction. In *International School of Physics, Enrico Fermi, Course XLVI: Physics with Intersecting Storage Rings Varenna, Italy, June 16-26, 1969*, volume C6906161, pages 257–411, 1969.

21. H. W. Koch and J. W. Motz. Bremsstrahlung cross-section formulas and related data. *Rev. Mod. Phys.*, 31:920–955, Oct 1959.

22. Martin J. Berger and Stephen M. Seltzer. Bremsstrahlung and photoneutrons from thick tungsten and tantalum targets. *Phys. Rev. C*, 2:621–631, Aug 1970.

23. G Zschornack. *Handbook of X-Ray Data*. Springer, 2007.

24. A. F. Bielajew. *Fundamentals of the Monte Carlo method for neutron and charged particle transport*. University of Michigan, 2001.

25. B Nordell and A Brahme. Angular distribution and yield from bremsstrahlung targets (for radiation therapy). *Physics in Medicine and Biology*, 29(7):797–810, 1984.

26. Martin Stearns. Mean square angles of bremsstrahlung and pair production. *Phys. Rev.*, 76:836–839, Sep 1949.

27. I. Frank and Ig. Tamm. *Coherent visible radiation of fast electrons passing through matter*, pages 29–35. Springer Berlin Heidelberg, Berlin, Heidelberg, 1991.

7

Multi-Particle Motion

In the preceding chapters we described the three most important aspects of the behaviour of charged particles in accelerators and the devices associated with them: the effects of electric fields upon charges; the effects of magnetic fields upon charges; how the charges produce electromagnetic radiation. In those chapters each charge was considered to be moving independently of the other charges that may accompany it; the effect of having many charges moving together in a bunch being merely additive. For example, the beam loading in a cavity is proportional to the amount of passing charge, as is the intensity of synchrotron radiation. Moreover, we ignored that the moving charges might influence each other. This is true in situations where the bunch charge is sufficiently low. However, there are many circumstances where we must take account of the beam intensity – expressed either in terms of the bunch charge (or current) or in terms of the bunch density (which depends upon the bunch volume). There are a variety of phenomena that can manifest themselves – too many for the scope of this book – so in this chapter we describe the principles underlying them, and give a few of the most important examples that the reader may encounter. We divide our discussion of these self-fields in terms of i) the effect of moving charges upon each other (intra-beam forces, space charge and scattering), ii) The effect of bunches upon the vacuum system and the consequent effect onto the bunches (wakefields and instabilities) and iii) the enhancement of radiation by coherent effects.

DOI: 10.1201/9781351007962-7

7.1   Intra-Beam Forces and Space Charge

We begin by recalling that a slowly-moving (i.e. non-relativistic) charge exerts both an electric and magnetic field; we consider now two charges moving together at the same velocity v and separated by a distance r (where $\mathbf{v} \perp \mathbf{r}$). The electric force seen by each charge (due to the other) is $F_E = -e^2/4\pi\epsilon_0 r^2$ and does not depend on v. The magnetic force varies with velocity; the magnetic field experienced by one charge due to the motion of the other is $B = \mu_0 q v/4\pi r^2$, and the force $F_B = q v B = q^2 v^2 \mu_0 B/4\pi r^2$. When $v \to c$ we have $F_E + F_B = 0$; the forces cancel*. We may state this equivalently in terms of time dilation: in the rest frame of the charges there is only an electrostatic force, but when moving at v with respect to a (stationary) observer the overall Lorentz force $F = qE_c/\gamma$ where E_c is the electric field experienced by one charge due to the other in their mutual rest frame. We see therefore that slowly-moving bunches will experience a mutual repulsion, known as *space charge* that becomes much less strong as the average bunch velocity approaches c. We expect also that the strength of the space charge force is greater if we have more charges (i.e. more particles) in the bunch.

The self-forces within a bunch give rise to a number of phenomena, many of them unwanted. These include defocusing (leading to a change in the betatron tunes), energy loss, and so on. It is conventional to divide the space-charge forces into collisional interactions – those in which particles collide individually – and the overall smooth space-charge force. The boundary between these regimes is given by the Debye length, which describes the distance over the field of a single particle is screened. The Debye length is

$$\lambda_D = \frac{\sqrt{\langle v^2 \rangle}}{\omega_p}, \tag{7.1}$$

where $\langle v^2 \rangle$ is the average (thermal) RMS velocity of each charge and $\omega_p = \sqrt{e^2 n/m\epsilon_0}$ is the plasma frequency for unit-charge particles with a number density n. If the whole bunch is moving with some energy and relativistic factor γ and we have an (equilibrium) distribution of velocities (i.e. a Maxwell-Boltzmann distribution), then the Debye length can be given as

$$\lambda_D = \sqrt{\frac{\epsilon_0 \gamma k_B T_b}{e^2 n}}; \tag{7.2}$$

T_b is the RMS thermal temperature of the bunch charges with respect to their average velocity such that $\gamma m \tilde{v}^2 = k_B T_b/\gamma$. If the Debye length is large compared to the bunch size (radius) then individual collisions will dominate; if the Debye length is small then collective, smooth forces will dominate. For example, a typical relativistic electron bunch with radius 100 μm and length 6 ps may have an effective temperature $k_b T_b = 0.2$ eV, for which $\lambda_D \simeq 5$ μm but an inter-particle distance of, say, less than 1 μm. In most situations the collisional forces are therefore small compared to the smoothed forces, except that it is the collisions that lead to there being an equilibrium 'thermal' beam distribution and that also contribute to there being an equilibrium beam size. We now consider some practical examples.

7.1.1   Space-Charge Forces

In a rapidly-moving bunch ($\gamma \gg 1$) the electric field from any given charge is only felt in a plane that is co-moving with that charge. Hence we may calculate the space-charge force

---

*This is still true when we include the effects of length contraction on the electric and magnetic fields.

on a particle in a bunch by considering only the two-dimensional charge distribution at the same z location as that particle; often the distribution will be Gaussian in all three dimensions x, y, and z and we can approximate the electric field seen by a given ('test') charge as

$$E_x \simeq \frac{e\lambda}{2\pi\epsilon_0} \frac{x}{\sigma_x(\sigma_x + \sigma_y)}, \quad E_y \simeq \frac{e\lambda}{2\pi\epsilon_0} \frac{y}{\sigma_y(\sigma_x + \sigma_y)}; \tag{7.3}$$

here, $e\lambda$ is the longitudinal charge density and σ_x, σ_y are the 1-σ sizes of the bunch in the x and y dimensions. The force on each charge in the bunch appears as an effective defocusing effect in both planes; one consequence is that the overall betatron tune is reduced (really, a tune spread is induced and hence this effect is called an incoherent tune shift). In a circular accelerator the tune shift (for example in the vertical plane) is

$$\Delta\nu_y = \frac{1}{2\pi} \oint \beta_y k_y \mathrm{d}s, \quad k_y = -\frac{2r_e\lambda}{\beta^2\gamma^3\sigma_y(\sigma_x + \sigma_y)}, \tag{7.4}$$

where here β_y is the vertical β-function (integrated around the ring) and $\beta = v/c$. Hence for N particles in a Gaussian bunch of length σ_z we may re-cast this tune shift as

$$\Delta\nu_y = -\frac{2r_eN}{(2\pi)^{3/2}\beta^2\gamma^3} \oint \frac{\beta_y}{\sigma_y(\sigma_x + \sigma_y)} \mathrm{d}s. \tag{7.5}$$

As one would expect, the tune shift is proportional to the bunch charge (eN), and is larger for smaller-sized bunches. Also, since σ_y is often much smaller than σ_x the vertical tune shift is generally more important (hence why the formulae were given for $\Delta\nu_y$). Since the effect is to give a tune spread rather than a single tune change for all particles, it cannot be compensated by changing the strengths of the magnetic lattice quadrupoles; at a sufficiently-large value of $\Delta\nu$ particles will be driven onto resonances and be lost, causing a beam lifetime reduction. As an example, we consider one design of the TESLA 5 GeV, 17 km long, damping ring [1]; here the final emittance after damping is $\gamma\epsilon_x = 9~\mu$m, $\gamma\epsilon_y = 2~\mu$m with $N = 2\times10^{10}$ a bunch length of 6 mm. The tune shifts are $\Delta\nu_x \simeq -0.02$, $\Delta\nu_y \simeq -0.3$. $\Delta\nu_y$ is very large; to reduce the tune shift one must do one or more of: decrease the circumference; increase the electron energy; increase the (specified) emittance; or decrease the circulating bunch charge.

7.1.2 Space-Charge Dominated Beams

An intense beam generates mutual repulsion in a moving bunch that gives rise to an effective defocusing force between the charges; in general this force is nonlinear. However, there is a special (transverse) distribution, known as the Kapchinsky-Vladimirksy (KV) distribution, for which this space-charge force is linear [2]. Particles with a KV distribution are uniformly-distributed in any two phase-space coordinates (x, x', y, y') such that the RMS size of the beam is exactly half the actual beam radius. Since the space-charge force is linear, its effect on any particle may be determined by modifying the ordinary Hill's equations to give

$$x'' = -k_x x + \frac{2I}{\beta^3\gamma^3 I_A} \frac{x}{\tilde{\sigma}_x(\tilde{\sigma}_x + \tilde{\sigma}_y)}, \tag{7.6}$$

$$y'' = -k_y y + \frac{2I}{\beta^3\gamma^3 I_A} \frac{x}{\tilde{\sigma}_y(\tilde{\sigma}_x + \tilde{\sigma}_y)}, \tag{7.7}$$

where $\tilde{\sigma}_x$, $\tilde{\sigma}_y$ are the beam extents in each plane ($\tilde{\sigma}_x \neq \tilde{\sigma}_y$ means of course that the beam is transversely elliptical). $I = N\beta$ is the beam current at any point along the bunch for a (linear) bunch density N, and I_A is the Alfvén current, which for electrons is

$$I_A = 4\pi\epsilon_0 \frac{m_e c^3}{e} \simeq 17.045 \text{ kA}. \tag{7.8}$$

Bunches with a KV distribution are said to be stably transported [3] while other distributions can tend to a KV distribution; however, this is a complex topic and the reader is referred in particular to Lund's review [4], earlier work by Hoffman et al. [5] and Reiser's more specialist text [6].

The envelope equations (known as the KV equations) for the transport of a KV distribution can be found as

$$\tilde{\sigma}_x'' + k_x \tilde{\sigma}_x + \frac{2K_{\text{sc}}}{\tilde{\sigma}_x + \tilde{\sigma}_y} - \frac{\epsilon_x^2}{\tilde{\sigma}_x^3} = 0, \tag{7.9}$$

$$\tilde{\sigma}_y'' + k_y \tilde{\sigma}_y + \frac{2K_{\text{sc}}}{\tilde{\sigma}_x + \tilde{\sigma}_y} - \frac{\epsilon_y^2}{\tilde{\sigma}_y^3} = 0, \tag{7.10}$$

where the normalised space-charge perveance $K_{\text{sc}} = 2Nr_e/\beta^2\gamma^3$ is equivalent to the ratio $I/\beta^3\gamma^3 I_A$ (we have here defined $\epsilon_{x,y} = \tilde{\sigma}_{x,y}^2/\beta_{x,y}$, so that $\epsilon_{x,y} = 4\epsilon_{x\text{RMS},y\text{RMS}}$). The three focusing terms are from the lattice (terms in k_x and k_y), from space-charge defocusing (terms in K_{sc}, a perveance effect), and from so-called thermal defocusing (terms in $\epsilon_{x,y}$). A beam is said to be laminar if

$$\frac{\epsilon_x}{\beta_x} \ll \frac{2I}{\beta^3\gamma^3 I_A}. \tag{7.11}$$

Defining a laminarity parameter

$$\rho = \frac{1}{2}\frac{I}{I_A}\frac{\sigma_x^2}{\gamma\epsilon_{xn}^2} \tag{7.12}$$

for a normalised emittance $\epsilon_{xn} = \gamma\epsilon_x$ and (RMS) beam size σ_x, the condition $\rho \gg 1$ means that a beam is space-charge dominated; particles move on trajectories that do not cross and the emittance will grow. $\rho \ll 1$ is the condition for a so-called thermal beam where space-charge forces can be neglected, and in which individual particle trajectories do cross. As a bunch is accelerated, for example in a linac, ρ gradually reduces and there is a transition energy with (relativistic) γ

$$\gamma_\rho = \frac{1}{2}\frac{I}{I_A}\frac{\sigma_x^2}{\epsilon_{xn}^2} \tag{7.13}$$

above which the beam changes from being laminar to being thermal. A typical example might be an electron bunch driving an X-ray free-electron laser, for which $\sim$kA peak currents are required to obtain laser gain. Taking $I = 1$ kA, $\epsilon_{xn} = 1$ mm-mrad and a beam size of $\sigma_x = 300$ μm, the transition energy is 1350 MeV. Hence, space-charge forces can be significant in electron linacs driving free-electron lasers since they utilise electron bunches with these sorts of parameters.

7.1.3 Emittance Growth and Compensation

Particle bunches typically do not have ideal (KV) transverse distributions, but are often Gaussian both transversely and longitudinally; having a Gaussian longitudinal distribution (i.e. in the s direction) means that the current I varies from one end of the bunch to the other. The consequence of this is that the space-charge force also varies, for different longitudinal slices through the bunch. We saw above that for certain distributions the space-charge force is linear – it looks like a focusing term – and hence can be reversed by a suitable (external) focusing force; in other words, the correlated emittance effect can be compensated for using quadrupoles. However, this is only true within a single slice of the bunch; the centre of a bunch will have a larger defocusing than the bunch ends and there will be an *uncorrelated* phase space dilution for the bunch as a whole. In some circumstances this can be alleviated by instead using solenoidal fields in a technique known as emittance compensation, first described in 1989 by Carlsten [7] and later developed by others [8, 9].

Practically, the estimation of space-charge effects is carried out using one of many available codes that take inter-particle forces into account. These codes loosely fall into one of three categories, depending on the number of dimensions that are used to describe the particle density within the bunch. One-dimensional codes such as HOMDYN [10] treat a bunch as independent longitudinal slices, and within each slice effectively use a single (circular) size and density which is used to evolve the particle distribution using an approximation to the envelope equations; a bunch is followed ('tracked') incrementally in small steps along s (or in time t) through an accelerator lattice to obtain an estimate of the resulting particle distribution. Two-dimensional algorithms such as in ASTRA [11] divide each longitudinal slice further into concentric, cylindrically-symmetric rings with varying charge density. Fully three-dimensional algorithms follow a reduced but representative subset of particles – known as macroparticles – and calculate their individual motion due to the integrated effect of all the others; GPT [12, 13] is a widely-used example of such a code. In these codes, the electric field is generally calculated by solving Poisson's equation to obtain the fields on a finite mesh of points (a particle-in-cell method) [14]. As the number of dimensions used increases there are more inter-particle calculations to be performed each step; this was the original motivation for using lower-dimension approximations and for using macroparticles for three-dimensional simulations. Limborg et al. have studied the relative accuracy and speeds of a number of codes [15], and Neveu at al. have performed a more recent analysis [16]. Today, fast space-charge solvers and parallelisations of several codes exist (for example OPAL [17] and IMPACT-T [18, 19]) that allow more accurate calculations to be performed in a reasonable time.

7.2 Scattering Processes

7.2.1 Intrabeam Scattering

Intrabeam scattering (IBS) is a collective effect that occurs within particle bunches; multiple Coulomb (elastic) scattering events occur between pairs of electrons, giving rise to a transfer of energy between them; this process was originally called multiple Touschek scattering (see below). The process can be thought of in terms of there being an effective temperature in each of the x, y, s bunch directions due to the different particle momenta, and IBS gives rise to an equilibrium being formed between those directions; in addition, there is net energy being given to the particles from the RF acceleration and in the case of electrons there is also radiation damping. Thus, for hadron (e.g. proton) beams we expect a steady and unbounded growth of the beam emittance – so that we must keep the IBS growth rate small – and for electrons we expect an equilibrium emittance to be obtained which is larger than the natural (i.e. 'zero-current') emittance. Originally formulated by Piwinski [20] and developed by Bjorken and Mtingwa [21], there are today also more convenient approximate formulae from Kubo and Wolski [22], or from Bane [23]. In an electron storage ring, we start by writing an evolution of the emittance with time as

$$\frac{\mathrm{d}\epsilon_x}{\mathrm{d}t} = -\frac{1}{\tau_x}\left(\epsilon_x - \epsilon_{x0}\right) + \frac{2}{T_x}\epsilon_x \tag{7.14}$$

where T_x is the growth time due to IBS and τ_x is the synchrotron radiation damping time giving the natural emittance ϵ_{x0} (see Chapter 6); similar expressions may be written for the

y and z directions. The equilibrium emittance including IBS, ϵ_x^* is given by

$$\epsilon_x^* = \frac{T_x}{T_x - \tau_x}\epsilon_{x0}.$$
(7.15)

The general form of the growth rate ($1/T_u$ in each plane $u = x, y, s$) is

$$\frac{1}{T_u} = 4\pi A(\log)\left\langle \int f_u(\ldots) \right\rangle,$$
(7.16)

where $\langle \rangle$ denotes an average of f_u around the ring lattice (f_u being a function of the lattice and bunch parameters), (log) is the co-called Coulomb logarithm and

$$A = \frac{N_0 r_e c}{64\pi^2 \beta^3 \gamma^4 \epsilon_x \epsilon_y \sigma_\delta \sigma_s};$$
(7.17)

here, $\beta = v/c$, γ is the relativistic factor, σ_δ the energy spread, and σ_s is the bunch length. Immediately we see that the scaling $\propto 1/\gamma^4$ means that IBS is only relevant at lower electron energies, which for practical purposes in electron storage rings is around 3 GeV (or perhaps somewhat higher for some damping ring designs). The Coulomb logarithm describes the ratio between the maximum and minimum impact parameters relevant for an electron-electron collision, and it is conventional to use the classical electron radius r_e as the minimum and the vertical beam size as the maximum. For example, taking a modern 3rd-generation ring we might have $\epsilon_y \sim 10$ pm and $\langle \beta_y \rangle \sim 5$ m, so that

$$(\log) = \ln\left[\frac{\sqrt{5 \times 10^{-12}}}{r_e}\right] \approx 20.$$
(7.18)

This is the original value assumed by Bjorken and Mtingwa in their analysis. However, there is no clear consensus as to the correct value of (log) and values between 10 and 20 have been proposed; similarly, modifications to account for the non-Gaussian nature of some beams have also been proposed. Hence, calculated IBS growth rates should be checked against measured values as has been done in the Japanese ATF ring. A useful approximation to IBS is given by the CIMP (Completely Integrated Modified Piwinski) formulae [22], which are

$$\frac{1}{T_s} \simeq 2\pi^{3/2}A(\log)\left\langle \frac{\sigma_H^2}{\sigma_\delta^2}\left(\frac{g\left(\frac{b}{a}\right)}{a} + \frac{g\left(\frac{a}{b}\right)}{b}\right) \right\rangle,$$

$$\frac{1}{T_x} \simeq 2\pi^{3/2}A(\log)\left\langle -ag\left(\frac{b}{a}\right) + \frac{\mathcal{H}_x\sigma_H^2}{\epsilon_x}\left(\frac{g\left(\frac{b}{a}\right)}{a} + \frac{g\left(\frac{a}{b}\right)}{b}\right) \right\rangle,$$

$$\frac{1}{T_y} \simeq 2\pi^{3/2}A(\log)\left\langle -bg\left(\frac{a}{b}\right) + \frac{\mathcal{H}_y\sigma_H^2}{\epsilon_y}\left(\frac{g\left(\frac{b}{a}\right)}{a} + \frac{g\left(\frac{a}{b}\right)}{b}\right) \right\rangle,$$
(7.19)

where

$$g(\alpha) = \sqrt{\frac{\pi}{\alpha}}\left[P^0_{-1/2}\left(\frac{\alpha^2 + 1}{2\alpha}\right) - \frac{3}{2}P^{-1}_{-1/2}\left(\frac{\alpha^2 + 1}{2\alpha}\right)\right] \qquad \text{if } \alpha < 1,$$
(7.20)

$$g(\alpha) = \sqrt{\frac{\pi}{\alpha}}\left[P^0_{-1/2}\left(\frac{\alpha^2 + 1}{2\alpha}\right) + \frac{3}{2}P^{-1}_{-1/2}\left(\frac{\alpha^2 + 1}{2\alpha}\right)\right] \qquad \text{if } \alpha > 1.$$
(7.21)

P^m_n are the associated Legendre functions, and we define also

$$\frac{1}{\sigma_H^2} = \frac{1}{\sigma_\delta^2} + \frac{\mathcal{H}_x}{\epsilon_x} + \frac{\mathcal{H}_y}{\epsilon_y}$$
(7.22)

and the scaled bunch dimensions

$$a = \frac{\sigma_H}{\gamma}\sqrt{\frac{\beta_x}{\epsilon_x}}, \quad b = \frac{\sigma_H}{\gamma}\sqrt{\frac{\beta_y}{\epsilon_y}}. \tag{7.23}$$

These are fairly involved expressions, and there are established (and validated) codes such as `elegant` [24] that will determine IBS growth rates. To give an idea of the effect of IBS, consider again the 3 GeV, 561.6 m Diamond storage ring with a coupling (i.e. emittance ratio) $\kappa = 0.01$. With $\epsilon_{x0} = 2746$ nm-rad, and 936 bunches with zero-current bunch length $\sigma_{s0} = 3.83$ mm (13 ps), with 300 mA circulating current there is only a very small increase in emittance from IBS (2754 nm-rad). Conversely, one multi-bend achromat design gives $\epsilon_{x0} = 44$ pm-rad; even with a larger $\kappa = 0.1$ to reduce the growth rate the equilibrium emittance grows to 79 pm-rad – an increase of 80%, with similar increases in the vertical emittance and bunch length. Whilst IBS is predominantly considered for circular machines, it may also occur in sufficiently-long linacs when the electron emittance is small [25].

7.2.2 Touschek Scattering

Touschek scattering (first explained in 1963 by Bruno Touschek and collaborators [26]) is related to intrabeam scattering, but here we are concerned with those scattering events in which an appreciable transfer of momentum occurs between two particles – perhaps 1% of the momentum or more. After such a scattering event, one particle has an energy $+\Delta p$ above the mean bunch energy and the other has an energy $-\Delta p$ below (both Δp values are the same magnitude). Again we consider here an electron ring, and in this case the initial momentum change from the scattering will persist for a time $\sim \tau_s$ before the electrons damp back into the 'core' of the bunch; there is therefore a diffuse 'halo' of Touschek-scattered particles, with density above that from the synchrotron radiation, continuously being excited and damped. However, the RF accelerating voltage gives a bounded energy acceptance ϵ_{RF} that is typically between 1% and 3% (see Chapter 6); electrons that are Touschek-scattered outside this limit are lost, and this leads to a finite beam lifetime given by

$$\frac{1}{\tau} = -\frac{1}{N_b}\frac{dN}{dt} = \frac{r_e^2 c N_b}{8\pi\sigma_x\sigma_y\sigma_s}\frac{1}{\gamma^2\epsilon_{RF}^3}D(\epsilon), \tag{7.24}$$

where $\sigma_{x,y,s}$ are the bunch dimensions (in metres), ϵ is a scaled parameter defined as

$$\epsilon = \left(\frac{\epsilon_{RF}\beta_x}{\gamma\sigma_x}\right)^2, \tag{7.25}$$

and the function $D(\epsilon)$ is defined as

$$D(\epsilon) = \sqrt{\epsilon}\left[-\frac{3}{2}e^{-\epsilon} + \frac{\epsilon}{2}\int_\epsilon^\infty \frac{\ln u}{u}e^{-u}du + \frac{1}{2}(3\epsilon - \epsilon\ln\epsilon + 2)\int_\epsilon^\infty \frac{e^{-u}}{u}du\right]. \tag{7.26}$$

This function is shown in Fig 7.1, and arises from consideration of the electron-electron (Möller) scattering rate into a given energy deviation (see for example the detailed derivation by Le Duff [27]). Often we have $\epsilon \ll 1$, and hence the lifetime $\tau \propto \epsilon_{RF}^2$. In modern-day electron storage rings and damping rings the Touschek lifetime is often measured in terms of hours and is the shortest beam lifetime encountered; Touschek scattering determines how long the beam can survive for. One may increase the RF voltage to increase ϵ_{RF}, and generally despite the bunch shortening caused by the higher voltage the Touschek lifetime is increased. For a large enough voltage the RF acceptance becomes larger than the

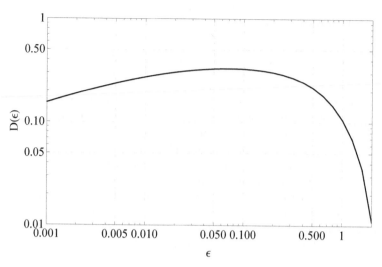

FIGURE 7.1 Behaviour of the function $D(\epsilon)$.

energy acceptance given by other limitations; these are the physical momentum acceptance – since the Touschek-scattered particles move laterally because of dispersion such that $\Delta x = \eta_x(\Delta p/p)$ – and the dynamic momentum acceptance. The dynamic momentum acceptance is here defined as the largest momentum deviation that remains stable (commonly found to be no more than 3 – 4%), and the limitation arises because of non-linear effects upon the electron motion (see Chapter 5 for a fuller discussion of dynamic aperture and momentum acceptance).

A number of approximations have been used to determine the Touschek lifetime, starting from Brück's [28] and including the method of Völkel [29]; some older codes such as ZAP use these approximations [30]. These may be used in certain circumstances depending mainly upon the relative horizontal and vertical beam velocities with respect to the particle energy, and the reader is advised to check the limits of validity for those approximations before using such calculations. More modern codes such as elegant exist that are generally reliable across different electron parameter regimes.

7.3 Wakefields

As we have just seen, a charged particle moving at $v = c$ through a vacuum – or equivalently moving through a smooth beam pipe whose walls have zero resistance – does not see the fields generated by other particles unless they have the same longitudinal coordinate s; this is because the electromagnetic field takes the form of a thin disk travelling with the particle. However, a finite-conductivity vacuum boundary near the bunch contains charges that will be attracted under the effect of the passing charge's field, creating an image current. Since the image current will dissipate energy due to the finite conductivity, there is therefore energy loss from the charge and the beampipe acts effectively as an impedance to slow the charge. Due to the retarded time $\tau = t - r/c$ between when the charge passes and when the charge's field is seen at the beam pipe (for a distance r between charge and pipe boundary) there will remain an electromagnetic field in the wake of the passing charge. This wakefield may then act back upon charges that follow the first; high-velocity charges cannot exert a wakefield in front of them – a form of the principle of causality. We introduced and briefly discussed wakefields in Chapter 3 from an RF cavity perspective and now we discuss it in more detail.

When a particle bunch reaches a beampipe discontinuity – such as an insulating gap or a cross-section change such as a cavity – the (free) particles in the bunch continue to travel unimpeded, but the image current must go around the discontinuity through the conduction path. There will be a decelerating force on the bunch as it moves away from the image charge/current, and the beam will lose energy in the form of radiation to the electromagnetic fields driven by these surface currents and charges. This radiation will remain and can interact with later bunches, or indeed later charges within the same bunch and can in some cases cause significant beam disruption. Energy in unwanted modes can adversely affect the shape and trajectory of later bunches in a bunch train through beam-beam instabilities. If the wakefield from the head (front) of a bunch interacts with the particles in the tail of the same bunch, we call it an intra-bunch, or short-range, wakefield; if the wakefield from one bunch interacts with a later bunch, it is referred to as an inter-bunch, or long-range, wakefield.

If we have a beam propagating in a perfectly-conducting beampipe of constant cross section, the phase velocity of all the waveguide modes of the beampipe are greater than the speed of light at all frequencies and there is no power loss from the beam's space-charge fields to the perfectly conducing walls due to the image charge, so there is no net interaction of these modes with the beam. Hence an ultra-relativistic bunch ($\gamma \gg 1$) in a perfectly-conducting smooth-walled beampipe generates no wakefield; to induce a wakefield, one of two things must occur: either there is an obstacle to reflect the fields to slow down the phase velocity/localise the fields (known as a geometric wake), or there is a finite conductivity causing the image current to lose energy (known as resistive-wall wake).

In geometric wakes, a discontinuity scatters the field. However the discontinuity needn't only be a cavity: it may be a coupler, a corrugation (such as a vacuum bellows), a surface imperfection (small features such as flanges, diagnostic instrumentation or pumping ports), surface roughness, or any similar change in the beampipe's otherwise constant cross section. Where the beampipe or cavity cross section is reduced or increased, the wakefield is strongly dependent on the smallest aperture size. Typically the aperture size is proportional to the wavelength of an RF cavity, hence higher-frequency cavities tend to have higher wakefields. In resistive walls the longitudinal wavenumber, k_z, becomes complex due to the finite conductivity of the walls, causing a transfer of energy between the particle and the EM wave.

A *driving* charge q' traversing a discontinuity induces a wakefield that will persist for some time. At a later time, a *test* charge q, a distance s behind the driving charge, will experience that wakefield. The wake function $w_z(s)$ describes the effect of the driving charge on the test charge, where w_z is the voltage per unit drive charge as a function of the distance between the two particles, given as a superposition of the voltage from all modes in the system. If we consider a particle trailing the driving charge by a distance, s, and we integrate the electric field, E_z, seen by that particle along the beam path in z – divided by the driving charge – we obtain the wakefield

$$w_z(s) = -\frac{1}{q} \int_0^L E_z(z, t = (z+s)/c) \mathrm{d}z, \tag{7.27}$$

where q is the charge of the driving bunch, and the field extends for a distance, L. The wakefield is normally stated in V/nC or V/pC. It is often more useful to represent the wakefields in the frequency domain. This is known as the coupling impedance and is the relationship between the decelerating voltage and the beam current. The coupling impedance, $Z_\parallel$ is calculated by performing a Fourier transform on the wake potential, and is measured in Ohms [31]; we have

$$Z_{\parallel}(\omega) = \int_{-\infty}^{\infty} w_z \exp(-i\omega t)dt. \tag{7.28}$$

For an RF cavity or other discontinuity with an electromagnetic resonance at some frequency (where there are two discontinuities between which radiation can reflect), the impedance plot has 3 distinct regimes, namely i) cut-off, ii) narrowband impedance, and iii) broadband impedance. This is illustrated in Fig 7.2.

Below the beampipe cut-off frequency the impedance is close to zero as there are no propagating modes with a real k_z, although there is a small impedance due to the bandwidth of modes above cut-off extending to lower frequencies and the evanescent modes. The resonant discontinuity is connected to the rest of the accelerator by a beampipe, which is normally a circular waveguide. Below the cut-off frequency of the beampipe any modes which are resonant are trapped and only exist as certain frequencies. Each of these modes is a narrowband impedance as it has a small bandwidth and the modes do not overlap greatly. In between the resonances we see regions that have very low impedance over a very narrow band. These are known as *Fano resonances*, where the reactance of a mode above its resonant frequency cancels out the reactance of a mode below its resonant frequency. Above the cut-off of the beam-pipe the modes are travelling waves and can propagate out of the system through the beampipes. If the bandwidth of each mode is very large, at any given frequency the fields will be a superposition of several modes. This causes a continuous broad impedance spectrum. The beam may interact with the beam at the discontinuity or elsewhere in the machine if the radiation has propagated away.

The narrowband regime is normally treated in the frequency domain. The impedance can also be given from the equivalent circuit impedance as seen in Chapter 3,

$$Z = \frac{R_{s,circuit}}{1 + iQ_L\left(\frac{\omega}{\omega_0} - \frac{\omega_0}{\omega}\right)}. \tag{7.29}$$

where R_s is the shunt impedance of the cavity, and Q_L is the loaded Q factor. If we transform the impedance back into the time domain, we get a wave that is described by

$$w_z(t) = \frac{\omega_0 R}{2Q} s^{-t/\tau} \left(\cos \omega_0 \sqrt{1 - \frac{1}{Q_L^2}}t - \frac{1}{2Q_L\sqrt{1 - \frac{1}{4Q_L^2}}} \sin \omega_0 \sqrt{1 - \frac{1}{Q_L^2}}t\right), \tag{7.30}$$

where τ is the decay time given as $2Q_L/\omega_0$, and Q_L is the loaded Q factor of the cavity. The damping causes a phase shift between the beam and the excited RF wave, as well as a small shift in the wake's frequency from the resonant frequency of the mode, ω_0. Normally the Q factor is sufficiently large that this reduces to

$$w_z(t) = \frac{\omega R}{2Q} s^{-t/\tau} (\cos \omega_0 t). \tag{7.31}$$

It is commont to define the loss parameter, k, for each mode to quantify the strength of interaction where,

$$k = \frac{\omega R}{2Q}. \tag{7.32}$$

The effect of the beam's longitudinal profile on the wakefield is found by multiplying the frequency spectrum of the beam by the impedance, which can then be converted back to a wake potential by using an inverse Fourier transform. The beam spectrum, $\tilde{I}(\omega)$, for a

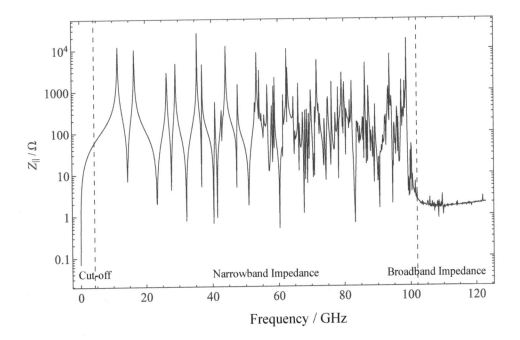

FIGURE 7.2 Illustration of the behaviour of coupling impedance as a function of frequency, showing typical impedance spectrum in the cut-off, narrowband and broadband impedance regimes.

bunch of charge q with a Gaussian charge density distribution in space, is also a Gaussian in the frequency domain and is given by

$$\tilde{I}(\omega) = q \exp -\frac{\omega^2 \sigma_z^2}{2c^2} \tag{7.33}$$

where σ_z is the (Gaussian) standard deviation of the bunch in the z direction.

7.3.1 Short-Range Wakefields

In order to obtain the wakefield of an entire bunch, W_z, it is necessary to convolute the wakefield of a single point charge, w_z, with the charge density of the bunch.

$$W_z(s) = \int_{-\infty}^{\infty} w_z(z)\lambda_q(s-z)\mathrm{d}z \tag{7.34}$$

where λ_q is the longitudinal charge density profile of the bunch.

The total energy lost by a bunch due to its own wakefield is given by the loss factor (ΔK), in electron-volts per unit drive charge, which integrates the wakefield multiplied by the charge density along the length of the interaction

$$\Delta K = -\frac{1}{q} \int_{-\infty}^{\infty} W_z(z)\lambda_q(z)\mathrm{d}z. \tag{7.35}$$

Most bunches do not have Gaussian distributions, instead having ripple in the charge spectrum, but it is taken as the standard distribution for wakefield calculations due to its smooth roll-off with frequency. If the beam has a Gaussian spectrum then the standard deviation

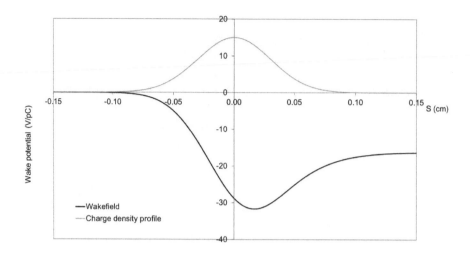

FIGURE 7.3   The short-range wakefield for a Gaussian bunch with $\sigma_z = 0.3$ mm from a 9-cell RF cavity with a 30 mm aperture along with the line charge density. The head of the bunch is to the left of the plot.

will be inversely proportional to the beam's time duration. Long-range wakefields are dominated by the narrow-band impedance region as the broadband impedance quickly decays as the energy propagates down the beampipe. For short bunches, which hence have a large spectrum the short-range wakefield is dominated by the broadband region, as while the impedance is lower than the narrowband region it extends over a wide frequency spectrum for short bunches. The short-range wakefield for a Gaussian bunch with $\sigma_z = 0.3$ mm from a 9-cell RF cavity with a 30 mm aperture is shown in Fig 7.3. The electrons at the head of the bunch do not see any deceleration due to the wakefield, while electrons later see the full wakefield. For very long bunches, the higher-frequency components in the wake may oscillate causing the electrons in the tail to be accelerated.

7.3.2   Long-Range Wakefields

If we integrate the real part of the impedance in frequency around a mode and divide by the resonant frequency of the mode, we obtain the geometric shunt impedance (R/Q) of that mode. The longitudinal wakefield induced in a single mode in the narrowband region by a single bunch is given by

$$W_z = 2k \cos(\omega t)e^{-\omega t/2Q_L}. \tag{7.36}$$

Due to superposition we can simply sum the wake from each mode (with index m) together to obtain the total wakefield

$$W_z = \sum_{m=1}^{\infty} 2k_m \cos(\omega_m t)e^{-\omega_m t/2Q_m}. \tag{7.37}$$

It can be seen that the longitudinal wake is given as a sum of damped cosine waves, each with different amplitude and frequency, which are all initially in phase at $t = 0$.

7.3.3 Transverse Wakes

As well as accelerating beams, wakefields can also deflect bunches. Dipole modes have transverse electric and magnetic fields which can kick the beam if excited. In order to excite a dipole mode there must be an energy exchange between the bunch and the dipole mode, which requires electrons to be decelerated by a longitudinal electric field. TE dipole modes do not have longitudinal electric fields hence they cannot be excited by the beam. The longitudinal electric field of a TM dipole mode is given by

$$E_z = E_0 J_1 \left(r \frac{\zeta_{mn}}{a} \right) \cos(\phi) \cos \left(p\pi \frac{z}{L} \right) e^{i\omega t} \tag{7.38}$$

where ζ_{mn} is the n-th root of the m-th Bessel function (in this case the 1st Bessel function J_1), L is the cavity length, and a is the cavity radius. A beam travelling along the central axis ($r = 0$) will not excite a dipole mode, but a beam that is offset will excite a dipole wakefield, referred to as the transverse wake. Once excited by an offset bunch, future bunches will be deflected even if travelling along the central axis. It is also possible to excite a transverse wake if the cavity is asymmetric due to couplers, or asymmetric cavity geometries which cause the dipole mode to gain a longitudinal electric field at $r = 0$. The transverse Lorentz force in the direction of the x-axis, F_x, is given by

$$F_x = e(E_x + vB_y), \tag{7.39}$$

where v is the beam velocity. If the beam is highly relativistic then the transverse movement is small over the cavity length, L, and hence the transverse momentum change, Δp_x, is given by

$$\Delta p_x = e \int_0^L (E_x + cB_y)\mathrm{d}z. \tag{7.40}$$

It is useful to define a transverse voltage, V_x; it is not strictly a voltage as part of the force comes from the magnetic field. However the electric and magnetic forces are very similar if the beam is very relativistic due to the longitudinal momentum being orders of magnitude greater than the transverse momentum.

$$V_x = \int_0^L (E_x + cB_y)\mathrm{d}z \tag{7.41}$$

and

$$V_y = \int_0^L (E_y + cB_x)\mathrm{d}z. \tag{7.42}$$

Applying Maxwell's equations to these equations and integrating along the cavity with the limits in the zero field region, we can derive a relation between the transverse and longitudinal voltage of a dipole mode

$$V_\perp = -\frac{ic}{\omega}\nabla_\perp V_z \tag{7.43}$$

where $V_\perp$ is a vector sum of V_x and V_y. By applying this to the equation for the longitudinal wakefield, we can derive the transverse wakefield, $W_\perp$, as

$$W_\perp = \frac{c}{2}\frac{R}{Q}\sin(\omega t)e^{-\omega t/2Q_L}. \tag{7.44}$$

The effect of the dipole wake is to deflect the beam and the beam offset downstream is given by

$$\Delta x = M_{12}\frac{V_\perp}{K}, \tag{7.45}$$

where M_{12} is the transport matrix element (described in Chapter 5), relating the beam offset at a point downstream to the divergence at the cavity and K is the beam's kinetic energy. The transverse wake is more sensitive to aperture size than the longitudinal wakefield. The CLIC-G 12 GHz structure has a peak transverse wakefield of -250 V/pC/mm/m which is damped to -2 V/pC/mm/m for a particle 0.15 m behind the driving bunch.

The excitation of a dipole-mode wakefield in a cavity is commonly used for measuring beam position in accelerators, as the wakefield is proportional to the beam offset. This can be performed either by using wakefield monitors in accelerating RF cavities or by making special cavity beam position monitors as bespoke devices specifically for measuring the beam position. Dipole modes are also used as deflecting cavities in accelerators as bunch separators, or to give the head and tail equal and opposite kicks to *rotate* a bunch as a diagnostic or to align a bunch for collision at a crossing angle (known as a *crab* cavity).

7.3.4 Multiple Bunches

As well as using superposition to sum the wake from each mode, we can also use superposition to sum the wakes from multiple bunches. We must include the energy lost by an electron bunch due to its own wake. As the wake grows in time as the beam passes the finite cavity length, on average the deceleration voltage experienced by a bunch is half of the wakefield voltage it induces, known as the *fundamental theorem of beam loading*. The value of half can be derived by considering energy conservation of the deceleration of an electron bunch by a cavity voltage, and considering both the energy gained by the cavity and the energy lost by the beam [32]. Taking this into account the total wakefield, W_z, over several modes, m, and N bunches is given by [33]

$$W_z = \sum_{m=1}^{\infty} \left(k_m + \sum_{n=1}^{N} 2k_m \cos(\omega_m n\tau) e^{-\omega_m n\tau/2Q_{Lm}} \right) \qquad (7.46)$$

for the longitudinal wake, where τ is the bunch spacing in time and is equal to the reciprocal of the bunch repetition frequency, f_{rep}, ($\tau = 1/f_{rep}$). While for the transverse wake, $W_\perp$, it is

$$W_\perp = \sum_{n=1}^{N} \sum_{m=1}^{\infty} \frac{c}{2} \frac{R_m}{Q} \sin(\omega_m n\tau) e^{-\omega_m n\tau/2Q_{Lm}}. \qquad (7.47)$$

The longitudinal wake is given as a sum of damped cosine waves above, which converges to a finite value – known as the sum wake – after the wake from the first bunch decays. If a given mode has a resonant frequency at a harmonic of the bunch repetition frequency, the wakes have constructive interference driving a larger wake, while if they are a half integer the wake cancels every second mode. The harmonics of the bunch repetition frequency are known as *machine lines*, and special care must be taken in designing an accelerator with modes close to them. For example, in ESS the specification requires no HOM is within 3 MHz of a machine line. The higher the Q factor of a HOM, the higher the sum wake as the field excited may remain in the cavity for several times the bunch seperation. To reduce the sum wake it is usually necessary in multi-bunch machines to damp the HOMs using HOM couplers as discussed in Chapter 3

For the transverse wake, the sum wake is the sum of damped sine waves. In this case the maximum wakefield isn't at a machine line, as while this will drive a large dipole field in the cavity, the transverse force is zero as the transverse and longitudinal wakes are 90 degrees out of phase. Instead, the largest wake occurs at a frequency slightly off the machine line, with the exact frequency depending on the Q factor of the dominant dipole mode. If we take an example of a bunch separation of 10 ns and a wake dominated by three modes

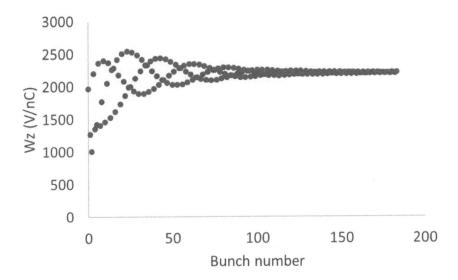

FIGURE 7.4 The total wakefield for each bunch in a train of 183 bunches, for a bunch separation of 10 ns and a wake dominated by three modes at frequencies of 3.050 GHz, 4.000 GHz and 8.035 GHz, with (R/Q) of 120, 20 and 100 Ω and loaded Q factors of 50, 1000 and 10,000 respectively.

at frequencies of 3.050 GHz, 4.000 GHz and 8.035 GHz, with (R/Q) of 60, 10 and 50 Ω and loaded Q factors of 50, 1000 and 10,000 respectively, we get the total wake shown in Fig 7.4. The first mode at 3.050 GHz is strongly damped, so it provides almost the same wake to every bunch. The second mode at 4.000 GHz is at a machine line and hence has a wake which increases every bunch over the first 50 bunches but then remains constant; this mode has the largest effect on the sum wake, even though it has the lowest impedance. The third mode is not at a bunch harmonic and hence the wake oscillates every bunch. Hence for very short bunch trains we are most concerned with modes with high R/Q, but a mode that is dominant over short timescales may not be dominant over longer timescales, hence for longer bunch trains we are most concerned with modes with high Q factors that are close to machine lines.

7.3.5 Wakefield-Driven Instabilities

Wakefields/impedances lead to energy spread and emittance growth in accelerators, but also cause instabilities that drive the beam offset and can cause beam loss. As we have seen, the transverse wake is zero if the driving charge is on the beam axis, but increases linearly with beam offset. The force on a trailing particle, where a wake has previously been induced, is independent of offset to first order. This leads to two feedback mechanisms, known as beam-breakup instabilities (BBU). The first is cumulative BBU, where an offset bunch at the start of a linac will cause subsequent bunches to be deflected, independent of their offset. These subsequent bunches will enter the next cavity at an offset, driving a larger wake which will in turn deflect the following bunches; this induces still larger wakes in the third cavity, and so on. In long linacs, this can lead to significant beam loss at the end of the linac. The second feedback mechanism occurs in energy recovery linacs (ERLs), and is known as *regenerative BBU*. In ERLs each bunch will have one or more accelerating passes of a linac, before passing the same cavity again for an equal number of decelerating passes before being dumped. This means the beam loading will cancel out between bunches being accelerated and decelerated, allowing very high beam currents. The beam current in an ERL is instead limited by regenerative BBU. Here an offset bunch again deflects subsequent bunches, which then return to the same cavity with an offset. If the offset bunch drives a wake such that the next bunch will return to the cavity with a larger offset than the first, and hence the wakefield amplitude grows, then the bunch offset will increase with each subsequent bunch until beam loss occurs. The BBU start current, I_{st} – where the offset starts to grow – for each mode is given by [34]

$$I_{st} = -\frac{2c^2}{e(R/Q)Q_L\omega M_{12}\sin(\omega t_r)}, \tag{7.48}$$

where t_r is the revolution time, i.e. the time it takes an electron bunch to make a single loop of the ERL and return to the same cavity. The regenerative BBU start current is the lowest start current for all modes. It can be seen that the start current is inversely proportional to $\sin \omega t_r$, hence for the highest impedance modes we can design the revolution time to be a harmonic of the RF frequency to increase the start current for that mode. The start current can also be increased with careful design of M_{12} or by using strong HOM damping.

Short-range wakefields are also an issue in linacs, with the head of an offset bunch driving a wake which deflects the tail creating so-called *banana* bunches due to their curved transverse profile. This effect can be reduced by an approach known as Balakin, Novokhatsky and Smirnov (BNS) damping after the originators [35]; they suggested that if the head and tail had different kinetic energies then their betatron motion would be different for the head and the tail causing them to oscillate in position and transverse momentum at different frequencies. As such, as the tail oscillates back and forward, it would be out of phase with the wake causing cancellation over the length of the linac.

In circular machines the transverse wakefield leads to tune shifts, which if sufficiently large can lead to transverse instabilities as discussed in Chapter 5. Like with space charge these lead to tune spreads which therefore cannot be compensated with quadrupoles. We can also use octupoles to stabilise the beam against external excitations, inducing what is known as Landau damping; for more details see the discussion in Lee [36].

7.4   Coherent Synchrotron Radiation (CSR)

In chapter 6 we looked at synchrotron radiation emitted from moving charges in accelerators. We saw that this radiation depends linearly on the number of charges in a bunch, i.e. the emitted power $P \propto N_b$. However, looking again at the Larmor formula,

$$P = \frac{q^2a^2\gamma^4}{6\pi\epsilon_0c^3}, \tag{7.49}$$

we note the factor q^2 in the numerator; this would imply that two charges (say, electrons) in very close proximity should radiate with a power $\propto q^2$ rather than $\propto 2q$ – i.e. they should radiate *coherently*. This is in fact true, and charges do radiate coherently if they are close enough together. In the case of synchrotron radiation, a bunch of electrons emits either incoherent synchrotron radiation (ISR) or coherent synchrotron radiation (CSR) depending upon the separation of the electrons; we expect coherent radiation for those emitted wavelengths λ which are comparable to the size of the electron bunch $\sigma_{x,y}$, as described by Schiff in 1946 [37] and first observed in 1989 [38]. Since the number of electrons in a bunch can be quite large – perhaps $N_b \sim 10^{10}$ to 10^{11} – the coherent enhancement of the radiation power can be huge. We should distinguish here the (coherent) enhancement of the radiated power – due to the proximity of the electrons to each other – from the coherence of the photon output. Incoherently-emitted photons at wavelength λ may be coherent with each other, if the source size of the electrons is small. The condition for *photon* coherence is that the emittance in each plane

$$\epsilon_{x,y} < \frac{\lambda}{4\pi};$$ (7.50)

such an (ISR) source is known as diffraction-limited, but even for emittances $\epsilon \sim \lambda/4\pi$ or larger there will be some partial coherence of the emitted radiation.

To illustrate the power enhancement due to CSR, we consider an electron bunch with total charge $Q = 1$ nC circulating with kinetic energy 3 GeV in the Diamond electron storage ring, choosing a rather short buch duration of 1 ps (in other words, the electrons are grouped together in a length $l = ct \simeq 0.3$ mm); these are realistic values for the length and charge*. As we saw in Chapter 6, the total incoherent power radiated by each electron from the dipoles is 86 nW (over all wavelengths), and therefore the total incoherent power radiated by the electron bunch is 537 W – already a large value. However, wavelengths $\lambda > 0.3$ mm (i.e. in the microwave part of the spectrum) radiate coherently, and so there is an enhancement of the radiated power by a factor of $Q/e \sim 6 \times 10^9$ (although emission at wavelengths larger than the size of the vacuum vessel are suppressed); the coherent enhancement in the power emitted at those wavelength is absolutely enormous. Remembering that relativistic electrons already see a power enhancement $\propto \gamma^4$, we see that coherent radiation from relativistically-moving charges can generate huge numbers of photons at long wavelengths since we can produce significant numbers of electrons in a small bunch. This enhancement is shown in Fig 7.5.

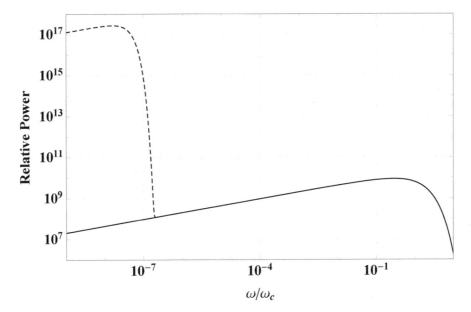

FIGURE 7.5 The solid line shows the incoherent synchrotron radiation power for an electron bunch charge of 1 nC, i.e. a number of electrons $N_b \simeq 6.24 \times 10^9$; the vertical scale is the radiated power compared to that radiated power of 1 electron at the critical frequency $\omega_c = 1.28 \times 10^{19}$ s^{-1} (8.4 keV photons). The dotted line shows the coherent enhancement up to another factor N_b that occurs for long-wavelength (low-frequency) photons for an electron bunch duration of 1 ps; N_b^2 is an enormous factor even for modest bunch charges of 1 nC.

---

*The transverse size of the bunch is much smaller than the length.

A complete treatment of CSR is complex and we refer the reader to many excellent works in the literature [39, 40, 41]. The analysis begins with the Liènard-Wiechert interaction between two electrons; we consider a 'steady-state' regime in which all electrons within a bunch experience a constant magnetic field, for example as found in the middle region of a bending magnet. The complete calculation for two electrons can be found in Saldin [39] and gives the power emitted in terms of the charge and the separation. The extension to a larger number of particles requires some approximation; a common one is to assume a one-dimensional charge distribution (i.e. all electrons follow the same orbit). The validity of such 1D CSR models is typically given in terms of the Derbenev criterion [42], stated in terms of bending radius ρ and the bunch size σ_z as

$$\frac{\sigma_x}{(\rho\sigma_z^2)^{1/3}} \ll 1; \tag{7.51}$$

this must be satisfied for the 1D model to be valid in the horizontal plane. A similar condition can be made in the vertical plane. For practical calculations of CSR numerical methods are generally employed, for example using `elegant` [40] or `GPT` [7, 8, 41]. A recent paper by Brynes et al. gives a useful review [41].

One practical consequence of CSR is that the emitted radiation (in a dipole) gradually overtakes the electrons themselves, and produces an effective wakefield; in contrast to the conventional wakefields described above, the CSR wake occurs in front of the emitting electrons rather than behind. CSR emission causes an average loss in energy for any given electron, but electrons towards the front of a bunch are generally accelerated with respect to those at the rear, which are relatively decelerated. This in turn can result in a number of phenomena that may degrade the quality of the electron bunch [43]. One consequence is that the overall emittance may be increased [44], but may also be compensated using a suitable beam-optics arrangement [45]. Another is that CSR can give rise to a longitudinal modulation of the electron density over distances shorter than the bunch length, a so-called microbunching; this is particularly important in free-electron laser (FEL) design, where unwanted microbunching may interfere with the similar process for FEL gain. The microbunching instability was described by Heifets et al. [46] with useful formulae given by Huang and Kim [47].

The FEL itself is the premier source of coherent radiation from undulators. They are able to generate tuneable laser-like output and can operate up to X-ray wavelengths. An excellent review of the physics of FELs and their output properties is given in [48].

Exercises

1. Estimate the Debye length for the following situations. i) a typical fluorescent lamp plasma, in which the electron density is around 10^{16} m^{-3} and the electron temperature around 1 eV, ii) a tokamak plasma for which the (fully-ionised) plasma density is around 10^{20} m^{-3} and the ion/electron temperature $\sim$10 keV, iii) a low-energy electron bunch with transverse size 0.1 mm, length 10 ps and a charge of 100 pC.

2. Estimate the space-charge-induced tune shift in a proton ring in which a smooth 1 A current of 10 MeV protons moves in a radius of 1 metre. You may take the emittance in each plane to be 10 mm-mrad and the β-function to be 1 metre.

3. The ALICE energy-recovery linac circulated electron bunches with a charge of 80 pC, a normalised emittance (discussed in chapter 5) of around 3 mm-mrad and a typical bunch length of 4 ps. Determine for what energy range the bunches will experience significant space charge; is space charge important for the ALICE energy range from 350 keV (injector) to 35 MeV (FEL operation)?

4. The 6 GeV PETRA-III storage ring at DESY has a very small natural emittance of 1.3 nm-rad and operates with a small coupling around 0.6%. Making sensible assumptions about the β-functions and bunch length, estimate the Touschek lifetime for a stored current of 100 mA and a momentum acceptance of 1.6%. Is the Touschek lifetime significant when compared to a beam-gas scattering lifetime of 14 hours?

5. The dominant higher-order mode in a cavity, the TM_{020} mode, has a resonant frequency of 2 GHz, a geometric shunt impedance of 100 Ω and a loaded Q factor of 50. Calculate the frequency shift, and phase shift at $t = 0$ between the resonant frequency of the cavity and the frequency of the wake caused by the low Q factor.

6. For the same cavity a continuous train of bunches of 10 nC charge, separated in time by by 15 ns, traverses the cavity. Calculate the sum wake for the TM_{020} mode, assuming the bunches are very short compared to the mode's wavelength.

7. If the same cavity is placed in an ERL, with $M_{12} = 1$ and a revolution time of 500 ns, at what current does regenerative BBU begin?

8. What is the diffraction-limited emittance for 8 keV photon emission?

9. The 6 GeV European Synchrotron Radiation Facility has recently been upgraded from a natural emittance of 4 nm-rad to an emittance of 0.13 nm-rad. Making sensible assumptions about the insertion device field and coupling, determine whether the ESRF output is diffraction-limited either with its old or new design. You may assume a vertical emittance of 10 pm-rad in both configurations.

10. Using the same ALICE parameters as in the problem given earlier and assuming a dipole field of 0.2 T, determine for what wavelengths you would expect CSR to be important and how much enhancement of the power there would be. What is the Derbenev criterion in this situation?

References

1. A. Wolski and W. Decking. Damping ring designs and issues. In *Proc. 2003 Particle Accelerator Conference, p. 652*, 2003.

2. P. M. Kapchinksy and V. V. Vladimirsky. In *Proc. 2nd Int. Conference on High Energy Accelerators, CERN, Geneva, p. 274*, 1959.

3. R. L. Gluckstern, R. Chasman, and K. Crandall. In *Proc. Proton Linear Accelerator Conference (Linac'70), Batavia, Illinois, p. 823*, 1970.

4. Steven M. Lund, Takashi Kikuchi, and Ronald C. Davidson. Generation of initial kinetic distributions for simulation of long-pulse charged particle beams with high space-charge intensity. *Phys. Rev. ST Accel. Beams*, 12:114801, Nov 2009.

5. I. Hoffman, L. J. Laslett, L. Smith, and I. Haber. Stability of the kapchinskij-vladimirskij (k-v) distribution in long periodic transport systems. *Particle Accelerators*, 13:145, 1983.

6. Reiser. *Theory and Design of Charged Particle Beams*. Wiley, 2007.

7. B.E. Carlsten. New photoelectric injector design for the los alamos national laboratory xuv fel accelerator. *Nuclear Instruments and Methods in Physics Research Section A: Accelerators, Spectrometers, Detectors and Associated Equipment*, 285(1):313 – 319, 1989.

8. Luca Serafini and James B. Rosenzweig. Envelope analysis of intense relativistic quasilaminar beams in rf photoinjectors:ma theory of emittance compensation. *Phys. Rev. E*, 55:7565–7590, Jun 1997.

9. Klaus Floettmann. Emittance compensation in split photoinjectors. *Phys. Rev. Accel. Beams*, 20:013401, Jan 2017.

10. Clendenin J E Palmer D T Rosenzweig J B Ferrario, M and L Serafini. Homdyn study for the lcls rf photo-injector. In *SLAC-PUB-8400*, 2000.

11. K. Floettmann. Astra: A space charge tracking algorithm. In *DESY, March 2017, Version 3.2.*

12. S. B. van der Geer and M. J. de Loos. *General Particle Tracer User Manual.* Pulsar Physics. Program Version 3.35.

13. Pulsar physics and general particle tracer code. http://www.pulsar.nl/gpt.

14. G. Poplau, U. van Rienen, B. van der Geer, and M. de Loos. Multigrid algorithms for the fast calculation of space-charge effects in accelerator design. *IEEE Transactions on Magnetics*, 40(2):714–717, March 2004.

15. C. Limborg et al. Code comparison for simulations of photo-injectors. In *Proc. 2003 Particle Accelerator Conference, p. 3548*, 2003.

16. N. R. Neveu et al. Benchmark of rf photoinjector and dipole using astra, gpt, and opal. In *Proc. 2016 North American Particle Accelerator Conference, p. 1194*, 2016.

17. Andreas Adelmann, Pedro Calvo, Matthias Frey, Achim Gsell, Uldis Locans, Christof Metzger-Kraus, Nicole Neveu, Chris Rogers, Steve Russell, Suzanne Sheehy, Jochem Snuverink, and Daniel Winklehner. Opal a versatile tool for charged particle accelerator simulations, 2019.

18. J. Qiang, R. D. Ryne, M. Venturini, A. A. Zholents, and I. V. Pogorelov. High resolution simulation of beam dynamics in electron linacs for x-ray free electron lasers. *Phys. Rev. ST Accel. Beams*, 12:100702, Oct 2009.

19. Ji Qiang. Fast 3d poisson solvers in elliptical conducting pipe for space-charge simulation. *Phys. Rev. Accel. Beams*, 22:104601, Oct 2019.

20. A Piwinski. Intra-beam-scattering. In *In Proc. 9th Int. Conference on High Energy Accelerators, Stanford, CA, p. 405*, 1974.

21. James D. Bjorken and Sekazi K. Mtingwa. Intrabeam scattering. *Particle Accelerators*, 13:115, 1983.

22. Kiyoshi Kubo, Sekazi K. Mtingwa, and Andrzej Wolski. Intrabeam scattering formulas for high energy beams. *Phys. Rev. ST Accel. Beams*, 8:081001, Aug 2005.

23. Karl L. F. Bane. An accurate, simplified model of intrabeam scattering, 2002.

24. Michael Borland. Features and Applications of the Program ELEGANT. In *Proceedings, 4th International Particle Accelerator Conference (IPAC 2013): Shanghai, China, May 12-17, 2013*, page THPPA02, 2013.

25. S. Di Mitri. Intrabeam scattering in high brightness electron linacs. *Phys. Rev. ST Accel. Beams*, 17:074401, Jul 2014.

26. C. Bernardini, G. F. Corazza, G. Di Giugno, G. Ghigo, J. Haissinski, P. Marin, R. Querzoli, and B. Touschek. Lifetime and beam size in a storage ring. *Phys. Rev. Lett.*, 10:407–409, May 1963.

27. J. Le Duff. Single and multiple touschek effects. In *CERN CAS Report 89-01*, 1989.

28. Henri Bruck. *Circular particle accelerators (translated by Ralph McElroy Co.* 1974.

29. Uta Volkel. Particle loss by Touschek effect in a storage ring. In *DESY Report DESY-67-5*, 1967.

30. M.S. Zisman, S. Chattopadhyay, and J.J. Bisognano. Zap user's manual.

31. SA Heifets and SA Kheifets. Coupling impedance in modern accelerators. *Reviews of Modern Physics*, 63(3):631, 1991.

32. T. P. Wangler. *RF Linear Accelerators, Second Edition.* Wiley, 2008.

33. R Wanzenberg. Monopole, dipole and uadrupole passbands of the tesla-cell cavity.

34. GH Hoffstaetter and IV Bazarov. Beam-breakup instability theory for energy recovery linacs. *Physical Review Special Topics: Accelerators and Beams*, 7:054401, May 2004.

35. M Ferrario, M Migliorati, and L Palumbo. Wake fields and instabilities in linear accelerators. 2006.

36. S.Y. Lee. *Accelerator Physics, 3rd edn.* World Scientific, 2011.

37. L. I. Schiff. Production of particle energies beyond 200 mev. *Review of Scientific Instruments*, 17(1):6–14, 1946.

38. T. Nakazato, M. Oyamada, N. Niimura, S. Urasawa, O. Konno, A. Kagaya, R. Kato, T. Kamiyama, Y. Torizuka, T. Nanba, Y. Kondo, Y. Shibata, K. Ishi, T. Ohsaka, and M. Ikezawa. Observation of coherent synchrotron radiation. *Phys. Rev. Lett.*, 63:1245–1248, Sep 1989.

39. E.L. Saldin, E.A. Schneidmiller, and M.V. Yurkov. On the coherent radiation of an electron bunch moving in an arc of a circle. *Nuclear Instruments and Methods in Physics Research Section A: Accelerators, Spectrometers, Detectors and Associated Equipment*, 398(2):373 – 394, 1997.

40. M. Borland. Simple method for particle tracking with coherent synchrotron radiation. *Phys. Rev. ST Accel. Beams*, 4:070701, Jul 2001.

41. A D Brynes, P Smorenburg, I Akkermans, E Allaria, L Badano, S Brussaard, M Danailov, A Demidovich, G De Ninno, D Gauthier, G Gaio, S B van der Geer, L Giannessi, M J de Loos, N S Mirian, G Penco, P Rebernik, F Rossi, I Setija, S Spampinati, C Spezzani, M Trov, P H Williams, and S Di Mitri. Beyond the limits of 1d coherent synchrotron radiation. *New Journal of Physics*, 20(7):073035, jul 2018.

42. Ya. S. Derbenev, J. Rossbach, E. L. Saldin, and V. D. Shiltsev. Microbunch radiative tail - head interaction. 1995.

43. Alexander Wu Chao, Karl Hubert Mess, Maury Tigner, and Frank Zimmermann, editors. *Handbook of Accelerator Physics and Engineering 2nd Edition.* World Scientific, 2013.

44. Ryoichi Hajima. A first-order matrix approach to the analysis of electron beam emittance growth caused by coherent synchrotron radiation. *Japanese Journal of Applied Physics*, 42(Part 2, No. 8A):L974–L976, aug 2003.

45. S. Di Mitri, M. Cornacchia, and S. Spampinati. Cancellation of coherent synchrotron radiation kicks with optics balance. *Phys. Rev. Lett.*, 110:014801, Jan 2013.

46. S. Heifets, G. Stupakov, and S. Krinsky. Coherent synchrotron radiation instability in a bunch compressor. *Phys. Rev. ST Accel. Beams*, 5:064401, Jun 2002.

47. Zhirong Huang and Kwang-Je Kim. Formulas for coherent synchrotron radiation microbunching in a bunch compressor chicane. *Phys. Rev. ST Accel. Beams*, 5:074401, Jul 2002.

48. E A Seddon, J A Clarke, D J Dunning, C Masciovecchio, C J Milne, F Parmigiani, D Rugg, J C H Spence, N R Thompson, K Ueda, S M Vinko, J S Wark, and W Wurth. Short-wavelength free-electron laser sources and science: a review. *Reports on Progress in Physics*, 80(11):115901, oct 2017.

Index

T - #0630 - 101024 - C0 - 254/178/18 - PB - 9781032399843 - Gloss Lamination